# Cancer Metabolomics 2018

# Cancer Metabolomics 2018

Special Issue Editors

**Paula Guedes De Pinho**
**Márcia Carvalho**
**Joana Pinto**

MDPI • Basel • Beijing • Wuhan • Barcelona • Belgrade

*Special Issue Editors*

Paula Guedes De Pinho
University of Porto
Portugal

Márcia Carvalho
University Fernando Pessoa
Portugal

Joana Pinto
University of Porto
Portugal

*Editorial Office*
MDPI
St. Alban-Anlage 66
4052 Basel, Switzerland

This is a reprint of articles from the Special Issue published online in the open access journal *Metabolites* (ISSN 2218-1989) from 2018 to 2019 (available at: https://www.mdpi.com/journal/metabolites/special_issues/cancer_metabolomics_2018)

For citation purposes, cite each article independently as indicated on the article page online and as indicated below:

LastName, A.A.; LastName, B.B.; LastName, C.C. Article Title. *Journal Name* **Year**, *Article Number*, Page Range.

ISBN 978-3-03921-345-0 (Pbk)
ISBN 978-3-03921-346-7 (PDF)

Cover image courtesy of Filipa Amaro.

# Contents

# About the Special Issue Editors

**Paula Guedes de Pinho** is graduated in Pharmaceutical Sciences by the Faculty of Pharmacy of University of Porto (FFUP) and completed a PhD degree in Medical and Biological Sciences at the University of Bordeaux II, France. During her PhD work, she specialized in biochemical mechanisms of living organisms, by using analytical methodologies such as Chromatography and Mass Spectrometry. She is presently Coordinator Researcher at UCIBIO-REQUIMTE/FFUP and Invited Associate Professor (with Habilitation) at the Faculty of Science of University of Porto (Department of Biology). She published more than 200 papers (*h*-index 33-Scopus) in international peer review journals, 13 book chapters, and 5 prizes or awards. She is responsible at the Laboratory of Toxicology of FFUP for GC-MS analysis. She implemented the research line of metabolomics applied to cancer biomarker discovery at the Laboratory of Toxicology of FFUP, getting financed projects for its development.

**Márcia Carvalho** is graduated in Pharmaceutical Sciences by the Faculty of Pharmacy, University of Porto (FFUP), Portugal. She holds a Master in Science degree in Quality Control—Scientific Area in Drug Substances and Medicinal Plants, and a PhD degree in Toxicology from the same University. Márcia Carvalho is presently Associate Professor of Toxicology at the Faculty of Health Sciences, University Fernando Pessoa (Porto, Portugal), researcher at the CEBIMED/FP-ENAS (UFP, Portugal) and at the Laboratory of Toxicology of the Associated Laboratory UCIBIO-REQUIMTE (FFUP, Portugal). Her main areas of research are Toxicology and Metabolomics, with a special interest in mechanistic studies to elucidate the toxicity of drugs of abuse and metabolomics studies for the identification of biomarkers and metabolic pathways altered in urological cancers. She has published over 70 papers in international scientific journals / book chapters in these fields, and holds an *h*-index of 27.

**Joana Pinto** holds a BSc degree in Biochemistry (2008), a MSc in Biomolecular Methods (2010) and a PhD in Biochemistry (2015) from the University of Aveiro, Portugal. Her PhD thesis was focused on the investigation of potential biomarkers of several prenatal disorders through NMR-based metabolomic analysis of maternal blood plasma. She is currently a researcher and co-PI in a project aiming the definition of a volatile signature characteristic of renal cell carcinoma based on GC-MS metabolomics and electronic nose analysis, at the Associated Laboratory UCIBIO@REQUIMTE, Faculty of Pharmacy, University of Porto, Portugal. Joana Pinto has published 1 book chapter, 22 articles in international peer-reviewed journals and 3 scientific conference proceedings (*h*-index 9). Her current research interests are biomarker discovery using GC-MS and NMR based metabolomics for diagnosis, prognosis and treatment improvement and the development of new non-invasive diagnostic tools for cancer detection based on the profile of urinary volatile compounds.

*Review*

# Metabolomics Contributions to the Discovery of Prostate Cancer Biomarkers

Nuria Gómez-Cebrián [1,2,3], Ayelén Rojas-Benedicto [1,2], Arturo Albors-Vaquer [1,2], José Antonio López-Guerrero [3], Antonio Pineda-Lucena [1,2] and Leonor Puchades-Carrasco [2,*]

[1]  Drug Discovery Unit, Instituto de Investigación Sanitaria La Fe, Valencia 46026, Spain; ngomez@cipf.es (N.G.-C.); ayelen_rojas@iislafe.es (A.R.-B.); arturo_albors@iislafe.es (A.A.-V.); pineda_ant@gva.es (A.P.-L.)

[2]  Joint Research Unit in Clinical Metabolomics, Centro de Investigación Príncipe Felipe/Instituto de Investigación Sanitaria La Fe, Valencia 46012, Spain

[3]  Laboratory of Molecular Biology, Fundación Instituto Valenciano de Oncología, Valencia 46009, Spain; jalopez@fivo.org

*  Correspondence: leonor.puchadescarrasco@icr.ac.uk; Tel.: +34-96-124-6713

Received: 31 January 2019; Accepted: 4 March 2019; Published: 8 March 2019

**Abstract:** Prostate cancer (PCa) is one of the most frequently diagnosed cancers and a leading cause of death among men worldwide. Despite extensive efforts in biomarker discovery during the last years, currently used clinical biomarkers are still lacking enough specificity and sensitivity for PCa early detection, patient prognosis, and monitoring. Therefore, more precise biomarkers are required to improve the clinical management of PCa patients. In this context, metabolomics has shown to be a promising and powerful tool to identify novel PCa biomarkers in biofluids. Thus, changes in polyamines, tricarboxylic acid (TCA) cycle, amino acids, and fatty acids metabolism have been reported in different studies analyzing PCa patients' biofluids. The review provides an up-to-date summary of the main metabolic alterations that have been described in biofluid-based studies of PCa patients, as well as a discussion regarding their potential to improve clinical PCa diagnosis and prognosis. Furthermore, a summary of the most significant findings reported in these studies and the connections and interactions between the different metabolic changes described has also been included, aiming to better describe the specific metabolic signature associated to PCa.

**Keywords:** metabolomics; metabolism; prostate cancer; biomarker; early diagnosis; prognosis

---

## 1. Introduction

Prostate cancer (PCa) is the second most frequently diagnosed cancer and represents the fifth leading cause of death in men [1]. In 2018, new cases of PCa were estimated to account for over 1.3 million, and 359.000 PCa-associated deaths were expected worldwide [1]. PCa is a hormone-dependent tumor characterized by an extremely variable clinical course, ranging from an indolent condition to a rapid progression into an aggressive phenotype that disseminates and metastasizes to the lymph nodes and bones. Moreover, there is a current lack of reliable and reproducible assays to identify tumors destined to remain indolent. Thus, stratifying PCa patients into different risk phenotypes at time of diagnosis is still a major clinical challenge.

Nowadays, PCa screening tests rely on the determination of prostate-specific antigen (PSA) serum levels and digital rectal examination (DRE). Based on the results of these screening tests, trans-rectal ultrasound (TRUS)-guided prostate biopsy is performed to confirm diagnosis when necessary. However, these tests suffer from a number of limitations and do not provide enough information to enable a precise discrimination between indolent and aggressive tumors. While PSA provides high sensitivity and low specificity for PCa diagnosis, (TRUS)-guided prostate biopsy has been associated with high false negative rates due to the high degree of PCa inter- and intra-heterogeneity [2]. Moreover, even the recently updated histopathology-based estimation of the Gleason Score (GS), the current clinical gold standard for assessing the risk of PCa metastasis and prognosis, exhibits limitations [3]. During the last years, many research studies have focused on the identification of molecular biomarkers that could help to improve early diagnosis and risk stratification of PCa patients [4–7]. Among them, a potential biomarker, that has been evaluated in combination with PSA levels, is the non-coding transcript PCA3 (overexpressed in >95% of PCa). The quantification of PCA3 levels in urine has shown improvement, when combined with PSA, in PCa detection [8], although no optimal cut-off for urinary PCA3 levels has been established for maximizing clinical benefit while avoiding overdiagnosis [9]. Another potential biomarker is the TMPRSS2:ERG fusion transcript [10], that is being evaluated as a potential diagnostic and therapeutic target associated with PCa invasion [11]. Despite being 100% indicative of PCa [12], it is only detected in 50% of PCa cases [13]. In summary, although intense efforts have been devoted to the discovery and development of new PCa biomarkers, there still exists an unmet clinical need to identify accurate PCa biomarkers for early diagnosis, prognosis and monitoring of PCa patients, both in terms of sensitivity and specificity [14,15].

Moreover, additional clinically robust biomarkers able to differentiate between indolent and aggressive PCa are urgently needed. In this context, several metabolomics studies have been carried out to attempt the characterization of a specific PCa metabolic profile, with the ultimate goal of identifying potential metabolic biomarkers that could improve the clinical management of PCa patients [16–19].

## 2. Cancer and Metabolic Reprogramming: Metabolomics Opportunities

The metabolic profile is closely associated with the pathophysiological condition of an individual. In particular, the metabolic composition can be strongly influenced, both from a qualitative and quantitative point of view, as a result of pathological processes or in the presence of specific drug treatments [20]. These changes can provide useful clues for the characterization of biomarkers associated with the onset and progression of diseases, as well as with the prediction of the response to therapeutic interventions.

Different studies, linking significant metabolic alterations and cancer onset and progression, have been extensively described since Warburg's pioneering studies [21]. The metabolic rewiring associated with the neoplastic processes is the result of mutations in specific oncogenes and tumor suppressors, leading to the activation of different signaling pathways and transcriptional networks [22]. Furthermore, it is well known that neoplastic processes have a strong influence on gene expression, cellular differentiation and tumor microenvironment [23,24]. Metabolites represent the end products of biochemical pathways, and the concentrations of these compounds are extremely sensitive to different alterations. At the molecular level, the progression of cancer involves multiple alterations in metabolic pathways that are specifically required for cancer cells to survive [23]. Interestingly, cancer cells exhibit different metabolic phenotypes [25,26]. Thus, some tumors preferentially use aerobic glycolysis to proliferate [27], while others rely on glutaminolysis [28], or one-carbon metabolism [29]. There are also tumors that benefit from the utilization of several of these metabolic routes at the same time [25,26,28].

In this context, metabolomics, that relies on the systematic analysis of low-molecular-weight metabolites present in biological samples, provides an accurate and complementary approach for getting a better understanding of the biochemical alterations responsible for the onset and progression of neoplastic processes, thus offering new opportunities for biomarker discovery in complex diseases [30]. Metabolomics studies offer a holistic view of the biochemical processes that could contribute to getting a deeper insight into the molecular alterations underlying pathological processes. This information could significantly improve the opportunities to identify clinically relevant biomarkers for the diagnosis and prognosis of different pathological processes, including PCa.

## 3. Metabolomics and PCa

The ultimate goal of metabolomics is to measure and identify as many metabolites as possible, ideally obtaining a complete overview of the metabolome. Metabolomics can provide an accurate description of the phenotype of an individual because it represents the final step of the omics cascade. The analysis of metabolic changes associated with specific biochemical pathways offers unprecedented opportunities for identifying the molecular mechanisms of complex diseases. Taken into consideration the limitations of current diagnostic procedures, this information could result in the characterization of specific and novel disease biomarkers [31].

At the analytical level, these studies are extremely challenging [32,33]. The complexity of the matrix to be examined (e.g., osmolarity, the presence of proteins, and inorganic salt concentration), the dynamic range of metabolites concentrations, and the vast chemical diversity of metabolite types (e.g., acidic, neutral, basic, lyophilic, and hydrophilic) greatly complicate the choice of analytical modality. However, a number of technical improvements have been introduced over the last few years. This has led to the development of a wide variety of analytical platforms that are currently used to characterize the metabolic content of biological samples [34–36]. The selection of the appropriate approach usually depends on the experimental objectives and the biological matrix. The detection of metabolites in cells, tissues or biofluids is usually carried out by either Nuclear Magnetic Resonance (NMR) spectroscopy or mass spectrometry (MS). In general, NMR spectroscopy, mostly $^1$H-NMR, and MS, particularly liquid chromatography (LC)-MS, are the two most important analytical platforms used in metabolomics studies.

PCa is a disease of great interest from a metabolomics perspective. A number of studies, focused on the characterization of the specific PCa metabolic phenotype using different experimental approaches, have been reported recently [37–61]. These studies have shown that healthy prostate cells are characterized by a decreased citrate oxidation and metabolism within the tricarboxylic acid (TCA) cycle, resulting in citrate accumulation [62] and the reliance on glucose oxidation for energy production [63]. Benign prostate cells accumulate zinc, resulting in the inhibition of the m-aconitase (ACO), the enzyme that catalyzes the isomerization of citrate in the TCA cycle [62]. However, when prostate cells undergo malignant transformation, their characteristic ability to accumulate zinc is lost, leading to the TCA activation. Furthermore, it has been shown that early PCa does not exhibit the Warburg effect [64], relying on lipids and other energetic molecules for energy production, but not on aerobic respiration [65,66]. In this context, it should be noted that several metabolic alterations have also been identified in PCa tissue compared with normal tissue, including an increase of choline [67] and sarcosine [68], and a decrease of polyamine and citrate levels [69,70]. Nevertheless, the clinical relevance of some of these changes remains controversial due to the contradictory results reported in different studies (e.g., alterations in sarcosine levels–further discussed in the following section).

Overall, the possibility to directly evaluate the metabolic phenotype of PCa patients offers a great potential from a clinical perspective. To this end, many metabolomics projects, based on the analysis of different biological samples, have been conducted over the last few years with a focus on the discovery of new biomarkers that could improve the clinical management of PCa patients (Table 1).

## 4. PCa Metabolic Biomarkers in Biofluids

Changes in the concentration of metabolites in biofluids are reflective of alterations in the physiological status of an individual. The metabolome, that is, the set of all metabolites present on a particular biological sample, represents the downstream end product of the omics cascade, and a closer approach to the phenotype. Therefore, metabolite signatures obtained from biofluids can be a useful approach for identifying non-invasive biomarkers and characterizing the molecular mechanisms associated with pathological conditions. The most widely used biofluids in PCa studies have been urine, serum and seminal fluid.

### 4.1. Urine Biomarkers

Urine samples offer some advantages for carrying out metabolomics studies since they can be collected non-invasively and have a less complex composition compared with other biofluids, thus facilitating the discovery of novel biomarkers [71]. However, the analysis of this biofluid has several limitations, including the presence of diluted urinary constituents and interferences between molecules [37,71], that can result in failing to detect underrepresented metabolites or to correctly identify the molecules. Despite these problems, different studies have discovered metabolic alterations in urine samples from PCa patients and evaluated their clinical utility as biomarkers for this neoplastic process.

Urine is anatomically close to the prostate, which explains why it has been extensively studied for metabolic biomarker discovery in PCa [37]. As shown in Table 1, most of these studies have aimed to identify metabolic dysregulations that could provide clinically relevant PCa biomarkers. Most of these studies focused on the characterization of the metabolic differences between urine samples from healthy individuals [38–43] or benign prostate hyperplasia (BPH) patients [37,44,45] and PCa patients. In general, they were performed using mass spectrometry (MS)-based metabolomics as an analytical platform ($n = 8$), and only one study was performed using NMR spectroscopy for the analysis of urine samples [44].

The study conducted by Liang et al., including the analysis of 233 healthy individuals and 236 PCa patients, highlighted the clinical utility of three metabolites: 5-hydroxy-L-tryptophan, hippurate, and glycocholic acid, as potential metabolic biomarkers for the early diagnosis of PCa (area under the curve, (AUC) > 0.95) [38]. A metabolite called 5-hydroxy-L-tryptophan is involved in tryptophan metabolism, a pathway that has been associated with the ability of several tumors to evade the antitumor immune response [72,73]. Another metabolite involved in this pathway, kynurenic acid, also exhibited a moderate diagnostic value (AUC = 0.62) in a study conducted by Gkotsos et al. for the detection of PCa using urine samples obtained after prostatic massage [39].

Another metabolite that has been extensively investigated as a potential biomarker of PCa is sarcosine. Sarcosine is an intermediate product in the synthesis and degradation of glycine. In 2009, Sreekumar et al. identified sarcosine as a promising PCa biomarker, being highly correlated with PCa progression and more detectable in the urine of PCa patients when compared with healthy individuals [68]. Similarly, Khan et al. reported in 2013 markedly elevated sarcosine levels in the urine sediments of PCa patients compared with controls [74]. In serum, Kumar et al. [46,47] also found increased sarcosine levels in PCa samples compared with healthy individuals. In these studies, it was shown that sarcosine, in combination with other metabolites, could accurately differentiate PCa patients from healthy individuals (accuracy = 90.2%) [47] and PCa from BPH patients (87.7% sensitivity and 85.5% specificity) [46]. Furthermore, the authors showed that metabolomics provided better predictions than serum PSA levels for the discrimination between PCa patients and healthy individuals as well as between PCa and BPH patients. However, the role of sarcosine as a metabolic biomarker for PCa diagnosis and prognosis remains controversial due to the contradictory results reported in further studies. In a case-control study conducted by Ankerst et al., the use of sarcosine as a biomarker for early PCa detection was investigated in serum samples of matched-age controls and PCa patients [75]. These authors reported no differences in sarcosine levels when comparing both groups. Furthermore, in another pilot study by Dereziński et al., where higher serum sarcosine levels were found in PCa patients when compared with the control group, no statistically significant differences were observed in urine samples [76]. Similarly, Pérez-Rambla et al. found elevated sarcosine levels in PCa patients when compared with BPH patients, although these alterations were not found to be statistically significant [44].

Beyond the alteration in sarcosine levels, Pérez-Rambla et al. also identified alterations in the urine levels of six metabolites that facilitated the discrimination of the metabolomic profile of PCa and BPH patients [44]. Among the characteristic changes, PCa patients showed decreased concentration of glycine, a metabolite involved in one-carbon metabolism and associated with cell transformation and tumorigenesis [77]. Interestingly, Struck-Lewicka et al. reported lower levels of this metabolite in urine samples from PCa patients when compared with a control group [40]. The overall results of this study showed alterations in the urine levels of metabolites associated with TCA cycle, purine, glucose, amino acid and urea metabolism in PCa patients. These findings are in agreement with those obtained by Fernández-Peralbo et al., where variations in the levels of 28 metabolites involved in amino acid, purine and pyrimidine, and tryptophan metabolism were also identified [41] when comparing PCa patients and healthy individuals. The results of this study led to a predictive model of high quality for the discrimination of these two groups (sensitivity = 88.4% sensitivity, specificity = 92.9%).

Metabolic changes have also been identified when comparing urine samples from low and high risk PCa patients. Heger et al. performed a study focused on the characterization of differences in protein expression levels between two different risk groups of PCa patients after radical prostatectomy (RP) [48]. The two experimental cohorts were divided based on the presence of positive ($n = 15$) or negative ($n = 15$) surgical margins. The analysis led to the identification of three proteins with different expression levels. Among them, the glycolytic enzyme lactate dehydrogenase C (LDHC), that plays a key role in metabolism, was detected at higher expression levels in PCa patients with positive surgical margins [48]. Beyond PCa, increased LDHC expression has also been observed in melanoma, lung and breast cancer [78]. Moreover, this enzyme has been shown to be involved in tumor invasion and migration in breast cancer [79].

A complementary approach, that has also been the focus of recent studies in the context of urinary alterations associated with PCa, is the analysis of extracellular vesicles (EV). The analysis of these particles still requires the optimization of methods for isolation and storage of urinary EV, as well as for the normalization of metabolite levels [80]. Nevertheless, in a preliminary study, Puhka et al. analyzed urine EV samples from three controls and three PCa patients, obtained before and after prostatectomy [42]. After normalization tests, decreased levels of glucuronate, D-ribose 5-phosphate and isobutyryl-L-carnitine were observed in pre-prostatectomy samples when compared with the healthy individuals and post-prostatectomy samples. In agreement with these results, Clos-García et al. also reported variations in carnitine-related metabolites when comparing urine EV samples from PCa ($n = 31$) and BPH ($n = 14$) patients [37]. In this study, changes in the expression levels of seven enzymes related to fatty acid, steroid biosynthesis, creatine, and cAMP metabolism were also observed [37]. Increased levels of another enzyme involved in fatty acid metabolism (fatty acid binding protein 5, FABP5) were also found in urinary EVs from PCa patients collected after prostatic massage [43]. In this study, the AUC for the prediction of PCa with GS $\geq$ 6 based on FABP5 was 0.757 (confidence interval 0.570–0.994, $p$-value = 0.027), whereas the AUC value for the prediction based on serum PSA was 0.593 (confidence interval 0.372–0.815, $p$-value = 0.42). FABP5 is an enzyme involved in the uptake and transport of fatty acids, that has been previously found to be overexpressed in PCa tissues [81]. Increased levels of this enzyme have been described in serum and tissue samples from PCa patients with lymph node metastasis [82].

Overall, these studies show that the urine metabolic phenotype of PCa patients is significantly different from that of healthy individuals and BPH patients. Taken together, alterations in the levels of metabolites involved in TCA cycle, tryptophan, amino acid, fatty acid, nucleotide, and carbon metabolism have been reported. In general, a significant limitation of these studies has been the sample size, except for the study carried out by Liang et al. where a total of 469 urine samples were analyzed [38]. Therefore, further analyses and validation studies will be necessary to assess the clinical utility of these findings.

## 4.2. Serum Biomarkers

Metabolic dysregulations in TCA cycle, fatty acid, amino acid, purine, histidine, creatine, glycine, and serine, and threonine metabolism have been described when analyzing serum metabolic profile of PCa patients. Particularly, a study conducted by Giskeødegård et al., comparing the serum metabolic profile of 21 BPH and 29 PCa patients, revealed significant changes in fatty acid, choline and amino acid metabolism [49]. In this study, different metabolomics analytical platforms were used to perform the analysis. The combination of the most relevant metabolites identified using the different platforms provided the best classification results, enabling the discrimination of PCa patients and BPH controls with a sensitivity and specificity of 81.5% and 75.2%, respectively. In a different study, Kumar et al. reported a metabolic signature of three metabolites (pyruvate, glycine, and sarcosine) that classified 90.2% of PCa samples ($n = 70$) with 84.8% sensitivity and 92.9% specificity compared with healthy controls ($n = 32$) [47]. Furthermore, Kumar et al., using filtered serum samples ($n = 210$), obtained a model based on five metabolites (alanine, sarcosine, creatinine, glycine, and citrate) that enabled the discrimination of BPH and PCa patients with high accuracy (88.3%) [46]. Finally, Zhao et al., analyzing the metabolic profile of plasma samples from 32 control cases and 32 PCa patients, reported alterations in different metabolic pathways, including amino acid, propanoate, butanoate, and nucleotide metabolism [50]. After evaluation of the predictive value of individual changes, a predictive model combining sarcosine, acetylglycine, and coreximine was reported. However, although a discrete increase in the diagnostic performance (AUC = 0.941; confidence interval 0.812–1) was found when compared with PSA levels (AUC = 0.926; confidence interval 0.851–0.978), this model partially relied on changes in the levels of coreximine, a compound belonging to a family of alkaloids and derivatives, probably from exogenous origin.

Regarding PCa biomarkers associated with disease progression and outcome, different studies, focused on the analysis of PCa serum samples, have been performed trying to identify metabolic alterations that could be useful from this clinical perspective [47,51]. These studies revealed alterations in TCA cycle, lipids, and amino acids metabolism. Lin et al. investigated the correlation between the plasma lipidome and the outcome of 96 castration-resistant PCa (CRPC) patients [51]. A three-lipid signature, comprising ceramide d18:1/24:1, sphingomyelin d18:2/16:0 and phosphatidylcholine 16:0/16:0, was found to be associated with poor prognosis in this study and further validated in an independent cohort of 63 CRPC patients. The results also revealed an association between the lipid signature in the serum of the patients and the overall survival time. Eleven out of the 63 patients of the validation cohort exhibited the three-lipid signature, and their median overall survival time was significantly shorter than those not displaying that signature (11.3 vs. 21.4 months). In another study performed in serum samples, Kumar et al. described a model consisting of three metabolites (alanine, pyruvate and glycine) that allowed the discrimination of low- ($n = 40$) from high-grade ($n = 30$) PCa serum samples with 92.5% sensitivity and 93.3% specificity [47]. Alanine and glycine can be metabolized to a common end product, pyruvate. Increased levels of these two metabolites have also been observed in urine [83] and tissue [84] from PCa patients. Tissue levels of both metabolites have also shown a statistically significant correlation with the GS [85]. Finally, in a study performed by Mondul et al., 200 matched-controls and 200 PCa patients (100 aggressive) were analyzed [52]. The authors reported inverse associations between the risk of aggressive PCa and the levels of glycerophospholipids and fatty acids, inositol-1-phosphate showing the strongest inverse association. On the contrary, aggressive PCa risk was correlated with the levels of $\alpha$-ketoglutarate, thyroxine, TMAO, and erucoyl-sphingomyelin, while metabolites involved in the metabolism of nucleotides, steroid hormones and tobacco were associated with non-aggressive PCa [52]. In this particular study, although levels of two known nicotine-derived metabolites (cotinine and hydroxycotinine) were found to be associated with non-aggressive PCa, the authors argued that it was unlikely that these changes were related to tobacco smoking as all individuals included in the study were smokers at the time of sample collection. Furthermore, results remained unchanged when adjusting for cigarettes smoked per day, suggesting that cigarette smoking did not strongly influence the results.

Additionally, some of the most recent PCa metabolomics studies based on the analysis of serum samples have aimed to identify metabolic alterations that could provide insights into the risk of developing PCa. These studies were carried out with a significant number of samples in each experimental cohort compared with those focused on the identification of biomarkers for PCa diagnosis and/or prognosis. Thus, Kühn et al. evaluated the association between the levels of pre-diagnostic metabolites and the risk of developing different cancers, including PCa [53]. Serum samples of 310 PCa patients with a median follow-up of 6.83 years were included in the study. High levels of lysophosphatidylcholines were found to be positively correlated to lower PCa risk, while high levels of phosphatidylcholines were associated with increased risk of developing the disease [53]. Schmidt et al. analyzed 1077 healthy and PCa serum samples to assess the risk of developing PCa [54]. In this study, higher citrulline levels were associated with a 27% decreased risk of PCa in the first five years of follow-up but not after longer periods of time [54]. The authors also reported inverse associations between 12 glycerophospholipids and advanced stage disease. In another study, Huang et al. analyzed serum samples from controls ($n$ = 200) and PCa patients classified according to their tumor stage (T2: $n$ = 71, T3: $n$ = 51, T4: $n$ = 15), and identified metabolites associated with the risk of being diagnosed with each stage [55]. Histidine and uridine-related metabolites were associated with risk of T2 stage. Glycerophospholipids and primary bile acid lipids showed inverse correlations with T3 stage, while sphingomyelins were positively associated with risk of T3. Secondary bile acid, sex steroids, histamine, and BCAA were associated with T4 risk, while citrate and fumarate were inversely correlated. Finally, a recent study carried out by Andras et al. used serum samples to identify variations in the metabolite levels that could be useful for predicting PCa before biopsy [56]. These authors analyzed 90 samples from patients with suspicion of PCa and derived a predictive score based on six metabolites, that was validated using a subgroup of patients. A cut-off value of 0.528 for the derived score showed good accuracy for PCa prediction before biopsy (AUC = 0.779; confidence interval 0.625–0.876), although not statistically significantly higher than the predictive ability of PSA levels (AUC = 0.793; confidence interval 0.665–0.889). In PCa patients with PSA levels < 10 ng/mL, this score had 80.95% sensitivity and 64.52% specificity for PCa detection at biopsy.

### 4.3. Seminal Fluid Biomarkers

Seminal fluid has a number of advantages over blood and urine in terms of its potential as a source of PCa specific biomarkers. Prostatic constituents are highly enriched in seminal fluid compared with other biofluids. In the last few years, several metabolomics studies have been performed aiming to analyze the metabolic profile of seminal fluid samples from either healthy individuals [57–59] or BPH patients [60] and PCa patients to discover metabolic alterations that could be useful for discriminating between both groups. In general, these studies were performed using NMR spectroscopy ($n$ = 4) and the sample size of the different cohorts was relatively small. Most of the metabolic alterations identified included changes in the TCA cycle, amino acid, and lipid metabolism. In a preliminary study, Averna et al. found decreased concentrations of citrate in PCa ($n$ = 3) compared to BPH ($n$ = 1) samples [60]. Similarly, Kline et al. also observed lower citrate levels in PCa samples both when analyzing seminal fluid samples and expressed prostatic secretions (EPS) from 33 healthy volunteers and 28 PCa patients [57]. In this study, authors reported good values for predicting PCa in patients (AUC = 0.81 in seminal fluid, confidence interval 0.60–0.92 and AUC = 0.73 in EPS, confidence interval 0.38–0.90), outperforming the predictive ability of PSA (AUC = 0.61, confidence interval 0.44–0.74) in these samples. Furthermore, using an ELISA assay, Etheridge et al. identified alpha methylacyl A coenzyme racemase (AMACR) as a promising biomarker for PCa diagnosis [58]. Higher levels of this enzyme were detected in seminal fluid samples of PCa patients ($n$ = 28) compared with age-matched controls ($n$ = 15). AMACR, a key regulator of lipid metabolism, is involved in the peroxisomal and mitochondrial β-oxidation of branched-chain fatty acids. This enzyme had been previously described as an immunohistological marker for PCa diagnosis [86,87], associated with poor prognosis in patients with localized PCa [88] and found to be overexpressed in PCa tissues [89]. Interestingly, AMACR has also been identified as a promising prognostic indicator in other cancer types, including gastric cancer [90] and hepatocellular [91] and nasopharyngeal [92] carcinomas.

Besides seminal fluid, EPS is another biofluid enriched in prostatic material that has shown potential utility for the identification of new PCa disease-specific biomarkers. EPS is obtained in the first void following vigorous DRE or prostatic massage. Given the nature of this biofluid, metabolites present in EPS are usually found at lower concentrations than in seminal fluid, thus requiring the use of highly sensitive detection methods. In 2008, Serkova et al. analyzed EPS samples from 26 healthy volunteers and 52 PCa patients aiming to identify potential metabolites that could contribute to PCa risk assessment [59]. This study revealed that concentrations of citrate, myo-inositol, and spermine were inversely correlated with PCa risk (AUC values of 0.89, 0.87 and 0.79, respectively). However, in a more recent study attempting to validate the role of these metabolites as biomarkers for assessing PCa risk, Roberts et al. found that citrate, spermine, and myo-inositol had minimal predictive ability when analyzing seminal fluid samples [61]. Therefore, further studies using larger cohorts will be required to confirm the utility of seminal fluid and EPS derived biomarkers for PCa diagnosis and prognosis.

**Table 1.** Metabolomics studies focused on the analysis of biofluids to identify clinically relevant prostate cancer (PCa) biomarkers.

| Article | Sample | Experimental Approach | Research Aim | Sample Cohort | Main Findings | Validation Cohort |
|---|---|---|---|---|---|---|
| Clos-Garcia et al., 2018 [37] | Urine EVs | UHPLC-MS | Diagnosis | 31 × PCa; 14 × BPH | Statistically significant changes in 76 metabolites and 7 enzymes related to urea cycle, TCA cycle, and metabolism of steroid hormone biosynthesis, leukotriene, and prostaglandin, linoleate and purine, glycerophospholipid and tryptophan | No |
| Liang et al., 2017 [38] | Urine | FPLC-MS/MS | Diagnosis | 236 × PCa; 233 × HV | ↑ glycocholic acid; hippurate; chenodeoxycholic acid: PCa > HV<br>↓ taurocholic acid; 5-hydroxy-L-tryptophan: PCa < HV | No |
| Gkotsos et al., 2017 [39] | Urine | UPLC-MS/MS | Diagnosis | 52 × PCa, 49 × HV | ↓ kynurenic acid: PCa < HV | No |
| Struck-Lewicka et al., 2015 [40] | Urine | HPLC-TOF-MS; GC-QqQ-MS | Diagnosis | 32 × PCa; 32 × HV | Statistically significant changes in 82 metabolites related to amino acid, organic acids, sphingolipids, fatty acids, and carbohydrates metabolism | No |
| Fernández-Peralbo et al., 2016 [41] | Urine | LC-QTOF-MS/MS | Diagnosis | 43 × PCa; 29 × HV | ↑ 7-methylguanine: PCa > HV<br>↓ Statistically significant changes in 27 metabolites related to amino acid metabolism: PCa < HV | 19 × PCa; 13 × HV |
| Puhka et al., 2017 [42] | Urine EVs; Plasma EVs | UPLC-MS/MS | Diagnosis | 3 × PCa pre-prostatectomy; 3 × PCa post-prostatectomy; 3 × HV | ↓ glucoronate; isobutyryl-L-carnitine; D-Ribose-5-phosphate: pre- < post-prostatectomy and HV | No |
| Fujita et al., 2017 [43] | Urine EVs | iTRAQ; LC-MS/MS | Diagnosis and prognosis | 12 × PCa (6 × HG PCa; 6 × LG PCa); 6 × HV | ↑ FABP5: PCa > HV<br>↑ FABP5; GRN; AMBP; CHMP4A; CHMP4C associated with higher GS | 18 × PCa (6 × HG; 12 × LG); 11 × HV |
| Perez-Rambla et al., 2017 [44] | Urine | $^1$H-NMR | Diagnosis | 64 × PCa; 51 × BPH | ↑ BCAAs; glutamate; pseudouridine: PCa > BPH<br>↓ glycine; dimethylglycine; fumarate; 4-imidazole-acetate: PCa < BPH | No |
| Davalieva et al., 2015 [45] | Urine | 2D-DIGE-MS | Diagnosis | 8 × PCa; 16 × BPH | ↑ AMBP: PCa > BPH<br>↓ HP: PCa < BPH | 16 × PCa; 16 × BPH |
| Heger et al., 2015 [48] | Urine | 2D-DIGE; MALDI-TOF-MS | Diagnosis | 15 × HG PCa; 15 × LG PCa | ↑ CDK6; M2BP; LDHC: HG PCa > LG PCa | No |
| Kumar et al., 2016 [46] | Serum | $^1$H-NMR | Diagnosis | 75 × PCa; 70 × BPH; 65 x HV | ↑ alanine; sarcosine; creatine; creatinine: PCa > BPH and HV<br>↑ pyruvate; 3-methylhistidine; xanthine; hypoxanthine: BPH and PCa > HV<br>↓ glycine: PCa < HV<br>↓ citrate: PCa < BPH and HV | No |

**Table 1.** *Cont.*

| Article | Sample | Experimental Approach | Research Aim | Sample Cohort | Main Findings | Validation Cohort |
|---|---|---|---|---|---|---|
| Kumar et al., 2015 [47] | Serum | ¹H-NMR | Diagnosis and prognosis | 21 × HG PCa; 28 × LG PCa; 22 × HV | ↑ alanine; sarcosine: LG PCa > HG PCa and HV<br>↑ pyruvate: LG PCa and HG PCa > HV<br>↓ glycine: LG PCa and HG PCa < HV | 9 × HG PCa; 12 × LG PCa; 12 × HV |
| Giskeødegård et al., 2015 [49] | Plasma/Serum | ¹H-NMR; UPLC-MS/MS; GC-MS | Diagnosis | 29 × PCa; 21 × BPH | ↑ decanoylcarnitine (c10); tetradecenoylcarnitine (c14:1); octanoyl-carnitine (c8); dimethylsulfone; phenylalanine; lysine: PCa > BPH<br>↓ phosphatidylcholine diacyl (c34:4); lipid -(CH2)n-CH2-CH2-CO: PCa < BPH | No |
| Zhao et al., 2017 [50] | Plasma | UPLC-MS/MS | Diagnosis | 32 × PCa; 32 × HV | Statistically significant changes in 19 metabolites related to amino acid, nucleotide, butanoate and propionate metabolism | No |
| Lin et al., 2017 [51] | Plasma | LC-MS/MS | Prognosis | 96 × CRPC | ↑ ceramide d18:1/24:1; sphingomyelin d18:2/16:0; phosphatidylcholine 16:0/16:0 correlated with shorter overall survival | 63 × CRPC |
| Mondul et al., 2015 [52] | Serum | UHPLC-MS; GC-MS | PCa risk Prognosis | 100 × HG PCa; 100 × LG PCa; 200 × HV | Statistically significant changes in 22 metabolites related to lipid and amino acid metabolism associated with overall PCa risk<br>Statistically significant changes in 14 metabolites related to TCA cycle and lipid metabolism associated with HG PCa<br>Statistically significant changes in 34 metabolites related to lipid, amino acid and nucleotide metabolism associated with LG PCa | No |
| Kühn et al., 2016 [53] | Plasma | LC-MS/MS; FIA-MS/MS | PCa risk | 310 × PCa; 774 × HV | ↑ Phosphatidylcholine (PC) associated with higher risk of PCa<br>↑ lysoPC associated with lower risk of PCa | No |
| Schmidt et al., 2017 [54] | Plasma | QqQ-MS | PCa risk Prognosis | 1077 × PCa; 1077 × HV 208 × advanced PCa; 456 × localized PCa | Statistically significant changes in 14 metabolites related to lipid and amino acid metabolism associated with overall PCa risk<br>12 glycerophospholipids inversely associated with risk of advanced PCa | No<br>No |

**Table 1.** *Cont.*

| Article | Sample | Experimental Approach | Research Aim | Sample Cohort | Main Findings | Validation Cohort |
|---|---|---|---|---|---|---|
| Huang et al., 2017 [55] | Serum | UHPLC-MS; GC-MS | PCa risk | 71 × PCa T2 stage; 51 × PCa T3 stage; 15 × PCa T4 stage; 200 × HV | Statistically significant changes in 8 metabolites related to histidine and uridine metabolism associated with PCa T2 risk. Statistically significant changes in 12 metabolites related to fatty acid and primary bile acid metabolism associated with PCa T3 risk Statistically significant changes in 16 metabolites related to TCA, BCAA secondary bile acid, sex steroids and histamine metabolism associated with PCa T4 risk. | No |
| Andras et al., 2017 [56] | Serum | HPLC-ESI-QTOF-MS | Prediction | 59 × patients with high PSA levels | 6 metabolites involved in lipid, purine and tryptophan metabolism predictive for prostate biopsy outcome | 31 × patients with high PSA levels |
| Kline et al., 2006 [57] | Seminal fluid; Prostatic secretion | ¹H-NMR | Diagnosis | 28 × PCa; 33 × HV | ↓ citrate: PCa < HV | No |
| Etheridge et al., 2018 [58] | Seminal fluid | ELISA | Diagnosis | 28 × PCa; 15 × HV | ↑ AMACR: PCa > HV | No |
| Serkova et al., 2008 [59] | Prostatic secretion | ¹H-NMR | Prediction PCa risk | 52 × PCa; 26 × HV | ↓ citrate; myo-inositol; spermine shown highly predictive of PCa and inversely associated with PCa risk | No |
| Averna et al., 2005 [60] | Seminal fluid | ¹H-NMR | Diagnosis | 3 × PCa; 1 × BPH; 4 × HV | ↓ citrate: PCa < BPH | No |
| Roberts et al., 2017 [61] | Seminal fluid | ¹H-NMR | Prediction | 98 × PCa; 53 × HV | Statistically significant changes in choline, valine and leucine, | No |

¹H-NMR: Proton nuclear magnetic resonance spectroscopy; 2D-DIGE-MS: Two dimensional–difference gel electrophoresis–mass spectrometry; BCAA: Branched-chain amino acids; BPH: Benign Prostatic Hyperplasia; CRPC: Metastatic castration-resistant prostate cancer; ELISA: Enzyme-linked immunosorbent assay; EV: Extracellular vesicles; FIA-MS/MS: Flow injection analysis–tandem mass spectrometry; FPLC-MS: Fast ultra-high performance liquid chromatography–mass spectrometry; GC-MS: Gas chromatography–mass spectrometry; GC-QqQ-MS: Gas chromatography–triple quadrupole–mass spectrometry; HG: High-grade (GS ≥ 8); HPLC-ESI-QTOF-MS: High performance liquid chromatography–electrospray ionization–quadrupole time of flight–mass spectrometry; HPLC-TOF-MS: High performance liquid chromatography–time of flight–mass spectrometry; HV: Healthy Volunteers; iTRAQ: Isobaric tag for relative and absolute quantification; LC-MS: Liquid chromatography–mass spectrometry; LC-MS/MS: Liquid chromatography–tandem mass spectrometry; LG: Low-grade (GS ≤ 7); MALDI-TOF-MS: Matrix-assisted laser desorption ionization–time of flight–mass spectrometry; PCa: Prostate Cancer; PM: prostatic massage; QqQ-MS: Triple quadrupole–mass spectrometry; T: Stage; TCA: Tricarboxylic acid; UHPLC-MS: Ultra high performance liquid chromatography–mass spectrometry; UPLC-MS/MS: Ultra performance liquid chromatography–tandem mass spectrometry.

## 5. Conclusions and Future Perspectives

The identification and characterization of the metabolic changes accompanying the transformation of benign into malignant prostate cells has led to an increased interest, over the last few years, in the application of metabolomics for identifying clinically relevant biomarkers in this field. Omics approaches, including genomics, proteomics, transcriptomics, and metabolomics, are highly innovative areas of research. One of the major advantages of the omics approaches is their ability to provide information using unbiased large-scale approaches. Among them, metabolomics provides an unprecedented opportunity for understanding the pathophysiological condition of an individual. Metabolites represent the end products of biochemical pathways, and the concentrations of these compounds are extremely sensitive to different alterations. Thus, these metabolic fingerprints can provide useful clues for the characterization of biomarkers associated with the onset and progression of diseases. Furthermore, as metabolomics studies can be performed using biological fluids that could be easily accessible (e.g., serum, plasma, urine, and seminal fluid), it offers a high potential for clinical translatability when compared with other omics approaches.

In this manuscript, we aimed to review the main findings described in recent PCa metabolomics studies focused on the analysis of different biofluids (Table 1). Furthermore, a summary of the most significant findings reported in these studies and the connections and interactions between the different metabolic changes described has also been included, aiming to better describe the specific metabolic signature associated to PCa (Figure 1).

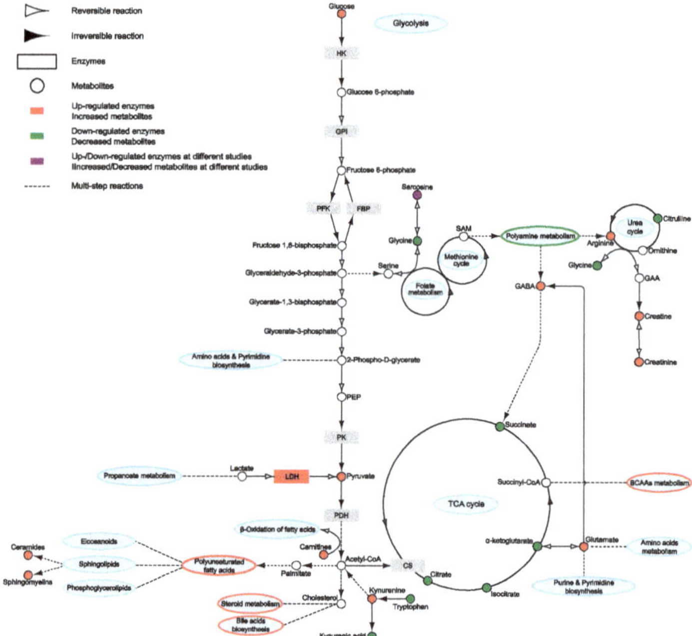


**Figure 1.** Overview of main metabolic changes described in metabolic-related studies of human biofluids applied to PCa biomarker discovery. BCAA: Branched-chain amino acids; CS: Citrate synthase; FBP1: Fructose-bisphosphatase; GAA: Guanidinoacetate; GABA: Gamma-aminobutyric acid; GPI: Glucose-6-phosphate isomerase; HK2: Hexokinase 2; LDH: Lactate dehydrogenase; PDH: Pyruvate dehydrogenase; PEP: Phosphoenolpyruvate; PFK: Phosphofructokinase; PK: Pyruvate kinase; SAM: S-Adenosyl methionine.

Most of the studies included in this review were based on the analysis of blood or urine samples, probably due to their easy accessibility and non-invasiveness. NMR and MS are the two most commonly used analytical platforms in these studies, though other analytical techniques have also been applied to the identification of PCa-related metabolic changes [58,93–95]. Although a significant number of studies focused on the identification of biomarkers for PCa diagnosis, some of them also explored the potential of metabolic biomarkers for patient prognosis and PCa risk evaluation.

Overall, these studies have revealed that alterations in TCA cycle, polyamines, glycolysis, one-carbon metabolism, nucleotide synthesis, amino acid, fatty acid, and lipid metabolism are associated with PCa onset and progression. Figure 1 illustrates the main alterations, in terms of metabolic pathways and metabolites, associated with PCa based on current literature.

The results of the different studies provide compelling evidence of the potential of metabolomics strategies for identifying new PCa biomarkers in biofluids that could be of interest from a clinical perspective. The potential of this approach for routine clinical diagnostics is significant since only minimal biological preparation is necessary. Despite the advances achieved in the field of PCa biomarker discovery, intense efforts are still required before metabolite profiling can be implemented in the clinic. So far, the variability in the metabolic alterations reported precludes consistent, universal signatures to be established, showing that a long path is still to be thread toward the full validation and clinical approval of putative new metabolic biomarkers. In this context, it is worth noting that although most of the reviewed studies included the internal validation of the statistical models developed during the study, either for PCa diagnosis or prognosis, a limited number of them included the assessment of the clinical utility of these findings using an external validation cohort of patients. Thus, future studies should include larger sample cohorts from adequately defined and matched groups of samples. In addition, statistical validation of multivariate models would benefit from full external validation. Finally, increased knowledge on the biological significance of potential PCa biomarkers should be assessed through the integration of metabolomics with other biochemical/biological approaches.

**Author Contributions:** Conceptualization, N.G.-C., A.P.-L. and L.P.-C.; Writing—Original Draft Preparation, N.G.-C. with input from L.P.-C.; Writing—Figure 1, A.R.-B.; Writing—Table 1, A.A.-V.; Clinical aspects and interpretation of literature data, J.A.L.-G.; Writing—Review and Editing, A.P.-L. and L.P.-C. All authors critically commented on and approved the final submitted version of the paper.

**Funding:** This research was funded by the Ministerio de Economía y Competitividad grant number [SAF2017-89229-R] and the Conselleria de Educación, Investigación, Cultura y Deporte grant number [GVA, PROMETEO/2016/103].

**Conflicts of Interest:** The authors declare no conflict of interest.

## Abbreviations

The following abbreviations are used in this manuscript:

| | |
|---|---|
| $^1$H-NMR | Proton nuclear magnetic resonance spectroscopy |
| 2D-DIGE-MS | Two dimensional–difference gel electrophoresis–mass spectrometry |
| AUC | Area under the curve |
| BCAA | Branched-chain amino acids |
| BPH | Benign prostatic hyperplasia |
| CRPC | Castration-resistant prostate cancer |
| CS | Citrate synthase |
| DRE | Digital rectal examination |
| ELISA | Enzyme-linked immunosorbent assay |
| EPS | Expressed prostatic secretions |
| EV | Extracellular vesicles |

| | |
|---|---|
| $^1$H-NMR | Proton nuclear magnetic resonance spectroscopy |
| FBP | Fructose-bisphosphatase |
| FIA-MS/MS | Flow injection analysis–tandem mass spectrometry |
| FPLC-MS | Fast ultra-high-performance liquid chromatography–mass spectrometry |
| GAA | Guanidinoacetate |
| GABA | Gamma-aminobutyric acid |
| GPI | Glucose-6-phosphate isomerase |
| GS | Gleason Score |
| GC-MS | Gas chromatography–mass spectrometry |
| GC-QqQ-MS | Gas chromatography–triple quadrupole–mass spectrometry |
| HG | High-grade (GS $\geq$ 8) |
| HK2 | Hexokinase 2 |
| HPLC-ESI-QTOF-MS | High performance liquid chromatography–electrospray ionization–quadrupole time of flight–mass spectrometry |
| HPLC-TOF-MS | High performance liquid chromatography–time of flight–mass spectrometry |
| HV | Healthy Volunteers |
| iTRAQ | Isobaric tag for relative and absolute quantification |
| LC-MS | Liquid chromatography–mass spectrometry |
| LC-MS/MS | Liquid chromatography–tandem mass spectrometry |
| LDH | Lactate dehydrogenase |
| LG | Low-grade (GS $\leq$ 7) |
| MALDI-TOF-MS | Matrix-assisted laser desorption ionization–time of flight–mass spectrometry |
| MS | Mass spectroscopy |
| NMR | Nuclear magnetic resonance |
| QqQ-MS: | Triple quadrupole–mass spectrometry |
| PCa | Prostate cancer |
| PDH | Pyruvate dehydrogenase |
| PEP | Phosphoenolpyruvate |
| PFK | Phosphofructokinase |
| PK | Pyruvate kinase |
| PM | Prostatic massage |
| PSA | Prostate specific antigen |
| SAM | S-Adenosyl methionine |
| T | Stage |
| TCA | Tricarboxylic acid |
| TMAO | Trimethylamine N-oxide |
| TRUS | Trans-rectal ultrasound |
| UHPLC-MS | Ultra-high-performance liquid chromatography–mass spectrometry |
| UPLC-MS/MS | Ultra performance liquid chromatography–tandem mass spectrometry |

## References


1. Bray, F.; Ferlay, J.; Soerjomataram, I.; Siegel, R.L.; Torre, L.A.; Jemal, A. Global cancer statistics 2018: GLOBOCAN estimates of incidence and mortality worldwide for 36 cancers in 185 countries. *CA Cancer J. Clin.* **2018**, *68*, 394–424. [CrossRef] [PubMed]
2. Schoenfield, L.; Jones, J.S.; Zippe, C.D.; Reuther, A.M.; Klein, E.; Zhou, M.; Magi-Galluzzi, C. The incidence of high-grade prostatic intraepithelial neoplasia and atypical glands suspicious for carcinoma on first-time saturation needle biopsy, and the subsequent risk of cancer. *BJU Int.* **2007**, *99*, 770–774. [CrossRef] [PubMed]
3. Offermann, A.; Hohensteiner, S.; Kuempers, C.; Ribbat-Idel, J.; Schneider, F.; Becker, F.; Hupe, M.C.; Duensing, S.; Merseburger, A.S.; Kirfel, J.; et al. Prognostic Value of the New Prostate Cancer International Society of Urological Pathology Grade Groups. *Front. Med.* **2017**, *4*, 157. [CrossRef] [PubMed]

4.  Chistiakov, D.A.; Myasoedova, V.A.; Grechko, A.V.; Melnichenko, A.A.; Orekhov, A.N. New biomarkers for diagnosis and prognosis of localized prostate cancer. *Semin. Cancer Biol.* **2018**, *52*, 9–16. [CrossRef] [PubMed]

5.  Gordetsky, J.; Epstein, J. Grading of prostatic adenocarcinoma: Current state and prognostic implications. *Diagn. Pathol.* **2016**, *11*, 25. [CrossRef] [PubMed]

6.  Foley, R.W.; Maweni, R.M.; Gorman, L.; Murphy, K.; Lundon, D.J.; Durkan, G.; Power, R.; O'Brien, F.; O'Malley, K.J.; Galvin, D.J.; et al. European Randomised Study of Screening for Prostate Cancer (ERSPC) risk calculators significantly outperform the Prostate Cancer Prevention Trial (PCPT) 2.0 in the prediction of prostate cancer: A multi-institutional study. *BJU Int.* **2016**, *118*, 706–713. [CrossRef]

7.  Nam, R.K.; Satkunasivam, R.; Chin, J.L.; Izawa, J.; Trachtenberg, J.; Rendon, R.; Bell, D.; Singal, R.; Sherman, C.; Sugar, L.; et al. Next-generation prostate cancer risk calculator for primary care physicians. *Can. Urol. Assoc. J.* **2017**, *12*, E64–E70. [CrossRef]

8.  Loeb, S.; Partin, A.W. Review of the Literature: PCA3 for Prostate Cancer Risk Assessment and Prognostication. *Rev. Urol.* **2011**, *13*, 191–195. [CrossRef]

9.  Sanhueza, C.; Kohli, M. Clinical and Novel Biomarkers in the Management of Prostate Cancer. *Curr. Treat. Options Oncol.* **2018**, *19*. [CrossRef]

10. Biomarkers PCA3 and TMPRSS2-ERG: Better together: Prostate cancer. *Nat. Rev. Urol.* **2014**, *11*, 129. [CrossRef]

11. Perner, S.; Mosquera, J.-M.; Demichelis, F.; Hofer, M.D.; Paris, P.L.; Simko, J.; Collins, C.; Bismar, T.A.; Chinnaiyan, A.M.; De Marzo, A.M.; et al. TMPRSS2-ERG Fusion Prostate Cancer: An Early Molecular Event Associated with Invasion. *Am. J. Surg. Pathol.* **2007**, *31*, 882–888. [CrossRef] [PubMed]

12. Barbieri, C.E.; Demichelis, F.; Rubin, M.A. Molecular genetics of prostate cancer: Emerging appreciation of genetic complexity. *Histopathology* **2012**, *60*, 187–198. [CrossRef] [PubMed]

13. Tomlins, S.A. Recurrent Fusion of TMPRSS2 and ETS Transcription Factor Genes in Prostate. *Cancer Sci.* **2005**, *310*, 644–648. [CrossRef] [PubMed]

14. Ferro, M.; Buonerba, C.; Terracciano, D.; Lucarelli, G.; Cosimato, V.; Bottero, D.; Deliu, V.M.; Ditonno, P.; Perdonà, S.; Autorino, R.; et al. Biomarkers in localized prostate cancer. *Future Oncol.* **2016**, *12*, 399–411. [CrossRef] [PubMed]

15. Hendriks, R.J.; van Oort, I.M.; Schalken, J.A. Blood-based and urinary prostate cancer biomarkers: A review and comparison of novel biomarkers for detection and treatment decisions. *Prostate Cancer Prostatic Dis.* **2017**, *20*, 12–19. [CrossRef] [PubMed]

16. Khan, A.; Choi, S.A.; Na, J.; Pamungkas, A.D.; Jung, K.J.; Jee, S.H.; Park, Y.H. Non-invasive Serum Metabolomic Profiling Reveals Elevated Kynurenine Pathway's Metabolites in Humans with Prostate Cancer. *J. Proteome Res.* **2019**. [CrossRef] [PubMed]

17. Andersen, M.K.; Rise, K.; Giskeødegård, G.F.; Richardsen, E.; Bertilsson, H.; Størkersen, Ø.; Bathen, T.F.; Rye, M.; Tessem, M.-B. Integrative metabolic and transcriptomic profiling of prostate cancer tissue containing reactive stroma. *Sci. Rep.* **2018**, *8*, 14269. [CrossRef] [PubMed]

18. Fujita, K.; Nonomura, N. Urinary biomarkers of prostate cancer. *Int. J. Urol.* **2018**, *25*, 770–779. [CrossRef] [PubMed]

19. Kumar, D.; Gupta, A.; Nath, K. NMR-based metabolomics of prostate cancer: A protagonist in clinical diagnostics. *Expert Rev. Mol. Diagn.* **2016**, *16*, 651–661. [CrossRef] [PubMed]

20. Holmes, E.; Wilson, I.D.; Nicholson, J.K. Metabolic phenotyping in health and disease. *Cell* **2008**, *134*, 714–717. [CrossRef] [PubMed]

21. Warburg, O. On the origin of cancer cells. *Science* **1956**, *123*, 309–314. [CrossRef] [PubMed]

22. Vander Heiden, M.G.; DeBerardinis, R.J. Understanding the Intersections between Metabolism and Cancer Biology. *Cell* **2017**, *168*, 657–669. [CrossRef] [PubMed]

23. Hanahan, D.; Weinberg, R.A. Hallmarks of cancer: The next generation. *Cell* **2011**, *144*, 646–674. [CrossRef] [PubMed]

24. Pavlova, N.N.; Thompson, C.B. The Emerging Hallmarks of Cancer Metabolism. *Cell Metab.* **2016**, *23*, 27–47. [CrossRef]

25. Levine, A.J.; Puzio-Kuter, A.M. The control of the metabolic switch in cancers by oncogenes and tumor suppressor genes. *Science* **2010**, *330*, 1340–1344. [CrossRef] [PubMed]

26. Ward, P.S.; Patel, J.; Wise, D.R.; Abdel-Wahab, O.; Bennett, B.D.; Coller, H.A.; Cross, J.R.; Fantin, V.R.; Hedvat, C.V.; Perl, A.E.; et al. The common feature of leukemia-associated IDH1 and IDH2 mutations is a neomorphic enzyme activity converting alpha-ketoglutarate to 2-hydroxyglutarate. *Cancer Cell* **2010**, *17*, 225–234. [CrossRef] [PubMed]

27. Hipp, S.J.; Steffen-Smith, E.A.; Patronas, N.; Herscovitch, P.; Solomon, J.M.; Bent, R.S.; Steinberg, S.M.; Warren, K.E. Molecular imaging of pediatric brain tumors: Comparison of tumor metabolism using 18F-FDG-PET and MRSI. *J. Neurooncol.* **2012**, *109*, 521–527. [CrossRef]

28. Zhan, H.; Ciano, K.; Dong, K.; Zucker, S. Targeting glutamine metabolism in myeloproliferative neoplasms. *Blood Cells Mol. Dis.* **2015**, *55*, 241–247. [CrossRef]

29. Sutinen, E.; Nurmi, M.; Roivainen, A.; Varpula, M.; Tolvanen, T.; Lehikoinen, P.; Minn, H. Kinetics of [(11)C]choline uptake in prostate cancer: A PET study. *Eur. J. Nuclear Med. Mol. Imaging* **2004**, *31*, 317–324. [CrossRef]

30. Srivastava, A.; Creek, D.J. Discovery and Validation of Clinical Biomarkers of Cancer: A Review Combining Metabolomics and Proteomics. *Proteomics* **2018**, 1700448. [CrossRef]

31. Zhang, A.; Sun, H.; Yan, G.; Wang, P.; Wang, X. Metabolomics for Biomarker Discovery: Moving to the Clinic. *Biomed. Res. Int.* **2015**, *2015*, 354671. [CrossRef] [PubMed]

32. Mirnaghi, F.S.; Caudy, A.A. Challenges of analyzing different classes of metabolites by a single analytical method. *Bioanalysis* **2014**, *6*, 3393–3416. [CrossRef] [PubMed]

33. Wolfender, J.-L.; Marti, G.; Thomas, A.; Bertrand, S. Current approaches and challenges for the metabolite profiling of complex natural extracts. *J. Chromatogr. A* **2015**, *1382*, 136–164. [CrossRef] [PubMed]

34. Alonso, A.; Marsal, S.; Julià, A. Analytical methods in untargeted metabolomics: State of the art in 2015. *Front. Bioeng. Biotechnol.* **2015**, *3*, 23. [CrossRef] [PubMed]

35. Bingol, K.; Brüschweiler, R. Two elephants in the room: New hybrid nuclear magnetic resonance and mass spectrometry approaches for metabolomics. *Curr. Opin. Clin. Nutr. Metab. Care* **2015**, *18*, 471–477. [CrossRef] [PubMed]

36. Fuhrer, T.; Zamboni, N. High-throughput discovery metabolomics. *Curr. Opin. Biotechnol.* **2015**, *31*, 73–78. [CrossRef] [PubMed]

37. Clos-Garcia, M.; Loizaga-Iriarte, A.; Zuñiga-Garcia, P.; Sánchez-Mosquera, P.; Rosa Cortazar, A.; González, E.; Torrano, V.; Alonso, C.; Pérez-Cormenzana, M.; Ugalde-Olano, A.; et al. Metabolic alterations in urine extracellular vesicles are associated to prostate cancer pathogenesis and progression. *J. Extracell. Vesicles* **2018**, *7*, 1470442. [CrossRef]

38. Liang, Q.; Liu, H.; Xie, L.; Li, X.; Zhang, A.-H. High-throughput metabolomics enables biomarker discovery in prostate cancer. *RSC Adv.* **2017**, *7*, 2587–2593. [CrossRef]

39. Gkotsos, G.; Virgiliou, C.; Lagoudaki, I.; Sardeli, C.; Raikos, N.; Theodoridis, G.; Dimitriadis, G. The Role of Sarcosine, Uracil, and Kynurenic Acid Metabolism in Urine for Diagnosis and Progression Monitoring of Prostate Cancer. *Metabolites* **2017**, *7*, 9. [CrossRef]

40. Struck-Lewicka, W.; Kordalewska, M.; Bujak, R.; Yumba Mpanga, A.; Markuszewski, M.; Jacyna, J.; Matuszewski, M.; Kaliszan, R.; Markuszewski, M.J. Urine metabolic fingerprinting using LC–MS and GC–MS reveals metabolite changes in prostate cancer: A pilot study. *J. Pharm. Biomed. Anal.* **2015**, *111*, 351–361. [CrossRef]

41. Fernández-Peralbo, M.A.; Gómez-Gómez, E.; Calderón-Santiago, M.; Carrasco-Valiente, J.; Ruiz-García, J.; Requena-Tapia, M.J.; Luque de Castro, M.D.; Priego-Capote, F. Prostate Cancer Patients–Negative Biopsy Controls Discrimination by Untargeted Metabolomics Analysis of Urine by LC-QTOF: Upstream Information on Other Omics. *Sci. Rep.* **2016**, *6*. [CrossRef] [PubMed]

42. Puhka, M.; Takatalo, M.; Nordberg, M.-E.; Valkonen, S.; Nandania, J.; Aatonen, M.; Yliperttula, M.; Laitinen, S.; Velagapudi, V.; Mirtti, T.; et al. Metabolomic Profiling of Extracellular Vesicles and Alternative Normalization Methods Reveal Enriched Metabolites and Strategies to Study Prostate Cancer-Related Changes. *Theranostics* **2017**, *7*, 3824–3841. [CrossRef] [PubMed]

43. Fujita, K.; Kume, H.; Matsuzaki, K.; Kawashima, A.; Ujike, T.; Nagahara, A.; Uemura, M.; Miyagawa, Y.; Tomonaga, T.; Nonomura, N. Proteomic analysis of urinary extracellular vesicles from high Gleason score prostate cancer. *Sci. Rep.* **2017**, *7*, 42961. [CrossRef] [PubMed]

44. Pérez-Rambla, C.; Puchades-Carrasco, L.; García-Flores, M.; Rubio-Briones, J.; López-Guerrero, J.A.; Pineda-Lucena, A. Non-invasive urinary metabolomic profiling discriminates prostate cancer from benign prostatic hyperplasia. *Metabolomics* **2017**, *13*. [CrossRef]

45. Davalieva, K.; Kostovska, I.M.; Kiprijanovska, S.; Markoska, K.; Kubelka-Sabit, K.; Filipovski, V.; Stavridis, S.; Stankov, O.; Komina, S.; Petrusevska, G.; et al. Proteomics analysis of malignant and benign prostate tissue by 2D DIGE/MS reveals new insights into proteins involved in prostate cancer: Proteomics Analysis of Prostate Cancer. *Prostate* **2015**, *75*, 1586–1600. [CrossRef] [PubMed]

46. Kumar, D.; Gupta, A.; Mandhani, A.; Sankhwar, S.N. NMR spectroscopy of filtered serum of prostate cancer: A new frontier in metabolomics: Metabolomics of Prostate Cancer. *Prostate* **2016**, *76*, 1106–1119. [CrossRef] [PubMed]

47. Kumar, D.; Gupta, A.; Mandhani, A.; Sankhwar, S.N. Metabolomics-Derived Prostate Cancer Biomarkers: Fact or Fiction? *J. Proteome Res.* **2015**, *14*, 1455–1464. [CrossRef] [PubMed]

48. Heger, Z.; Michalek, P.; Guran, R.; Cernei, N.; Duskova, K.; Vesely, S.; Anyz, J.; Stepankova, O.; Zitka, O.; Adam, V.; et al. Differences in urinary proteins related to surgical margin status after radical prostatectomy. *Oncol. Rep.* **2015**, *34*, 3247–3255. [CrossRef] [PubMed]

49. Giskeødegård, G.F.; Hansen, A.F.; Bertilsson, H.; Gonzalez, S.V.; Kristiansen, K.A.; Bruheim, P.; Mjøs, S.A.; Angelsen, A.; Bathen, T.F.; Tessem, M.-B. Metabolic markers in blood can separate prostate cancer from benign prostatic hyperplasia. *Br. J. Cancer* **2015**, *113*, 1712–1719. [CrossRef] [PubMed]

50. Zhao, Y.; Lv, H.; Qiu, S.; Gao, L.; Ai, H. Plasma metabolic profiling and novel metabolite biomarkers for diagnosing prostate cancer. *RSC Adv.* **2017**, *7*, 30060–30069. [CrossRef]

51. Lin, H.-M.; Mahon, K.L.; Weir, J.M.; Mundra, P.A.; Spielman, C.; Briscoe, K.; Gurney, H.; Mallesara, G.; Marx, G.; Stockler, M.R.; et al. A distinct plasma lipid signature associated with poor prognosis in castration-resistant prostate cancer: Prognostic lipid signature in metastatic prostate cancer. *Int. J. Cancer* **2017**, *141*, 2112–2120. [CrossRef] [PubMed]

52. Mondul, A.M.; Moore, S.C.; Weinstein, S.J.; Karoly, E.D.; Sampson, J.N.; Albanes, D. Metabolomic analysis of prostate cancer risk in a prospective cohort: The alpha-tocopherol, beta-carotene cancer prevention (ATBC) study: Serum Metabolomics Profiling of Prostate Cancer Risk. *Int. J. Cancer* **2015**, *137*, 2124–2132. [CrossRef] [PubMed]

53. Kühn, T.; Floegel, A.; Sookthai, D.; Johnson, T.; Rolle-Kampczyk, U.; Otto, W.; von Bergen, M.; Boeing, H.; Kaaks, R. Higher plasma levels of lysophosphatidylcholine 18:0 are related to a lower risk of common cancers in a prospective metabolomics study. *BMC Med.* **2016**, *14*. [CrossRef] [PubMed]

54. Schmidt, J.A.; Fensom, G.K.; Rinaldi, S.; Scalbert, A.; Appleby, P.N.; Achaintre, D.; Gicquiau, A.; Gunter, M.J.; Ferrari, P.; Kaaks, R.; et al. Pre-diagnostic metabolite concentrations and prostate cancer risk in 1077 cases and 1077 matched controls in the European Prospective Investigation into Cancer and Nutrition. *BMC Med.* **2017**, *15*. [CrossRef] [PubMed]

55. Huang, J.; Mondul, A.M.; Weinstein, S.J.; Karoly, E.D.; Sampson, J.N.; Albanes, D. Prospective serum metabolomic profile of prostate cancer by size and extent of primary tumor. *Oncotarget* **2017**, *8*. [CrossRef]

56. Andras, I.; Crisan, N.; Vesa, S.; Rahota, R.; Romanciuc, F.; Lazar, A.; Socaciu, C.; Matei, D.-V.; de Cobelli, O.; Bocsan, I.-S.; et al. Serum metabolomics can predict the outcome of first systematic transrectal prostate biopsy in patients with PSA <10 ng/mL. *Future Oncol.* **2017**, *13*, 1793–1800. [CrossRef]

57. Kline, E.E.; Treat, E.G.; Averna, T.A.; Davis, M.S.; Smith, A.Y.; Sillerud, L.O. Citrate Concentrations in Human Seminal Fluid and Expressed Prostatic Fluid Determined via 1H Nuclear Magnetic Resonance Spectroscopy Outperform Prostate Specific Antigen in Prostate Cancer Detection. *J. Urol.* **2006**, *176*, 2274–2279. [CrossRef]

58. Etheridge, T.; Straus, J.; Ritter, M.A.; Jarrard, D.F.; Huang, W. Semen AMACR protein as a novel method for detecting prostate cancer. *Urol. Oncol.* **2018**. [CrossRef]

59. Serkova, N.J.; Gamito, E.J.; Jones, R.H.; O'Donnell, C.; Brown, J.L.; Green, S.; Sullivan, H.; Hedlund, T.; Crawford, E.D. The metabolites citrate, myo-inositol, and spermine are potential age-independent markers of prostate cancer in human expressed prostatic secretions. *Prostate* **2008**, *68*, 620–628. [CrossRef]

60. Averna, T.; Kline, E.; Smith, A.; Sillerud, L. A decrease in 1h nuclear magnetic resonance spectroscopically determined citrate in human seminal fluid accompanies the development of prostate adenocarcinoma. *J. Urol.* **2005**, *173*, 433–438. [CrossRef]

61. Roberts, M.J.; Richards, R.S.; Chow, C.W.K.; Buck, M.; Yaxley, J.; Lavin, M.F.; Schirra, H.J.; Gardiner, R.A. Seminal plasma enables selection and monitoring of active surveillance candidates using nuclear magnetic resonance-based metabolomics: A preliminary investigation. *Prostate Int.* **2017**, *5*, 149–157. [CrossRef] [PubMed]

62. Eidelman, E.; Twum-Ampofo, J.; Ansari, J.; Siddiqui, M.M. The Metabolic Phenotype of Prostate Cancer. *Front. Oncol.* **2017**, *7*. [CrossRef] [PubMed]

63. Lima, A.; Araújo, A.; Pinto, J.; Jerónimo, C.; Henrique, R.; Bastos, M.; Carvalho, M.; Guedes de Pinho, P. GC-MS-Based Endometabolome Analysis Differentiates Prostate Cancer from Normal Prostate Cells. *Metabolites* **2018**, *8*, 23. [CrossRef] [PubMed]

64. Giunchi, F.; Fiorentino, M.; Loda, M. The Metabolic Landscape of Prostate Cancer. *Eur. Urol. Oncol.* **2018**, *2*, 28–36. [CrossRef]

65. Sadeghi, R.N.; Karami-Tehrani, F.; Salami, S. Targeting prostate cancer cell metabolism: Impact of hexokinase and CPT-1 enzymes. *Tumour Biol.* **2015**, *36*, 2893–2905. [CrossRef] [PubMed]

66. Twum-Ampofo, J.; Fu, D.-X.; Passaniti, A.; Hussain, A.; Siddiqui, M.M. Metabolic targets for potential prostate cancer therapeutics. *Curr. Opin. Oncol.* **2016**, *28*, 241–247. [CrossRef] [PubMed]

67. Awwad, H.M.; Geisel, J.; Obeid, R. The role of choline in prostate cancer. *Clin. Biochem.* **2012**, *45*, 1548–1553. [CrossRef] [PubMed]

68. Sreekumar, A.; Poisson, L.M.; Rajendiran, T.M.; Khan, A.P.; Cao, Q.; Yu, J.; Laxman, B.; Mehra, R.; Lonigro, R.J.; Li, Y.; et al. Metabolomic profiles delineate potential role for sarcosine in prostate cancer progression. *Nature* **2009**, *457*, 910–914. [CrossRef] [PubMed]

69. Giskeødegård, G.F.; Bertilsson, H.; Selnæs, K.M.; Wright, A.J.; Bathen, T.F.; Viset, T.; Halgunset, J.; Angelsen, A.; Gribbestad, I.S.; Tessem, M.-B. Spermine and Citrate as Metabolic Biomarkers for Assessing Prostate Cancer Aggressiveness. *PLoS ONE* **2013**, *8*, e62375. [CrossRef]

70. Zabala-Letona, A.; Arruabarrena-Aristorena, A.; Martín-Martín, N.; Fernandez-Ruiz, S.; Sutherland, J.D.; Clasquin, M.; Tomas-Cortazar, J.; Jimenez, J.; Torres, I.; Quang, P.; et al. mTORC1-dependent AMD1 regulation sustains polyamine metabolism in prostate cancer. *Nature* **2017**, *547*, 109–113. [CrossRef] [PubMed]

71. Wu, D.; Ni, J.; Beretov, J.; Cozzi, P.; Willcox, M.; Wasinger, V.; Walsh, B.; Graham, P.; Li, Y. Urinary biomarkers in prostate cancer detection and monitoring progression. *Crit. Rev. Oncol. Hematol.* **2017**, *118*, 15–26. [CrossRef] [PubMed]

72. Amobi, A.; Qian, F.; Lugade, A.A.; Odunsi, K. Tryptophan Catabolism and Cancer Immunotherapy Targeting IDO Mediated Immune Suppression. *Adv. Exp. Med. Biol.* **2017**, *1036*, 129–144. [PubMed]

73. Santhanam, S.; Alvarado, D.M.; Ciorba, M.A. Therapeutic targeting of inflammation and tryptophan metabolism in colon and gastrointestinal cancer. *Transl. Res.* **2016**, *167*, 67–79. [CrossRef] [PubMed]

74. Khan, A.P.; Rajendiran, T.M.; Bushra, A.; Asangani, I.A.; Athanikar, J.N.; Yocum, A.K.; Mehra, R.; Siddiqui, J.; Palapattu, G.; Wei, J.T.; et al. The Role of Sarcosine Metabolism in Prostate Cancer Progression. *Neoplasia* **2013**, *15*, 491–501. [CrossRef] [PubMed]

75. Ankerst, D.P.; Liss, M.; Zapata, D.; Hoefler, J.; Thompson, I.M.; Leach, R.J. A case control study of sarcosine as an early prostate cancer detection biomarker. *BMC Urol.* **2015**, *15*, 99. [CrossRef] [PubMed]

76. Dereziński, P.; Klupczynska, A.; Sawicki, W.; Pałka, J.A.; Kokot, Z.J. Amino Acid Profiles of Serum and Urine in Search for Prostate Cancer Biomarkers: A Pilot Study. *Int. J. Med. Sci.* **2017**, *14*, 1–12. [CrossRef] [PubMed]

77. Locasale, J.W. Serine, glycine and one-carbon units: Cancer metabolism in full circle. *Nat. Rev. Cancer* **2013**, *13*, 572–583. [CrossRef] [PubMed]

78. Koslowski, M.; Türeci, O.; Bell, C.; Krause, P.; Lehr, H.-A.; Brunner, J.; Seitz, G.; Nestle, F.O.; Huber, C.; Sahin, U. Multiple splice variants of lactate dehydrogenase C selectively expressed in human cancer. *Cancer Res.* **2002**, *62*, 6750–6755. [PubMed]

79. Kong, L.; Du, W.; Cui, Z.; Wang, L.; Yang, Z.; Zhang, H.; Lin, D. Expression of lactate dehydrogenase C in MDA-MB-231 cells and its role in tumor invasion and migration. *Mol. Med. Rep.* **2016**, *13*, 3533–3538. [CrossRef]

80. Merchant, M.L.; Rood, I.M.; Deegens, J.K.J.; Klein, J.B. Isolation and characterization of urinary extracellular vesicles: Implications for biomarker discovery. *Nat. Rev. Nephrol.* **2017**, *13*, 731–749. [CrossRef]

81. Myers, J.S.; von Lersner, A.K.; Sang, Q.-X.A. Proteomic Upregulation of Fatty Acid Synthase and Fatty Acid Binding Protein 5 and Identification of Cancer- and Race-Specific Pathway Associations in Human Prostate Cancer Tissues. *J. Cancer* **2016**, *7*, 1452–1464. [CrossRef] [PubMed]

82. Pang, J.; Liu, W.-P.; Liu, X.-P.; Li, L.-Y.; Fang, Y.-Q.; Sun, Q.-P.; Liu, S.-J.; Li, M.-T.; Su, Z.-L.; Gao, X. Profiling protein markers associated with lymph node metastasis in prostate cancer by DIGE-based proteomics analysis. *J. Proteome Res.* **2010**, *9*, 216–226. [CrossRef] [PubMed]

83. Wu, H.; Liu, T.; Ma, C.; Xue, R.; Deng, C.; Zeng, H.; Shen, X. GC/MS-based metabolomic approach to validate the role of urinary sarcosine and target biomarkers for human prostate cancer by microwave-assisted derivatization. *Anal. Bioanal. Chem.* **2011**, *401*, 635–646. [CrossRef] [PubMed]

84. Kami, K.; Fujimori, T.; Sato, H.; Sato, M.; Yamamoto, H.; Ohashi, Y.; Sugiyama, N.; Ishihama, Y.; Onozuka, H.; Ochiai, A.; et al. Metabolomic profiling of lung and prostate tumor tissues by capillary electrophoresis time-of-flight mass spectrometry. *Metabolomics* **2013**, *9*, 444–453. [CrossRef] [PubMed]

85. McDunn, J.E.; Li, Z.; Adam, K.-P.; Neri, B.P.; Wolfert, R.L.; Milburn, M.V.; Lotan, Y.; Wheeler, T.M. Metabolomic signatures of aggressive prostate cancer. *Prostate* **2013**, *73*, 1547–1560. [CrossRef] [PubMed]

86. Jiang, Z.; Woda, B.A. Diagnostic utility of alpha-methylacyl CoA racemase (P504S) on prostate needle biopsy. *Adv. Anat. Pathol.* **2004**, *11*, 316–321. [CrossRef]

87. Zhou, M.; Jiang, Z.; Epstein, J.I. Expression and diagnostic utility of alpha-methylacyl-CoA-racemase (P504S) in foamy gland and pseudohyperplastic prostate cancer. *Am. J. Surg. Pathol.* **2003**, *27*, 772–778. [CrossRef]

88. Box, A.; Alshalalfa, M.; Hegazy, S.A.; Donnelly, B.; Bismar, T.A. High alpha-methylacyl-CoA racemase (AMACR) is associated with ERG expression and with adverse clinical outcome in patients with localized prostate cancer. *Tumour Biol.* **2016**, *37*, 12287–12299. [CrossRef]

89. Alinezhad, S.; Väänänen, R.-M.; Ochoa, N.T.; Vertosick, E.A.; Bjartell, A.; Boström, P.J.; Taimen, P.; Pettersson, K. Global expression of AMACR transcripts predicts risk for prostate cancer—A systematic comparison of AMACR protein and mRNA expression in cancerous and noncancerous prostate. *BMC Urol.* **2016**, *16*, 10. [CrossRef]

90. Mroz, A.; Kiedrowski, M.; Lewandowski, Z. α-Methylacyl-CoA racemase (AMACR) in gastric cancer: Correlation with clinicopathologic data and disease-free survival. *Appl. Immunohistochem. Mol. Morphol.* **2013**, *21*, 313–317. [CrossRef]

91. Xu, B.; Cai, Z.; Zeng, Y.; Chen, L.; Du, X.; Huang, A.; Liu, X.; Liu, J. α-Methylacyl-CoA racemase (AMACR) serves as a prognostic biomarker for the early recurrence/metastasis of HCC. *J. Clin. Pathol.* **2014**, *67*, 974–979. [CrossRef] [PubMed]

92. Lee, Y.-E.; He, H.-L.; Lee, S.-W.; Chen, T.-J.; Chang, K.-Y.; Hsing, C.-H.; Li, C.-F. AMACR overexpression as a poor prognostic factor in patients with nasopharyngeal carcinoma. *Tumour Biol.* **2014**, *35*, 7983–7991. [CrossRef] [PubMed]

93. Da Costa, I.A.; Hennenlotter, J.; Stühler, V.; Kühs, U.; Scharpf, M.; Todenhöfer, T.; Stenzl, A.; Bedke, J. Transketolase like 1 (TKTL1) expression alterations in prostate cancer tumorigenesis. *Urol. Oncol.* **2018**, *36*, 472.e21–472.e27. [CrossRef] [PubMed]

94. Kojima, Y.; Yoneyama, T.; Hatakeyama, S.; Mikami, J.; Sato, T.; Mori, K.; Hashimoto, Y.; Koie, T.; Ohyama, C.; Fukuda, M.; et al. Detection of Core2 β-1,6-*N*-Acetylglucosaminyltransferase in Post-Digital Rectal Examination Urine Is a Reliable Indicator for Extracapsular Extension of Prostate Cancer. *PLoS ONE* **2015**, *10*, e0138520. [CrossRef] [PubMed]

95. Sato, T.; Yoneyama, T.; Tobisawa, Y.; Hatakeyama, S.; Yamamoto, H.; Kojima, Y.; Mikami, J.; Mori, K.; Hashimoto, Y.; Koie, T.; et al. Core 2 β-1, 6-N-acetylglucosaminyltransferase-1 expression in prostate biopsy specimen is an indicator of prostate cancer aggressiveness. *Biochem. Biophys. Res. Commun.* **2016**, *470*, 150–156. [CrossRef] [PubMed]

 *metabolites*

*Review*

# Function, Detection and Alteration of Acylcarnitine Metabolism in Hepatocellular Carcinoma

**Shangfu Li [1,2], Dan Gao [1,2,3],* and Yuyang Jiang [1,4]**

[1]  State Key Laboratory of Chemical Oncogenomics, Graduate School at Shenzhen, Tsinghua University, Shenzhen 518055, China; li.shangfu@sz.tsinghua.edu.cn (S.L.); jiangyy@sz.tsinghua.edu.cn (Y.J.)

[2]  National & Local United Engineering Lab for Personalized Anti-tumour Drugs, Graduate School at Shenzhen, Tsinghua University, Shenzhen 518055, China

[3]  Key Laboratory of Metabolomics at Shenzhen, Shenzhen 518055, China

[4]  School of Pharmaceutical Sciences, Tsinghua University, Beijing 100084, China

*  Correspondence: gao.dan@sz.tsinghua.edu.cn; Tel.: +86-755-26036035

Received: 11 January 2019; Accepted: 14 February 2019; Published: 21 February 2019

**Abstract:** Acylcarnitines play an essential role in regulating the balance of intracellular sugar and lipid metabolism. They serve as carriers to transport activated long-chain fatty acids into mitochondria for β-oxidation as a major source of energy for cell activities. The liver is the most important organ for endogenous carnitine synthesis and metabolism. Hepatocellular carcinoma (HCC), a primary malignancy of the live with poor prognosis, may strongly influence the level of acylcarnitines. In this paper, the function, detection and alteration of acylcarnitine metabolism in HCC were briefly reviewed. An overview was provided to introduce the metabolic roles of acylcarnitines involved in fatty acid β-oxidation. Then different analytical platforms and methodologies were also briefly summarised. The relationship between HCC and acylcarnitine metabolism was described. Many of the studies reported that short, medium and long-chain acylcarnitines were altered in HCC patients. These findings presented current evidence in support of acylcarnitines as new candidate biomarkers for studies on the pathogenesis and development of HCC. Finally we discussed the challenges and perspectives of exploiting acylcarnitine metabolism and its related metabolic pathways as a target for HCC diagnosis and prognosis.

**Keywords:** acylcarnitines; hepatocellular carcinoma; metabolite profiling; metabolomics

---

## 1. Introduction

Hepatocellular carcinoma (HCC) is the most common type of primary liver cancer. This intra-abdominal malignant tumours accounted for 90% of all cases of primary liver cancer [1]. HCC ranks as the second leading cause of cancer-related mortality in the world [2]. It has a very poor prognosis of malignant tumours, with prognosis less than 5% [3]. The main pathogenic factors of HCC are viruses, bacteria, alcohol, therapeutic drugs, and harmful substances [4]. Its occurrence is long-term, dynamic, and multi-stage with the complex regulation of multiple genes and factors [5]. Chronic liver damage and inflammation caused by chronic hepatitis B virus or hepatitis C virus (HBV, HCV) infections account for the majority of HCC cases [6]. The persistent inflammatory environment may promote simple hepatic steatosis to fibrosis, cirrhosis (CIR) and, ultimately, HCC [2,7]. Additionally, in the last 20 years, the rising rates of alcoholic liver disease, non-alcoholic fatty liver disease (NAFLD), and non-alcoholic steatohepatitis (NASH) increased the risk of HCC development in patients with viral hepatitis [8]. In fact, these liver metabolic disorders, including type II diabetes, obesity, and metabolic syndrome, are becoming emerging risk factors for the rapidly increasing incidence of HCC [9]. It has been reported that 4% to 27% of patients with NASH and CIR may have HCC [10]. However, the oncogenic mechanisms of these new metabolic risk factors that promote HCC are only beginning to be characterized [11]. In order to improve the early diagnosis of HCC and the prognosis of patients, investigation of the pathogenesis of HCC and exploration of high-sensitivity, high-specificity biomarkers are the hotspots for HCC research. The development of current research techniques provides a great deal of convenience to investigate HCC-related biomarkers [12,13]. In particular, the omics technologies, such as genomics, proteomics, and metabolomics, have greatly accelerated the progress in HCC research with its high-throughput technology advantages [14,15]. The investigations provide many sensitive and specific markers for early and accurate diagnosis of HCC [16].

Since the liver is an important organ of substance and energy metabolism, liver lesions, especially carcinogenesis, can strongly affect its metabolic process [17]. Therefore, quantitative and qualitative analysis of metabolic alteration in HCC samples can monitor the fluctuation of specified metabolic pathways, thus obtaining some important information for the diagnosis and pathogenesis studies of HCC [18,19]. These are the currently booming research scopes of metabolomics in recent years [20,21]. At present, targeted and non-targeted metabolomics studies on HCC have been widely reported [22–24]. However, due to the large variety of metabolites, there is currently no single prospecting technique that can fully cover all metabolites [25]. Generally, only some of the metabolites of interest can be detected by quantitative or qualitative methods or a mix of both. In this article, we do not attempt to summarize the changes of all metabolites in HCC as well, but rather focus on the acylcarnitines, which are a large class of substances closely related to HCC metabolism.

## 2. Function of Acylcarnitines in Cellular Metabolism

Acylcarnitines are esters of L-carnitine and fatty acids (Figure 1). They are a large class of metabolites that are members of the non-protein amino acid family. According to the Human Metabolome Database, there may be more than 1200 fatty acids in the human body [26,27]. Therefore, it is inferred that the number of acylcarnitines that may be formed with these fatty acids is very considerable. Similar to fatty acids, acylcarnitines are also differed by length of the acyl groups, often categorized as short, medium and long-chain acylcarnitines (simply marked as SCACs, MCACs and LCACs). Acylcarnitines are zwitterionic compounds, containing a carboxyl group and a quaternary ammonium group in the molecule (Figure 1).

**Figure 1.** The structure of L-carnitine and acylcarnitines.

The large number and special structure make acylcarnitines play an important role in cell physiological activities and become a key substance for cell metabolism [28]. The main function of acylcarnitines is involved in long-chain fatty acids (LCFAs) β-oxidation (Figure 2). They serve as carriers to transport activated LCFAs into mitochondria for subsequent β-oxidation to provide energy for cell activities [29]. The enzymes that regulate these processes are mainly long-chain acyl-coenzyme A synthetase (LACS), carnitine/acylcarnitine translocase (CACT), carnitine palmitoyl-transferase 1 and 2 (CPT1 and CPT2) [30]. The LCFAs are activated by linking to coenzyme A (CoA) via LACS, forming long-chain acyl-CoAs. The intermediaries are converted into LCACs catalysed by CPT1 which is located on the outer mitochondrial membrane [31]. Under catalysis of CACT, the LCACs are imported through the mitochondrial membranes into the mitochondrial matrix [32]. Then they are converted back to the corresponding long-chain acyl-CoAs in the presence of CPT2 for β-oxidation [33]. The end products, acetyl-CoAs, are converted to acetylcarnitines by carnitine O-acetyltransferase (CrAT). Finally, acetylcarnitines are exported from mitochondrion to cytoplasm by CACT [34]. The activities of the involved enzymes can be evaluated by ratios of LCACs/SCACs. For example, the activity of CPT1 can be estimated by (carnitine C16 + carnitine C18)/carnitine. Similarly, the changes of CPT2 can be estimated by (carnitine C16 + carnitine C18:1)/carnitine C2, long-chain Acyl-CoA dehydrogenase by carnitine C16/carnitine C8, and β-oxidation of even-carbon fatty acids by carnitine C2/carnitine [35].

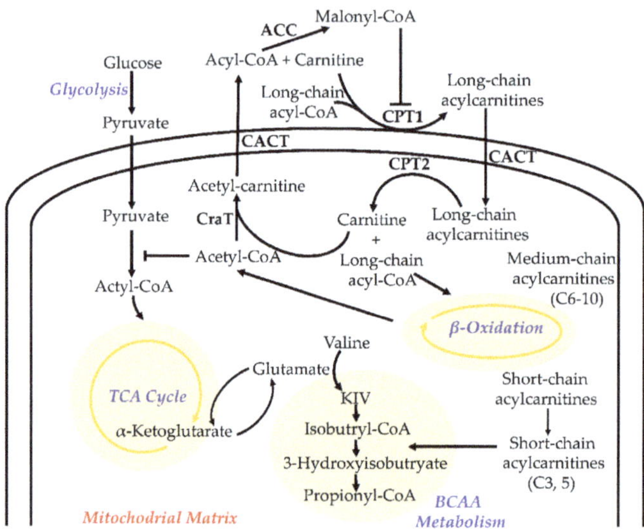


**Figure 2.** Overview of acylcarnitine-related cellular metabolism. For simplicity, not all intermediates and reversible processes are shown. Abbreviations: CPT1, carnitine-palmitoyl-transferase 1; CPT2, carnitine-palmitoyl-transferase 2; CACT, carnitine-acylcarnitine-translocase; CoA, coenzyme A; LACS, long-chain acyl-CoA synthetase; BCAA, branched-chain amino acid.

The metabolism of acylcarnitines is a key factor regulating the balance of intracellular sugar and lipid metabolism [36]. Acylcarnitine metabolism is involved in the metabolism of branched-chain amino acids [37]. They are also involved in maintaining the homeostasis of the mitochondrial acyl-CoA/CoA ratio. When the glucagon/insulin ratio is lowered, they stimulate the activity of pyruvate dehydrogenase to enhance the oxidation of pyruvate and enhance the aerobic oxidation of glucose [38]. Acetylcarnitine can be converted into malonyl-CoA in the cytosol to inhibit the activity of CPT1 and reduce the oxidation of fatty acids, which results in eliminating the adverse reactions caused by the accumulation of acyl-CoA metabolic intermediates in the mitochondria [39]. Acylcarnitines are also involved in other physiological processes such as peroxidation of fatty acids, and production of ketone bodies [37]. Therefore, the metabolism of acylcarnitines is not only related to the transport of fatty acids, but also plays a key role in regulating the balance of intracellular sugar and lipid metabolism (Figure 2) [36].

Acylcarnitines are closely related to many metabolic diseases [40]. Abnormal expression of enzymes involved in the metabolism of acylcarnitines may result in accumulation of acyl-CoA with a specific chain length [41]. These substances, if not removed by conversion to acylcarnitines, may have toxic effects on cells [42,43]. Since the levels of plasma acylcarnitines reflect the composition of the acylcarnitine pool within the cytoplasm, they are considered to be markers indicating the balance between acyl-CoA and acylcarnitine species [44]. Studies have shown that in organic acidemia, the content of acylcarnitines varies with the accumulation of organic acids. Therefore, acylcarnitines are clinically important parameters for organic acidemia diagnosis [45]. Acylcarnitines are also key indicators for screening genetic abnormalities in neonates [46]. In addition, changes of blood acylcarnitines also have significant correlation with type I diabetes and type II diabetes [47]. Mitochondrial fatty acid oxidation (FAO) disorders caused by gene mutations can lead to hereditary carnitine metabolism syndrome [48]. Secondary carnitine deficiency may be triggered by back of nutrition, absorption of gastrointestinal function, carnitine loss from blood and peritoneal dialysis [48].

Significant changes in acylcarnitine metabolism can also be observed in diseases such as coronary artery disease [49], heart failure [50], and dementia [51]. In cancer cells, acylcarnitine metabolism has been considered as a gridlock to finely trigger the metabolic flexibility of cancer cells on the basis of its fundamental role in tuning the switch between the glucose and fatty acid metabolism [34]. Metabolic reprogramming of cancer cells regulates the levels of acylcarnitines with varying chain lengths [52,53]. They intercalate acylcarnitines with other key metabolic pathways, factors and metabolites to create a balance between the production and consumption of energy and the synthesis of metabolic intermediates to meet rapid proliferation requirements [54,55]. For example, in prostate cancer cells (PCCs), carnitine cycle was a primary regulator of adaptive metabolic reprogramming, which was modulated by microRNAs (miRNAs) to deregulate mitochondrial fatty acid oxidation [56]. Results from the urine of kidney cancer patients and mouse models showed that urinary acylcarnitines are increased in a grade-dependent fashion. These compounds are likely emanating from the tumour tissue itself and have both cytotoxicity and immune modulatory properties which could be beneficial to the tumour in terms of growth and survival in situ [57]. Mass spectrometry images analysis found that in a human breast tumour xenograft model, two acylcarnitines, palmitoylcarnitine, and stearoylcarnitine, displayed the high percentage of overlap with hypoxic tumour regions, suggesting blockage of the β-oxidation process of fatty acids inside mitochondria [58]. In view of the importance of acylcarnitines in a variety of diseases, they are likely to be good biomarkers for clinical diagnosis. Therefore, studies on the function of acylcarnitines may help to deepen understanding of the disease mechanism, and it may also promote the development of disease diagnosis and treatment technology.

## 3. Detection of Acylcarnitines in Biological Samples

However, the difficulties encountered in the detection of acylcarnitines limit the study of their functions. The challenge for acylcarnitines detection is mostly attributed to the complex components of biological samples and the structural diversity of acylcarnitines caused by various acyl groups [37,59,60]. (1) The composition of biological samples is very complex. Matrix components will greatly interfere with the detection of acylcarnitines; (2) According to the different acyl groups attached, the acylcarnitines have a very long polarity span, covering the polarity from the polar SCACs to low-polar LCACs. SCACs have strong hydrophilicity due to the presence of quaternary ammonium groups and are difficult to retain on reversed-phase columns; (3) Due to a wide variety of species of fatty acids, the SCACs, MCACs and LCACs formed by the fatty acids constitute a large congener family of members, and the properties of some isomers are very close, leading to difficulty in chromatographic separation; (4) The limited number of commercial acylcarnitine standards affects the accurate identification of the specific structural composition of acyl groups. Due to the existence of these problems, the current detection methods can only focus on a few acylcarnitines that contain commercial standards, and the information of other acylcarnitines is still missed.

Biological samples usually contain macromolecules, such as proteins and nucleic acids, as well as small molecules such as phospholipids, amino acids, sugars, and inorganic salts. Therefore, the matrix effect caused by these ingredients cannot be ignored in the detection of acylcarnitines. Although there was a report that urinary acylcarnitines could be detected directly after the urine samples were subjected to simple centrifugation [61], the strong matrix effect still affected the sensitivity. Therefore, in order to efficiently detect acylcarnitines, appropriate sample preparation methods are necessary. The easiest way to handle the biological samples is the liquid–liquid extraction (LLE) method. Due to its convenient operation and low cost, it has been extensively used for sample preparation [62]. The usual procedure includes protein precipitation, centrifugation, and nitrogen drying [63]. The organic solvents used for deproteinization often are methanol (MeOH) or acetonitrile (ACN). Studies have shown that the choice of organic solvents has a great impact on the recovery of the methods, because acetonitrile itself is not a good solvent for all the acylcarnitine species [64]. Therefore, it is common to use MeOH or a mixture of MeOH and ACN for LLE [65–67]. In addition, using ACN containing 0.3% formic acid could also improve the extraction recovery, which was comparable to those approaches using

MeOH as solvent [66]. The limitation of LLE method is that, it requires to use potentially toxic organic solvents and high volume samples, and its sampling rate is low. In response to these problems, some new methods have been developed for the extraction of acylcarnitines. For example, using a continuous-flow microelectroextraction flow cell, acylcarnitines could be extracted from a large volume urine sample into a micro volume of stagnant acceptor phase under the electric field enhanced extraction [68]. The detection limit of the method was as low as 0.3–2 nM, which was appropriate for the detection of low concentration metabolites.

Solid-phase extraction (SPE) is another widely used method for the extraction of acylcarnitines. Its advantages are low cost, good selectivity, small solvent consumption, small sample amounts and high recovery [69]. Additionally, its disadvantages are long sample preparation times and multistep procedures. Despite these shortcomings, SPE is still extremely applied for the enrichment and isolation of acylcarnitines. For example, after purifying human urine using polymeric and weak cationic exchange cartridges, the matrix effect was significantly reduced as the urinary carnitine was analysed by UPLC-MS/MS [70]. An online SPE with an Oasis mixed-mode cation exchange (MCX) trapping column combined with LC-MS analysis was a rapid sample work-up method for the quantification of acylcarnitines with different polarity. The method required low sample consumption [71]. The sample preparation was more simplistic, and LLOQ was significantly lower than previously reported methods [72,73] after treating samples with semi-automatic microextraction by packed C2 of M1 (C8 + SCX) phase as a sorbent [74]. It was also reported that using a mixed-mode reversed-phase/strong cation-exchange 96-well SPE plate could achieve selective and accurate quantitation of C5 acylcarnitine in patients with isovaleric acidemia (IVA) by UHPLC–MS/MS [75]. With the same SPE plate isolation, 65 acylcarnitines were separated [76]. Although MCX SPE cartridges have been widely used in the extraction of acylcarnitines, it must be noted that the sulfo group on the packing could catalyse the carboxylic acid groups of about 40% dicarboxylic acylcarnitines reacting with MeOH in the elution solvent to form methylation products [77,78]. Therefore, Li et al. suggested that using ACN (containing 5% NH$_4$OH) instead of MeOH as elution may avoid the methylation problems [79].

The as prepared samples can be directly used for analysis [74,80], or analysed after derivatization. The derivatization procedure may introduce a chromophoric group to the targeted analytes, then it could be possibly detected by fluorescence or ultraviolet detector [81,82]. The derivatization procedure was also used to label water-soluble acylcarnitines to improve their retention on reversed-phase columns [83,84]. For example, the strategies have been used to detect L-carnitine and its chiral isomers D-carnitine [60]. And just recently, an isotope-labelling strategy with 3-nitrophenylhydrazine as derivatization reagents was employed for LC-MS-based quantitation of acylcarnitines in dried blood spots with good linearity, high precision and high accuracy [85]. One important aspect to note is that the application of derivatization methods requires systematically methodological evaluation of the chemical stabilities of acylcarnitines under various reaction conditions before they are used in the practical sample analysis. Since it has been confirmed that anhydrous n-butanol/HCl-based method, which was based on the acid-catalysed esterification and the most popular derivatization approach for acylcarnitines analysis at early stages [86], may cause part of the acylcarnitines hydrolysed and result in inaccuracies measurement from the hydrolysis of acylcarnitines [87].

Due to the complex composition of biological samples, proper separation means are beneficial for acylcarnitine analysis to obtain the maximum detection efficiency. As in the earlier study, chromatography separation coupled with fluorescence or UV detectors were commonly used methods [60,82]. There are also a small number of studies used GC-MS [88]. In recent years, LC-MS has become the most popular techniques for acylcarnitine detection [89–93]. Reversed-phase liquid chromatography (RPLC), hydrophilic interaction liquid chromatography (HILIC), ion chromatography, and capillary electrophoresis are different options for acylcarnitine separation. Due to their excellent separation ability and high sensitivity, dozens of acylcarnitines could be analysed simultaneously [94–97]. By using highly-selective scanning modes, such as selected reaction monitoring (SRM), multiple reaction monitoring (MRM) and parallel reaction monitoring (PRM), up to hundreds of acylcarnitines could be identified in plasma, urine and tissue samples [37,59,79]. These results provide significant reference value to annotation of acylcarnitines in biological samples. However, in these researches, especially for qualitative analysis, the identification of acylcarnitines relied primarily on accurate values of mass to charge ratio ($m/z$) and corresponding characteristic fragment ions obtained from high resolution mass spectrometry. Nevertheless, the majority of detected acylcarnitine structures cannot be actually verified because of not enough commercial standards for the diverse acylcarnitines. In some cases, such as the discovery of potential acylcarnitine biomarkers for clinical application, standards are still needed to be synthesized for their structures confirmation [79]. Although some of acylcarnitines can be obtained by conjugating the corresponding carboxyl compounds with L-carnitine, only a small fraction of currently known acylcarnitines can be synthesized. This is because the carboxyl compounds are also diverse and lack of sufficient standards. The un-authenticated acylcarnitines in biological samples limited the evaluation of these existing methods [59]. Therefore, the development of new approaches to high efficient and accurately identify acylcarnitines is still in urgent need.

## 4. Alteration of Acylcarnitine Metabolism in HCC

The liver is the most active organ for endogenous carnitine synthesis and metabolism [91,98]. Therefore, suffering from diseases may cause the liver to have a strong effect on the levels of acylcarnitines [99]. At different stages of liver disease, hepatocytes are stimulated by different risk factors, and the demand for glucose and lipids is not the same [100,101]. As a result, the disorder of acylcarnitine metabolism is also related to the stage of liver disease. The general rule is that as the condition worsens, the metabolic disturbance of acylcarnitines becomes more pronounced. Some research suggested that in non-alcoholic fatty liver disease (NAFLD) patients, the level of butyrylcarnitine was significantly elevated. When the disease progressed to more severe NASH, there was a significant increase in free carnitine, propionylcarnitine, butyrylcarnitine, and 2-methylbutyrylcarnitine [102]. In patients with liver fibrosis and CIR, both C16:1-acylcarnitine and C18:1-acylcarnitine have an increasing tendency, indicating reduced β-oxidation levels of these two fatty acids [103]. The changes of acylcarnitines caused by different pathogenic factors are also different. For example, serum LCAC levels in patients with CIR caused by viral hepatitis (HBV and HCV) showed an increasing trend, but in patients with CIR caused by alcohol consumption, both LCACs and SCACs were upward trend [35,104].

There are also many reports on the changes of acylcarnitines in HCC. Compared with human HCC clinical samples, cell and animal models are relatively easy to obtain and can perform knockout, silencing, high expression and other operations on genes of interest and, thus, are often used to study the disease mechanism of HCC. For example, Cheng et al. established SK-Hep1 cells underexpressing G6PD (Sk-Gi) to study the effect of a pharmacological dose of dehydroepiandrosterone on cellular metabolism. Compared with control cells (Sk-Sc), consumption of carnitine and its acyl derivatives was observed, suggesting the decline in fatty acid catabolism and mitochondrial malfunction and reduction in cellular ATP content [105]. Levels of acylcarnitines also enhanced the self-renewal of HCC cells. It was reported that Dih10 cells with CPT2 knockdown led to their resistance to lipotoxicity induced by the lipid-rich cellular environment via inhibiting the Src-mediated JNK activation. Simultaneously, by stimulating STAT 3, oleoylcarnitine may promote sphere formation in Dih10 cells [54]. In hepatitis B surface antigen (HBsAg) transgenic mouse model that mimics HBV carriers with and without AFB1 treatment, acylcarnitine concentration increased with increase in tumour growth in all HCC mouse models, indicating elevated metabolic activity and increased cell turnover. The results were consistent with a pilot study using human serum from HCC patients [106]. In addition to endogenous acylcarnitines, externally added carnitine may also affect HCC progression. It was reported that L-carnitine increased hepatic expression of genes related to long-chain fatty acid transport, mitochondrial β-oxidation, and antioxidant enzymes following suppression of hepatic oxidative stress markers and inflammatory cytokines in NASH, and mice treated with L-carnitine developed fewer liver tumours [107]. By using a non-targeted metabolomics method, Xu et al. analysed the diethylnitrosamine-induced rat HCC disease model. The level of palmityl-L-carnitine showed different trends at different ages. It decreased with age at the early stage. However, it increased significantly after 8 weeks between the two groups. The results of a stepwise histopathological progression indicated that the model was similar to human HCC, and it had potential practicality of HCC diagnosis at early stages [108]. This is also a rare report on the use of animal models to study the changes of acylcarnitines. The experimental results of cells and animals provide a good reference for the mechanism of HCC regulation of acylcarnitine metabolism.

However, because the pathogenesis of HCC is very complicated, cell culture or animal models cannot accurately simulate this process [109–111]. The results obtained from these two models still do not accurately reflect the regulation mechanism of HCC on acylcarnitines [105,108]. Therefore, most studies have focused on the analysis of clinical samples. Among them, blood and urine samples are dominant. Few reports are related to tissue samples because of the difficulty in sample collection. Since the development of HCC is closely related to CIR and hepatitis [112–114], there are also many literatures that compare these liver diseases together. It is hoped to uncover the relationship between hepatitis, CIR and HCC, and it is also hoped to discover some specific markers that can distinguish these diseases. For example, Du et al. detected 14 characteristic metabolites with significant differences in the homogenate of tissue samples obtained from HBV-related HCC patients. Five of these metabolites (beta-sitosterol, quinaldic acid, tetradecanal, oleamide, and arachidyl carnitine) were first discovered in HCC samples [115]. Differential acylcarnitines found between HCC and liver disease control groups were listed in Table 1. Some detailed examples are discussed below.

Xu et al. used the LC-MS combined with the random forest–recursive feature elimination method to compare the serum metabolic profiling of patients with chronic liver diseases (CLD) and HCC. The results demonstrated the accumulation of LCACs and the decline of free carnitine, MCACs and SCACs were associated with the severity of liver disease. A corresponding change was observed in the related enzyme activities. And HCC had less effect on the general changing extent of acylcarnitines than the non-malignant liver diseases. The authors speculated that this might be possible due to the special energy-expenditure mechanism of HCC cells [116]. The alteration of carnitine C16:1 and carnitine C18:1 was found to be consistent in another report, which was proposed by a mutual information-support vector machine-recursive feature elimination method to filter out noise and non-informative variables during data processing. The accumulation of the two LCACs demonstrated that compared with control, severe liver diseases (CIR and HCC) presented more notable implications of metabolic changes related to fatty acid β-oxidation. Moreover, HCC could be discriminated from CIR by SM (d18:0/22:2 (OH)), pimelylcarnitine and carnitine etc. [117]. The authors also used a pseudotargeted approach for further confirmation the changes of acylcarnitines. Serum metabolomic analysis of patients with HCC showed that the levels of MCACs (C8, C8:1, C10, and C10:1) reduced and LCACs (C18:1 and C18:2) levels were raised [118]. The study from Ong et al. also confirmed the similar difference between SCACs, MCACs and LCACs in HCC. In addition, they further verified that serum acetylcarnitine was a highly accurate biomarker for HCC diagnosis and progression, especially for AFP false-negative HCC patients [119]. However, the changes of LCACs appeared to be related to the type of hepatitis virus. Since it was reported that the level of octadecadienyl carnitine was higher in HBV-associated CIR group than in CHB and HBV-associated HCC groups [35].

Shariff et al. used a NMR system to analyse urine samples from hepatitis B surface antigen (HBsAg)-positive patients with HCC, HBsAg positive patients with CIR, and HBsAg negative healthy controls in Nigerian subjects. It was found that four metabolites, including creatinine, carnitine, creatine, and acetone, were strongly contributed to the grouping of the samples. The carnitine levels in the HCC group were meaningfully higher than in the healthy and CIR groups, reflecting that carnitine was excessive produced to meet the requirement of mitochondrial activity and rapid growth of cancer cells [120]. The method was then applied to detect HCV infected Egyptian patients with HCC. The metabolic profile presented similarity to that of Nigerian patients. It was firstly reported that metabolic alteration of HCC patients in two etiologically and ethnically distinct populations was similar, proposed that metabolic disorder caused by tumourogenesis did not rely on the two factors [121]. However, compared with the results of LC-MS analysis, the two studies also showed that except for carnitine, other acylcarnitines were difficult to be observed by NMR. The same results were also demonstrated in other studies [122]. By contrast, a LC-MS-based targeted and non-targeted study on the sera samples of 40 HCC and 49 CIR patients from Egypt detected more notably different metabolites. The results confirmed that the bile acid metabolites, LCACs and small peptide showed significant differences between the HCC and CIR control group. Of these remarkable metabolites, LCACs, oleoyl carnitine, palmitoyl carnitine, and linoelaidyl carnitine were down-regulated in HCC patients compared with CIR controls [123].

Combining multiple techniques to analyse disease samples can overcome the bias of a single method, improving the coverage of metabolite detection to obtain more comprehensive results. For example, Liu et al. reported on the use of NMR and LC-MS for global metabolomics analysis of serum of HCC cases [124]. GC-MS and LC-MS analyses were employed to investigate serum metabolic abnormalities in HBV-CIR and HCC patients [125,126]. Another report was used GC-MS and UPLC-MS/MS platforms to analyse the global serum metabolomes of HCC, hepatitis C CIR disease and healthy controls. The most significant altered metabolites included fatty acids, amino acids and acylcarnitines. Of these, SCACs and MCACs were highly overexpressed in HCC patients compared to disease controls, while LCACs trended downward [127]. To choose different stationary phases for sample separation was also an effective method to increase the detection rate of metabolites. For example, using both HILIC and RPLC to separate the urinary metabolites could found carnitine C10:1, carnitine C8:1, butylcarnitine, acetyl carnitine and cartinine, carnitine C9:1, carnitine C10:3, and carnitine C9:0 as potential biomarkers [128,129]. Zhang et al. sampled the HCC and CIR patients' blood samples on filter paper and dried at room temperature, then extracted using organic solvent and concentrated for mass spectrometry analysis. Using the detected amino acids, acylcarnitines and some of their relevant ratios as the evaluation criteria, it was found that in this model, in view of their individual odds ratios, C5-OH/C0, C3/Met, and Val/Phe seemed to be the most important risk factors for HCC, while Thr, C3DC/C10, and C18:1 seemed to be the risk factors for CIR [130].

The value of the area under the curve of the receiver operating characteristic curve (ROC) is commonly used as indicators for evaluating specificity and sensitivity of biomarker. A study on the urine samples of HCC and CIR discovered that combination of butyrylcarnitine and hydantoin-5-propionic acid could differentiate the two diseases. The area of the two metabolites under ROC curve was 0.786 and 0.773, respectively [23]. By comparing principal metabolic alteration obtained from 50 HCC tissue samples and 298 chronic hepatitis and CIR serum samples, it was found that betaine plus propionylcarnitine was efficient for distinguishing HCC from the two types of liver diseases with a 0.982 AUC value of the ROC curve, which was much better than that of a-fetoprotein (AFP, 0.697). The combination was useful for the diagnosis of both AFP false-positive and false-negative HCC patients [24]. Another study showed that undecanoyl- L-carnitine, whose level was lower in HCC than in HBVs and NCs, in combination with α-fetoprotein could provide highly sensitive and specific for HCC diagnosis. The values of the area under the curve of the ROC curve was 0.92 [131]. This finding demonstrated that the combination of differential biomarkers presented good diagnostic potential to HCC.

Nielsen et al. established a genome-scale hepatocyte metabolic model and used system biology to analyse the metabolic changing of different HCV progressions. The levels of acylcarnitines were disturbed markedly in the dysplastic nodule and early HCC stages. This was related to the up-regulated genes, including BCAT1, PLOD3 and six other methyltransferase genes, which influenced carnitine biosynthesis and S-adenosylmethionine metabolism. Meanwhile, acyl-CoA consumption was regulated by GNPAT and BCAP31 upregulated expression. These genes could be used as potential targets for the therapy of liver disorders related to HCV [132].

**Table 1.** Differential acylcarnitines between HCC and liver disease control groups.

| Reference | Sample | Platform | Main Findings |
|---|---|---|---|
| [23] | Urine<br>27 CIR<br>33 HCC<br>26 HC | LC-MS<br>Non-targeted | HCC *vs.* CIR: MCACs and SCACs increased.<br>CIR & HCC *vs.* HC: LCAC and MCAC decreased.<br>HCC *vs.* CIR: carnitine C4:0 elevated |
| [24] | Tissues<br>50 HCT<br>50 DNT<br>Serum<br>81 CH<br>78 CIR<br>139 HCC | LC-MS<br>CE-MS<br>Non-targeted<br>Targeted | HCT *vs.* DNT: carnitine (C2/C0) upregulated, propionylcarnitine to carnitine (C3/C0) downregulated, SCAC decreased, LCAC increased<br>HCC *vs.* CIR and CH: propionylcarnitine decreased, acetylcarnitine elevated |
| [35] | Sera<br>136 CHB<br>104 CIR<br>95 HCC | LC-MS<br>Targeted | CIR and HCC *vs.* CHB: carnitine, 2 SCAC, and 4 MCAC decreased.<br>CIR *vs.* CHB and HCC: AC C18:2 higher, AC C3-OH lower |
| [105] | SK-Hep1 cells underexpressing G6PD (Sk-Gi) and control cells (Sk-Sc) after dehydroepiandrosterone (DHEA) treatment | LC-MS<br>Non-targeted | Carnitine and acyl derivatives declined in DHEA-treated Sk-Gi cells |
| [108] | Rat sera<br>52 HCC<br>28 HCHuman sera<br>262 HCC<br>76 CIR<br>74 HBV | LC-MS Non-targeted | HCC *vs.* HC (rat): palmityl-L-carnitine decreased (with the age of the animals), increased in week 8 |
| [115] | HBV-related HCC tissue<br>10 CTT<br>10 ANT<br>10 DNT | LC-MSNon-targeted | DAT *vs.* CTT: arachidyl carnitine lower |
| [116] | Sera<br>30 HC<br>30 CHB<br>30 CIR<br>30 HCC | LC-MS<br>Non-targeted | CHB *vs.* CIR *vs.* HCC: C16:1-CN increased (with severity of chronic liver diseases), MCAC and SCAC decreased (including C10-CN, C10:1-CN, C8-CN, and C6-CN)<br>CIR and HCC *vs.* HC: C14-CN increased |

**Table 1.** *Cont.*

| Reference | Sample | Platform | Main Findings |
|---|---|---|---|
| [117] | Sera<br>30 CHB<br>30 CIR<br>30 HCC<br>30 HC | LC-MS<br>Non-targeted | HCC *vs.* CHB: pimelylcarnitine and acetylcarnitine rise<br>HCC *vs.* CIR: pimelylcarnitine and carnitine rise<br>CIR and HCC *vs.* CHB and HC: C16:1-CN and C18:1-CN accumulated |
| [119] | Tissue<br>50 HCC (38 males and 12 females).<br>Serum<br>18 HCC<br>20 CIR<br>20 HC | LC-MS<br>Non-targeted | CTT *vs.* ANT: LCAC accumulated, MCAC and SCAC decreased.<br>Acetylcarnitine reduced (gradually)<br>HCC *vs.* CIR and HC: serum acetylcarnitine reduced |
| [120] | Urine<br>18 HCC<br>10 CIR<br>15 HC | NMR<br>Non-targeted | HCC *vs.* CIR and HC: carnitine raised |
| [121] | Urine and serum<br>16 HCC<br>14 CIR<br>17 HC | NMR<br>Non-targeted | HCC *vs.* CIR and HC: urinary carnitine elevated |
| [122] | Urine<br>42 HCC<br>47 CIR<br>7 HC | NMR<br>Non-targeted | HCC *vs.* HC: carnitine increased |
| [123] | Sera<br>40 HCC<br>49 CIR | LC-MS<br>Targeted<br>Non-targeted | HCC *vs.* CIR: LCAC, oleoyl carnitine, palmitoyl carnitine, and linoelaidyl<br>carnitine down-regulation |
| [124] | Serum<br>71 HCC<br>80 CIR<br>26 HC | NMR and LC-MS<br>Non-targeted | HCC *vs.* CIR and HC: acylcarnitine increased, MCAC and LCAC decreased<br>(including decanoyl- -carnitine and palmitoylcarnitine)<br>CIR and HCC *vs.* HC: carnitine decreased (progressive) |

**Table 1.** *Cont.*

| Reference | Sample | Platform | Main Findings |
|---|---|---|---|
| [125] | Serum<br>39 HC<br>49 HBV-CIR<br>51 HCC | LC-MS Non-targeted<br>Targeted | HBV-CIR and HCC *vs.* HC: carnitine altered (stepwise) |
| [126] | Serum and urine<br>82 HCC<br>71 HC | LC-MS<br>Non-targeted | HCC *vs.* HC: serum carnitine 1.36 times |
| [127] | Serum<br>30 HCC,<br>27 CIR<br>30 HC | LC-MS<br>Non-targeted | HCC *vs.* CIR: acetylcarnitine, glutarylcarnitine and succinylcarnitine upregulated<br>CIR *vs.* HC: MCAC overexpression<br>HCC *vs.* CIR: LCAC downward |
| [128] | Urine<br>24 HC<br>21 HCC | LC-MS<br>Non-targeted | HCC *vs.* HC: carnitine C8:1, carnitine C9:0, carnitine C9:1, carnitine C10:1 and carnitine C10:3 reduced, acetylcarnitine increased |
| [129] | Serum<br>24 HCC<br>25 CIR<br>25 HC | LC-MS Non-targeted | HCC and CIR *vs.* HC: carnitine, carnitine fragment and acetylcarnitine increased |
| [130] | Serum<br>92 male and 44 female CIR<br>41 male and 9 female HCC | LC-MS<br>Targeted | HCC *vs.* CIR: decanoylcarnitine, decenoylcarnitine, propionylcarnitine/methionine, Methylmalonylcarnitine, 3-Hydroxy-isovalerylcarnitine/carnitine increased; octadecenoylcarnitine, malonylcarnitine/decanoylcarnitine, butyrylcarnitine/octanoylcarnitine decreased |
| [131] | Serum<br>267 HCC<br>48 HBV<br>272 HC | LC-MS<br>Non-targeted | HCC *vs.* HC: SCAC and MCAC decreased, LCAC increased |
| [133] | 22 HBV<br>6 HCV<br>38 HBV-associated HCC<br>31 HCV-associated HCC<br>31 HC | LC-MS<br>Targeted<br>Non-targeted | HCC *vs.* HC: C18:1-CN and C18:2-CN increased. |

**Table 1.** *Cont.*

| Reference | Sample | Platform | Main Findings |
|---|---|---|---|
| [134] | Serum<br>30 HC,<br>30 CHB,<br>30 CIR<br>30 HCC | LC-MS<br>Non-targeted | CIR and HCC *vs.* HC and CHB: C16:1-CN and C16:0-CN elevated<br>CHB, CIR and HCC *vs.* HC: C10-CN decreased, LCAC elevated, MCAC and SCAC exhibited the opposite trend |

Abbreviation: Cirrhosis (CIR), hepatocellular carcinoma (HCC), healthy control (HC), hepatitis B virus (HBV), hepatitis C virus (HCV), chronic liver disease (CLD), chronic hepatitis B (CHB), chronic hepatitis (CH), adjacent noncancerous tissue (ANT), distal noncancerous tissue (DNT), central tumour tissue (CTT), hepatocellular carcinoma tissue (HCT), medium-chain acylcarnitine (MCAC), short-chain acylcarnitine (SCAC), and long-chain acylcarnitine (LCAC).

Although the acylcarnitines with significant differences were not exactly the same in different reports due to the different samples, instruments and detection methods, the general rule of changes can still be in conclusion. It is certain that HCC can significantly regulate the metabolism of acylcarnitines. The levels of serum acylcarnitines in HCC patients showed specific patterns, mainly including increased levels of free carnitine, decreased levels of SCACs and MCACs, and increased levels of LCACs [116,118,133]. The main role of SCACs and MCACs is to remove organic acids from organelles such as mitochondria and excrete them to the urine and bile. The declining levels in serum indicated an increase in excretion rate of organelles or an obstacle of accumulation rate. Conversely, the promoted formation of LCACs in cells demonstrated increasing β-oxidation and producing more energy duo to the enhanced transport of LCFCs into the mitochondria [117]. However, it should be noted that there are low correlations between LCACs and SCACs/MCACs according to the results of acylcarnitine metabolic profiling from 80 pairs of HCC tissues and adjacent noncancerous tissues (ANTs) [135]. Among these significantly different acylcarnitines, some have been considered to have specificity of distinguish HCC from chronic hepatitis and CIR. It demonstrates the great potential of acylcarnitines as biomarkers for HCC diagnosis.

The mechanism of HCC-driven acylcarnitine changes has also been studied intensively. CPT1A and CPT2 are rate-limiting enzymes of LCFAs for β-oxidation [36,136]. Therefore, their expression is closely related to the changes of acylcarnitine levels. It was reported that in 66 post-operative liver tumour tissue from patients with resected HCCs, decreased expression of CPT1A was observed. And the expression changes appeared to correlate with risk factors for the prognosis of HCC patients, such as tumour size, histological grade, intrahepatic metastasis, and tumour–node–metastasis stage [137]. However, in another analysis based on the eighty pairs of HCC tissues and adjacent noncancerous tissues (ANTs), CPT1A expression was not significantly changed. These inconsistent findings suggest that the effect of CPT1A expression on the metabolism of acylcarnitines still requires further confirmation. In the later study, it was also found that downregulation of CPT2 was significantly associated with the presence of vascular invasion and poor tumour differentiation in HCC. And it caused low efficiency of the carnitine shuttle system, inducing the suppression of fatty acid β-oxidation in HCC. However, the downregulation of CPT2 could promote tumourigenesis, chemoresistance to cisplatin and lipogenesis [136]. Another independent study also identified CPT2 downregulation in HCC as a critical determinant in acylcarnitine accumulation. HCC cells presented resistance to lipotoxicity by the Src-mediated JNK inhibition after CPT2 was knocked out. In particular, oleoylcarnitine may act as an oncometabolite in hepatocarcinogenesis as it could promote HCC cell sphere formation by activating STAT3. Simultaneously, downregulation of CPT2 may mediate the metabolic reprogramming of HCC cells, which enables them to escape lipotoxicity and promotes hepatocarcinogenesis. These finding indicated that acylcarnitine accumulation was a surrogate marker of CPT2 downregulation [54]. These promising results offered mechanistic insights into acylcarnitine accumulation in HCC. As acylcarnitine metabolism is especially important for energy production in HCC, targeting this pathway is considered to be a potential strategy for cancer treatment.

## 5. Conclusions and Perspectives

Due to their special structure and function, the alteration of acylcarnitines in HCC has attracted significant attention. Certain acylcarnitines have been reported to present regular changes in HCC. These differential acylcarnitines have the potential to serve as biomarkers for HCC diagnosis. However, limited by the sensitivity of current detection techniques and the number of commercially available standards, only a small fraction of the known acylcarnitines could be accurately detected. The changes and function of these undetected acylcarnitines remain unknown in HCC. Therefore, it is urgent to develop new methods with high sensitivity and high selectivity to cover the detection of these metabolites as much as possible. Application of highly-selective sample preparation methods to enrich acylcarnitines and to reduce matrix interference may increase the probability of detection of low abundance acylcarnitines. In addition, the strategy of isotope labelling may be used for the relative quantification to overcome the problem of lack of standards. However, accurate structural identification and absolute quantification are still required in clinical detection. Therefore, obtaining the standards of acylcarnitines as many as possible is still the best way of approaching the problems.

In addition, the levels of acylcarnitines in HCC are affected by a number of factors, such as diet, renal dysfunction, biosynthesis rate, and other liver diseases. Therefore, in examining the regulation of HCC on acylcarnitines, the effects of these factors must also be carefully considered. In addition, acylcarnitine metabolism is an important node in the complex metabolic network of cells. Their levels are also affected by upstream and downstream changes in the metabolic pathways. To investigate the flux of acylcarnitines along the pathways may offer wonderful insight into the regulation mechanism of HCC on the acylcarnitine metabolism. The goals could be reached by accurate quantification of these metabolites using targeted metabolic profiling and metabolic flux analysis. With these comprehensive detection methods, some significantly differential acylcarnitines or their related metabolites may be discovered. They could be used as potential biomarkers for the subsequent study of HCC diagnosis or targets for drug development, which may supply a valuable reference for the pathogenesis and treatment investigation of HCC.

**Author Contributions:** S.L. and D.G. are responsible for preparing the manuscript. S.L., D.G. and Y.J. contributed to the writing of the article and have read and approved the final version.

**Funding:** This research was funded by the National Natural Science Foundation of China, grant number 21675096.

**Acknowledgments:** This work was supported by the grant from the National Natural Science Foundation of China.

**Conflicts of Interest:** The authors declare no conflict of interest.

## References


1. Gingold, J.A.; Zhu, D.; Lee, D.F.; Kaseb, A.; Chen, J. Genomic profiling and metabolic homeostasis in primary liver cancers. *Trends Mol. Med.* **2018**, *24*, 395–411. [CrossRef] [PubMed]
2. El-Serag, H.B.; Rudolph, K.L. Hepatocellular carcinoma: Epidemiology and molecular carcinogenesis. *Gastroenterology* **2007**, *132*, 2557–2576. [CrossRef] [PubMed]
3. Ferlay, J.; Shin, H.R.; Bray, F.; Forman, D.; Mathers, C.; Parkin, D.M. Estimates of worldwide burden of cancer in 2008: GLOBOCAN 2008. *Int. J. Cancer* **2010**, *127*, 2893–2917. [CrossRef] [PubMed]
4. Fujiwara, N.; Friedman, S.L.; Goossens, N.; Hoshida, Y. Risk factors and prevention of hepatocellular carcinoma in the era of precision medicine. *J. Hepatol.* **2018**, *68*, 526–549. [CrossRef] [PubMed]
5. Alavi, M.; Janjua, N.Z.; Chong, M.; Grebely, J.; Aspinall, E.J.; Innes, H.; Valerio, H.; Hajarizadeh, B.; Hayes, P.C.; Krajden, M.; et al. Trends in hepatocellular carcinoma incidence and survival among people with hepatitis C: An international study. *J. Viral Hepat.* **2018**, *25*, 473–481. [CrossRef] [PubMed]
6. Cohen, J.C.; Horton, J.D.; Hobbs, H.H. Human fatty liver disease: Old questions and new insights. *Science* **2011**, *332*, 1519–1523. [CrossRef] [PubMed]
7. Yu, L.X.; Ling, Y.; Wang, H.Y. Role of nonresolving inflammation in hepatocellular carcinoma development and progression. *NPJ Precis. Oncol.* **2018**, *2*, 6. [CrossRef]
8. Charrez, B.; Qiao, L.; Hebbard, L. Hepatocellular carcinoma and non-alcoholic steatohepatitis: The state of play. *World J. Gastroenterol.* **2016**, *22*, 2494–2502. [CrossRef]
9. Agosti, P.; Sabba, C.; Mazzocca, A. Emerging metabolic risk factors in hepatocellular carcinoma and their influence on the liver microenvironment. *Biochim. Biophys. Acta Mol. Basis Dis.* **2018**, *1864*, 607–617. [CrossRef]
10. Starley, B.Q.; Calcagno, C.J.; Harrison, S.A. Non-alcoholic fatty liver disease and hepatocellular carcinoma: A weighty connection. *Hepatology* **2010**, *51*, 1820–1832. [CrossRef]
11. Khan, F.Z.; Perumpail, R.B.; Wong, R.J.; Ahmed, A. Advances in hepatocellular carcinoma: Non-alcoholic steatohepatitis-related hepatocellular carcinoma. *World J. Hepatol.* **2015**, *7*, 2155–2161. [CrossRef] [PubMed]
12. Klingenberg, M.; Matsuda, A.; Diederichs, S.; Patel, T. Non-coding RNA in hepatocellular carcinoma: Mechanisms, biomarkers and therapeutic targets. *J. Hepatol.* **2017**, *67*, 603–618. [CrossRef] [PubMed]

13. Umeda, S.; Kanda, M.; Kodera, Y. Emerging evidence of molecular biomarkers in hepatocellular carcinoma. *Histol. Histopathol.* **2018**, *33*, 343–355. [CrossRef] [PubMed]
14. Kimhofer, T.; Fye, H.; Taylor-Robinson, S.; Thursz, M.; Holmes, E. Proteomic and metabonomic biomarkers for hepatocellular carcinoma: A comprehensive review. *Br. J. Cancer* **2015**, *112*, 1141–1156. [CrossRef] [PubMed]
15. Pineda-Solis, K.; McAlister, V. Wading through the noise of "multi-omics" to identify prognostic biomarkers in hepatocellular carcinoma. *Hepatobil. Surg. Nutr.* **2015**, *4*, 293–294. [CrossRef]
16. De Stefano, F.; Chacon, E.; Turcios, L.; Marti, F.; Gedaly, R. Novel biomarkers in hepatocellular carcinoma. *Dig. Liver Dis.* **2018**, *50*, 1115–1123. [CrossRef] [PubMed]
17. Rui, L. Energy metabolism in the liver. *Compr. Physiol.* **2014**, *4*, 177–197. [CrossRef] [PubMed]
18. Guo, W.; Tan, H.Y.; Wang, N.; Wang, X.; Feng, Y. Deciphering hepatocellular carcinoma through metabolomics: from biomarker discovery to therapy evaluation. *Cancer Manag. Res.* **2018**, *10*, 715–734. [CrossRef]
19. Wang, X.; Zhang, A.; Sun, H. Power of metabolomics in diagnosis and biomarker discovery of hepatocellular carcinoma. *Hepatology* **2013**, *57*, 2072–2077. [CrossRef]
20. Nicholson, J.K.; Lindon, J.C.; Holmes, E. 'Metabonomics': understanding the metabolic responses of living systems to pathophysiological stimuli via multivariate statistical analysis of biological NMR spectroscopic data. *Xenobiotica* **1999**, *29*, 1181–1189. [CrossRef]
21. Fiehn, O. Metabolomics—the link between genotypes and phenotypes. *Plant Mol. Biol.* **2002**, *48*, 155–171. [CrossRef] [PubMed]
22. Baniasadi, H.; Gowda, G.A.; Gu, H.; Zeng, A.; Zhuang, S.; Skill, N.; Maluccio, M.; Raftery, D. Targeted metabolic profiling of hepatocellular carcinoma and hepatitis C using LC-MS/MS. *Electrophoresis* **2013**, *34*, 2910–2917. [CrossRef] [PubMed]
23. Shao, Y.; Zhu, B.; Zheng, R.; Zhao, X.; Yin, P.; Lu, X.; Jiao, B.; Xu, G.; Yao, Z. Development of urinary pseudotargeted LC-MS-based metabolomics method and its application in hepatocellular carcinoma biomarker discovery. *J. Proteome Res.* **2015**, *14*, 906–916. [CrossRef] [PubMed]
24. Huang, Q.; Tan, Y.; Yin, P.; Ye, G.; Gao, P.; Lu, X.; Wang, H.; Xu, G. Metabolic characterization of hepatocellular carcinoma using non-targeted tissue metabolomics. *Cancer Res.* **2013**, *73*, 4992–5002. [CrossRef] [PubMed]
25. Dettmer, K.; Aronov, P.A.; Hammock, B.D. Mass spectrometry-based metabolomics. *Mass Spectrom. Rev.* **2007**, *26*, 51–78. [CrossRef]
26. Wishart, D.S.; Jewison, T.; Guo, A.C.; Wilson, M.; Knox, C.; Liu, Y.; Djoumbou, Y.; Mandal, R.; Aziat, F.; Dong, E.; et al. HMDB 3.0—The Human Metabolome Database in 2013. *Nucleic Acids Res.* **2013**, *41*, D801–D807. [CrossRef]
27. Wishart, D.S.; Feunang, Y.D.; Marcu, A.; Guo, A.C.; Liang, K.; Vazquez-Fresno, R.; Sajed, T.; Johnson, D.; Li, C.; Karu, N.; et al. HMDB 4.0: The human metabolome database for 2018. *Nucleic Acids Res.* **2018**, *46*, D608–D617. [CrossRef]
28. Batchuluun, B.; Al Rijjal, D.; Prentice, K.J.; Eversley, J.A.; Burdett, E.; Mohan, H.; Bhattacharjee, A.; Gunderson, E.P.; Liu, Y.; Wheeler, M.B. Elevated medium-chain acylcarnitines are associated with gestational diabetes mellitus and early progression to type 2 diabetes and induce pancreatic beta-cell dysfunction. *Diabetes* **2018**, *67*, 885–897. [CrossRef]
29. Tarasenko, T.N.; Cusmano-Ozog, K.; McGuire, P.J. Tissue acylcarnitine status in a mouse model of mitochondrial beta-oxidation deficiency during metabolic decompensation due to influenza virus infection. *Mol. Genet. Metab.* **2018**, *125*, 144–152. [CrossRef]
30. McCoin, C.S.; Knotts, T.A.; Adams, S.H. Acylcarnitines—old actors auditioning for new roles in metabolic physiology. *Nat. Rev. Endocrinol.* **2015**, *11*, 617–625. [CrossRef]
31. Hinder, L.M.; Figueroa-Romero, C.; Pacut, C.; Hong, Y.; Vivekanandan-Giri, A.; Pennathur, S.; Feldman, E.L. Long-chain acyl coenzyme A synthetase 1 overexpression in primary cultured Schwann cells prevents long chain fatty acid-induced oxidative stress and mitochondrial dysfunction. *Antioxid. Redox Signal.* **2014**, *21*, 588–600. [CrossRef] [PubMed]
32. Murthy, M.S.R.; Pande, S.V. Mechanism of carnitine acylcarnitine translocase-catalyzed import of acylcarnitines into mitochondria. *J. Biol. Chem.* **1984**, *259*, 9082–9089. [PubMed]
33. Pande, S.V.; Murthy, M.S. Carnitine-acylcarnitine translocase deficiency: Implications in human pathology. *Biochim. Biophys. Acta Mol. Basis Dis.* **1994**, *1226*, 269–276. [CrossRef]

34. Melone, M.A.B.; Valentino, A.; Margarucci, S.; Galderisi, U.; Giordano, A.; Peluso, G. The carnitine system and cancer metabolic plasticity. *Cell Death Dis.* **2018**, *9*, 228. [CrossRef] [PubMed]

35. Wu, T.; Zheng, X.; Yang, M.; Zhao, A.; Li, M.; Chen, T.; Panee, J.; Jia, W.; Ji, G. Serum lipid alterations identified in chronic hepatitis B, hepatitis B virus-associated cirrhosis and carcinoma patients. *Sci. Rep.* **2017**, *7*, 42710. [CrossRef]

36. Qu, Q.; Zeng, F.; Liu, X.; Wang, Q.J.; Deng, F. Fatty acid oxidation and carnitine palmitoyltransferase I: Emerging therapeutic targets in cancer. *Cell Death Dis.* **2016**, *7*, e2226. [CrossRef]

37. Xiang, L.; Wei, J.; Tian, X.Y.; Wang, B.; Chan, W.; Li, S.; Tang, Z.; Zhang, H.; Cheang, W.S.; Zhao, Q.; et al. Comprehensive analysis of acylcarnitine species in db/db mouse using a novel method of high-resolution parallel reaction monitoring reveals widespread metabolic dysfunction induced by diabetes. *Anal. Chem.* **2017**, *89*, 10368–10375. [CrossRef]

38. Seiler, S.E.; Koves, T.R.; Gooding, J.R.; Wong, K.E.; Stevens, R.D.; Ilkayeva, O.R.; Wittmann, A.H.; DeBalsi, K.L.; Davies, M.N.; Lindeboom, L.; et al. Carnitine acetyltransferase mitigates metabolic inertia and muscle fatigue during exercise. *Cell Metab.* **2015**, *22*, 65–76. [CrossRef]

39. Casals, N.; Zammit, V.; Herrero, L.; Fado, R.; Rodriguez-Rodriguez, R.; Serra, D. Carnitine palmitoyltransferase 1C: From cognition to cancer. *Prog. Lipid Res.* **2016**, *61*, 134–148. [CrossRef]

40. Kim, H.I.; Raffler, J.; Lu, W.; Lee, J.J.; Abbey, D.; Saleheen, D.; Rabinowitz, J.D.; Bennett, M.J.; Hand, N.J.; Brown, C.; et al. Fine mapping and functional analysis reveal a role of SLC22A1 in acylcarnitine transport. *Am. J. Hum. Genet.* **2017**, *101*, 489–502. [CrossRef]

41. Hagenbuchner, J.; Scholl-Buergi, S.; Karall, D.; Ausserlechner, M.J. Very long-/ and long Chain-3-Hydroxy Acyl CoA Dehydrogenase Deficiency correlates with deregulation of the mitochondrial fusion/fission machinery. *Sci. Rep.* **2018**, *8*, 3254. [CrossRef] [PubMed]

42. Vargas, C.R.; Ribas, G.S.; da Silva, J.M.; Sitta, A.; Deon, M.; de Moura Coelho, D.; Wajner, M. Selective Screening of fatty acids oxidation defects and organic acidemias by liquid chromatography/tandem mass spectrometry acylcarnitine analysis in Brazilian patients. *Arch. Med. Res.* **2018**, *49*, 205–212. [CrossRef] [PubMed]

43. Diekman, E.F.; Visser, G.; Schmitz, J.P.J.; Nievelstein, R.A.J.; de Sain-van der Velden, M.; Wardrop, M.; Van der Pol, W.L.; Houten, S.M.; van Riel, N.A.W.; Takken, T.; et al. Altered energetics of exercise explain risk of rhabdomyolysis in very long-chain acyl-coa dehydrogenase deficiency. *PLoS ONE* **2016**, *11*, e0147818. [CrossRef] [PubMed]

44. Makrecka-Kuka, M.; Sevostjanovs, E.; Vilks, K.; Volska, K.; Antone, U.; Kuka, J.; Makarova, E.; Pugovics, O.; Dambrova, M.; Liepinsh, E. Plasma acylcarnitine concentrations reflect the acylcarnitine profile in cardiac tissues. *Sci. Rep.* **2017**, *7*, 17528. [CrossRef] [PubMed]

45. Rizzo, C.; Boenzi, S.; Inglese, R.; la Marca, G.; Muraca, M.; Martinez, T.B.; Johnson, D.W.; Zelli, E.; Dionisi-Vici, C. Measurement of succinyl-carnitine and methylmalonyl-carnitine on dried blood spot by liquid chromatography-tandem mass spectrometry. *Clin. Chim. Acta* **2014**, *429*, 30–33. [CrossRef] [PubMed]

46. DiBattista, A.; McIntosh, N.; Lamoureux, M.; Al-Dirbashi, O.Y.; Chakraborty, P.; Britz-McKibbin, P. Temporal signal pattern recognition in mass spectrometry: a method for rapid identification and accurate quantification of biomarkers for inborn errors of metabolism with quality assurance. *Anal. Chem.* **2017**, *89*, 8112–8121. [CrossRef] [PubMed]

47. Sun, L.; Liang, L.; Gao, X.; Zhang, H.; Yao, P.; Hu, Y.; Ma, Y.; Wang, F.; Jin, Q.; Li, H.; et al. Early prediction of developing type 2 diabetes by plasma acylcarnitines: A population-based study. *Diabetes Care* **2016**, *39*, 1563–1570. [CrossRef]

48. Burkhardt, R.; Kirsten, H.; Beutner, F.; Holdt, L.M.; Gross, A.; Teren, A.; Tonjes, A.; Becker, S.; Krohn, K.; Kovacs, P.; et al. Integration of genome-wide snp data and gene-expression profiles reveals six novel loci and regulatory mechanisms for amino acids and acylcarnitines in whole blood. *PLoS Genet* **2015**, *11*, e1005510. [CrossRef]

49. Guasch-Ferre, M.; Zheng, Y.; Ruiz-Canela, M.; Hruby, A.; Martinez-Gonzalez, M.A.; Clish, C.B.; Corella, D.; Estruch, R.; Ros, E.; Fito, M.; et al. Plasma acylcarnitines and risk of cardiovascular disease: Effect of Mediterranean diet interventions. *Am. J. Clin. Nutr.* **2016**, *103*, 1408–1416. [CrossRef]

50. Ahmad, T.; Kelly, J.P.; McGarrah, R.W.; Hellkamp, A.S.; Fiuzat, M.; Testani, J.M.; Wang, T.S.; Verma, A.; Samsky, M.D.; Donahue, M.P.; et al. Prognostic implications of long-chain acylcarnitines in heart failure and reversibility with mechanical circulatory support. *J. Am. Coll. Cardiol.* **2016**, *67*, 291–299. [CrossRef]

51. Virmani, A.; Pinto, L.; Bauermann, O.; Zerelli, S.; Diedenhofen, A.; Binienda, Z.K.; Ali, S.F.; van der Leij, F.R. The carnitine palmitoyl transferase (CPT) system and possible relevance for neuropsychiatric and neurological conditions. *Mol. Neurobiol.* **2015**, *52*, 826–836. [CrossRef]

52. Nakagawa, H.; Hayata, Y.; Kawamura, S.; Yamada, T.; Fujiwara, N.; Koike, K. Lipid metabolic reprogramming in hepatocellular carcinoma. *Cancers (Basel)* **2018**, *10*, 447. [CrossRef] [PubMed]

53. Wang, Y.T.; Chen, Y.X.; Guan, L.H.; Zhang, H.Z.; Huang, Y.Y.; Johnson, C.H.; Wu, Z.M.; Gonzalez, F.J.; Yu, A.M.; Huang, P.; et al. Carnitine palmitoyltransferase 1C regulates cancer cell senescence through mitochondria- associated metabolic reprograming. *Cell Death Differ.* **2018**, *25*, 733–746. [CrossRef] [PubMed]

54. Fujiwara, N.; Nakagawa, H.; Enooku, K.; Kudo, Y.; Hayata, Y.; Nakatsuka, T.; Tanaka, Y.; Tateishi, R.; Hikiba, Y.; Misumi, K.; et al. CPT2 downregulation adapts HCC to lipid-rich environment and promotes carcinogenesis via acylcarnitine accumulation in obesity. *Gut* **2018**, *67*, 1493–1504. [CrossRef] [PubMed]

55. Wettersten, H.I.; Hakimi, A.A.; Morin, D.; Bianchi, C.; Johnstone, M.E.; Donohoe, D.R.; Trott, J.F.; Aboud, O.A.; Stirdivant, S.; Neri, B.; et al. Grade-dependent metabolic reprogramming in kidney cancer revealed by combined proteomics and metabolomics analysis. *Cancer Res.* **2015**, *75*, 2541–2552. [CrossRef] [PubMed]

56. Valentino, A.; Calarco, A.; Di Salle, A.; Finicelli, M.; Crispi, S.; Calogero, R.A.; Riccardo, F.; Sciarra, A.; Gentilucci, A.; Galderisi, U.; et al. Deregulation of MicroRNAs mediated control of carnitine cycle in prostate cancer: Molecular basis and pathophysiological consequences. *Oncogene* **2017**, *36*, 6030–6040. [CrossRef] [PubMed]

57. Ganti, S.; Taylor, S.L.; Kim, K.; Hoppel, C.L.; Guo, L.; Yang, J.; Evans, C.; Weiss, R.H. Urinary acylcarnitines are altered in human kidney cancer. *Int. J. Cancer* **2012**, *130*, 2791–2800. [CrossRef]

58. Chughtai, K.; Jiang, L.; Greenwood, T.R.; Glunde, K.; Heeren, R.M. Mass spectrometry images acylcarnitines, phosphatidylcholines, and sphingomyelin in MDA-MB-231 breast tumour models. *J. Lipid Res.* **2013**, *54*, 333–344. [CrossRef]

59. Yu, D.; Zhou, L.; Xuan, Q.; Wang, L.; Zhao, X.; Lu, X.; Xu, G. Strategy for comprehensive identification of acylcarnitines based on liquid chromatography-high-resolution mass spectrometry. *Anal. Chem.* **2018**, *90*, 5712–5718. [CrossRef]

60. Mansour, F.R.; Wei, W.; Danielson, N.D. Separation of carnitine and acylcarnitines in biological samples: A review. *Biomed. Chromatogr.* **2013**, *27*, 1339–1353. [CrossRef]

61. Horvath, T.D.; Stratton, S.L.; Bogusiewicz, A.; Owen, S.N.; Mock, D.M.; Moran, J.H. Quantitative measurement of urinary excretion of 3-hydroxyisovaleryl carnitine by LC-MS/MS as an indicator of biotin status in humans. *Anal. Chem.* **2010**, *82*, 9543–9548. [CrossRef] [PubMed]

62. Heinig, K.; Henion, J. Determination of carnitine and acylcarnitines in biological samples by capillary electrophoresis-mass spectrometry. *J. Chromatogr. B Biomed. Sci. Appl.* **1999**, *735*, 171–188. [CrossRef]

63. Vernez, L.; Wenk, M.; Krahenbuhl, S. Determination of carnitine and acylcarnitines in plasma by high-performance liquid chromatography/electrospray ionization ion trap tandem mass spectrometry. *Rapid Commun. Mass Spectrom.* **2004**, *18*, 1233–1238. [CrossRef] [PubMed]

64. Liu, A.; Pasquali, M. Acidified acetonitrile and methanol extractions for quantitative analysis of acylcarnitines in plasma by stable isotope dilution tandem mass spectrometry. *J. Chromatogr. B Anal. Technol. Biomed. Life Sci.* **2005**, *827*, 193–198. [CrossRef] [PubMed]

65. Pormsila, W.; Morand, R.; Krahenbuhl, S.; Hauser, P.C. Capillary electrophoresis with contactless conductivity detection for the determination of carnitine and acylcarnitines in clinical samples. *J. Chromatogr. B Anal. Technol. Biomed. Life Sci.* **2011**, *879*, 921–926. [CrossRef] [PubMed]

66. Costa, C.G.; Struys, E.A.; Bootsma, A.; tenBrink, H.J.; Dorland, L.; deAlmeida, I.T.; Duran, M.; Jakobs, C. Quantitative analysis of plasma acylcarnitines using gas chromatography chemical ionization mass fragmentography. *J. Lipid Res.* **1997**, *38*, 173–182. [PubMed]

67. Sun, D.; Cree, M.G.; Zhang, X.J.; Boersheim, E.; Wolfe, R.R. Measurement of stable isotopic enrichment and concentration of long-chain fatty acyl-carnitines in tissue by HPLC-MS. *J. Lipid Res.* **2006**, *47*, 431–439. [CrossRef]

68. Schoonen, J.W.; van Duinen, V.; Oedit, A.; Vulto, P.; Hankemeier, T.; Lindenburg, P.W. Continuous-flow microelectroextraction for enrichment of low abundant compounds. *Anal. Chem.* **2014**, *86*, 8048–8056. [CrossRef]

69. Qiu, C.L.; Raynie, D.E. The use of extraction technologies in food safety studies. *LC GC Eur.* **2017**, *30*, 662–669.

70. Isaguirre, A.C.; Olsina, R.A.; Martinez, L.D.; Lapierre, A.V.; Cerutti, S. Development of solid phase extraction strategies to minimize the effect of human urine matrix effect on the response of carnitine by UPLC-MS/MS. *Microchem. J.* **2016**, *129*, 362–367. [CrossRef]

71. Morand, R.; Donzelli, M.; Haschke, M.; Krahenbuhl, S. Quantification of plasma carnitine and acylcarnitines by high-performance liquid chromatography-tandem mass spectrometry using online solid-phase extraction. *Anal. Bioanal. Chem.* **2013**, *405*, 8829–8836. [CrossRef] [PubMed]

72. Vernez, L.; Thormann, W.; Krahenbuhl, S. Analysis of carnitine and acylcarnitines in urine by capillary electrophoresis. *J. Chromatogr. A* **2000**, *895*, 309–316. [CrossRef]

73. Kivilompolo, M.; Ohrnberg, L.; Oresic, M.; Hyotylainen, T. Rapid quantitative analysis of carnitine and acylcarnitines by ultra-high performance-hydrophilic interaction liquid chromatography-tandem mass spectrometry. *J. Chromatogr. A* **2013**, *1292*, 189–194. [CrossRef] [PubMed]

74. Magiera, S.; Baranowski, J. Determination of carnitine and acylcarnitines in human urine by means of microextraction in packed sorbent and hydrophilic interaction chromatography-ultra-high-performance liquid chromatography-tandem mass spectrometry. *J. Pharm. Biomed. Anal.* **2015**, *109*, 171–176. [CrossRef] [PubMed]

75. Minkler, P.E.; Stoll, M.S.K.; Ingalls, S.T.; Hoppel, C.L. Selective and accurate C5 acylcarnitine quantitation by UHPLC-MS/MS: Distinguishing true isovaleric acidemia from pivalate derived interference. *J. Chromatogr. B Anal. Technol. Biomed. Life Sci.* **2017**, *1061-1062*, 128–133. [CrossRef] [PubMed]

76. Minkler, P.E.; Stoll, M.S.; Ingalls, S.T.; Kerner, J.; Hoppel, C.L. Validated method for the quantification of free and total carnitine, butyrobetaine, and acylcarnitines in biological samples. *Anal. Chem.* **2015**, *87*, 8994–9001. [CrossRef] [PubMed]

77. Maeda, Y.; Ito, T.; Suzuki, A.; Kurono, Y.; Ueta, A.; Yokoi, K.; Sumi, S.; Togari, H.; Sugiyama, N. Simultaneous quantification of acylcarnitine isomers containing dicarboxylic acylcarnitines in human serum and urine by high-performance liquid chromatography/electrospray ionization tandem mass spectrometry. *Rapid Commun. Mass Spectrom.* **2007**, *21*, 799–806. [CrossRef]

78. Maeda, Y.; Nakajima, Y.; Gotoh, K.; Hotta, Y.; Kataoka, T.; Sugiyama, N.; Shirai, N.; Ito, T.; Kimura, K. Kinetic and molecular orbital analyses of dicarboxylic acylcarnitine methylesterification show that derivatization may affect the screening of newborns by tandem mass spectrometry. *Bioorg. Med. Chem. Lett.* **2016**, *26*, 121–125. [CrossRef]

79. Zuniga, A.; Li, L. Ultra-high performance liquid chromatography tandem mass spectrometry for comprehensive analysis of urinary acylcarnitines. *Anal. Chim. Acta* **2011**, *689*, 77–84. [CrossRef]

80. Peng, M.; Liu, L.; Jiang, M.; Liang, C.; Zhao, X.; Cai, Y.; Sheng, H.; Ou, Z.; Luo, H. Measurement of free carnitine and acylcarnitines in plasma by HILIC-ESI-MS/MS without derivatization. *J. Chromatogr. B Anal. Technol. Biomed. Life Sci.* **2013**, *932*, 12–18. [CrossRef]

81. Li, K.; Sun, Q. Simultaneous determination of free and total carnitine in human serum by HPLC with UV detection. *J. Chromatogr. Sci.* **2010**, *48*, 371–374. [CrossRef] [PubMed]

82. Cao, Q.R.; Ren, S.; Park, M.J.; Choi, Y.J.; Lee, B.J. Determination of highly soluble L-carnitine in biological samples by reverse phase high performance liquid chromatography with fluorescent derivatization. *Arch. Pharm. Res.* **2007**, *30*, 1041–1046. [CrossRef] [PubMed]

83. Ni, J.; Xu, L.; Li, W.; Wu, L. Simultaneous determination of thirteen kinds of amino acid and eight kinds of acylcarnitine in human serum by LC-MS/MS and its application to measure the serum concentration of lung cancer patients. *Biomed. Chromatogr.* **2016**, *30*, 1796–1806. [CrossRef] [PubMed]

84. Minkler, P.E.; Stoll, M.S.K.; Ingalls, S.T.; Hoppel, C.L. Correcting false positive medium-chain acyl-CoA dehydrogenase deficiency results from newborn screening; synthesis, purification, and standardization of branched-chain C8 acylcarnitines for use in their selective and accurate absolute quantitation by UHPLC-MS/MS. *Mol. Genet. Metab.* **2017**, *120*, 363–369. [CrossRef] [PubMed]

85. Han, J.; Higgins, R.; Lim, M.D.; Atkinson, K.; Yang, J.; Lin, K.; Borchers, C.H. Isotope-labelling derivatization with 3-nitrophenylhydrazine for LC/multiple-reaction monitoring-mass-spectrometry-based quantitation of carnitines in dried blood spots. *Anal. Chim. Acta* **2018**, *1037*, 177–187. [CrossRef] [PubMed]

86. Turgeon, C.; Magera, M.J.; Allard, P.; Tortorelli, S.; Gavrilov, D.; Oglesbee, D.; Raymond, K.; Rinaldo, P.; Matern, D. Combined newborn screening for succinylacetone, amino acids, and acylcarnitines in dried blood spots. *Clin. Chem.* **2008**, *54*, 657–664. [CrossRef] [PubMed]

87. Johnson, D.W. Inaccurate measurement of free carnitine by the electrospray tandem mass spectrometry screening method for blood spots. *J. Inherit. Metab. Dis.* **1999**, *22*, 201–202. [CrossRef]

88. Cyr, D.; Giguere, R.; Giguere, Y.; Lemieux, B. Determination of urinary acylcarnitines: A complementary aid for the high-risk screening of several organic acidurias using a simple and reliable GC/MS-based method. *Clin. Biochem.* **2000**, *33*, 151–155. [CrossRef]

89. Moreira, V.; Brasili, E.; Fiamoncini, J.; Marini, F.; Miccheli, A.; Daniel, H.; Lee, J.J.H.; Hassimotto, N.M.A.; Lajolo, F.M. Orange juice affects acylcarnitine metabolism in healthy volunteers as revealed by a mass-spectrometry based metabolomics approach. *Food Res. Int.* **2018**, *107*, 346–352. [CrossRef]

90. Kriisa, K.; Leppik, L.; Balotsev, R.; Ottas, A.; Soomets, U.; Koido, K.; Volke, V.; Innos, J.; Haring, L.; Vasar, E.; et al. Profiling of acylcarnitines in first episode psychosis before and after antipsychotic treatment. *J. Proteome Res.* **2017**, *16*, 3558–3566. [CrossRef]

91. Simcox, J.; Geoghegan, G.; Maschek, J.A.; Bensard, C.L.; Pasquali, M.; Miao, R.; Lee, S.; Jiang, L.; Huck, I.; Kershaw, E.E.; et al. Global analysis of plasma lipids identifies liver-derived acylcarnitines as a fuel source for brown fat thermogenesis. *Cell Metab.* **2017**, *26*, 509–522. [CrossRef] [PubMed]

92. Strand, E.; Pedersen, E.R.; Svingen, G.F.T.; Olsen, T.; Bjorndal, B.; Karlsson, T.; Dierkes, J.; Njolstad, P.R.; Mellgren, G.; Tell, G.S.; et al. Serum acylcarnitines and risk of cardiovascular death and acute myocardial infarction in patients with stable angina pectoris. *J. Am. Heart Assoc.* **2017**, *6*, e003620. [CrossRef] [PubMed]

93. Schooneman, M.G.; Houtkooper, R.H.; Hollak, C.E.; Wanders, R.J.; Vaz, F.M.; Soeters, M.R.; Houten, S.M. The impact of altered carnitine availability on acylcarnitine metabolism, energy expenditure and glucose tolerance in diet-induced obese mice. *Biochim. Biophys. Acta Mol. Basis Dis.* **2016**, *1862*, 1375–1382. [CrossRef] [PubMed]

94. Sarker, S.K.; Islam, M.T.; Bhuyan, G.S.; Sultana, N.; Begum, N.; Al Mahmud-Un-Nabi, M.; Howladar, M.A.A.N.; Dipta, T.F.; Muraduzzaman, A.K.M.; Qadri, S.K.; et al. Impaired acylcarnitine profile in transfusion-dependent beta-thalassemia major patients in Bangladesh. *J. Adv. Res.* **2018**, *12*, 55–66. [CrossRef] [PubMed]

95. Wang, F.; Sun, L.; Sun, Q.; Liang, L.; Gao, X.; Li, R.; Pan, A.; Li, H.; Deng, Y.; Hu, F.B.; et al. Associations of plasma amino acid and acylcarnitine profiles with incident reduced glomerular filtration rate. *Clin. J. Am. Soc. Nephrol.* **2018**, *13*, 560–568. [CrossRef]

96. Ruiz, M.; Labarthe, F.; Fortier, A.; Bouchard, B.; Thompson Legault, J.; Bolduc, V.; Rigal, O.; Chen, J.; Ducharme, A.; Crawford, P.A.; et al. Circulating acylcarnitine profile in human heart failure: A surrogate of fatty acid metabolic dysregulation in mitochondria and beyond. *Am. J. Physiol. Heart Circ. Physiol.* **2017**, *313*, H768–H781. [CrossRef]

97. Giesbertz, P.; Ecker, J.; Haag, A.; Spanier, B.; Daniel, H. An LC-MS/MS method to quantify acylcarnitine species including isomeric and odd-numbered forms in plasma and tissues. *J. Lipid Res.* **2015**, *56*, 2029–2039. [CrossRef]

98. Hassan, A.; Tsuda, Y.; Asai, A.; Yokohama, K.; Nakamura, K.; Sujishi, T.; Ohama, H.; Tsuchimoto, Y.; Fukunishi, S.; Abdelaal, U.M.; et al. Effects of oral L-carnitine on liver functions after transarterial chemoembolization in intermediate-stage HCC patients. *Mediat. Inflamm.* **2015**, *2015*, 608216. [CrossRef]

99. Flanagan, J.L.; Simmons, P.A.; Vehige, J.; Willcox, M.D.; Garrett, Q. Role of carnitine in disease. *Nutr. Metab. (Lond.)* **2010**, *7*, 30. [CrossRef]

100. Baffy, G.; Brunt, E.M.; Caldwell, S.H. Hepatocellular carcinoma in non-alcoholic fatty liver disease: An emerging menace. *J. Hepatol.* **2012**, *56*, 1384–1391. [CrossRef]

101. Ding, H.-r.; Wang, J.-l.; Ren, H.-z.; Shi, X.-l.J.B.R.I. Lipometabolism and Glycometabolism in Liver Diseases. *Biomed Res. Int.* **2018**, *2018*, 1287127. [CrossRef]

102. Kalhan, S.C.; Guo, L.; Edmison, J.; Dasarathy, S.; McCullough, A.J.; Hanson, R.W.; Milburn, M. Plasma metabolomic profile in non-alcoholic fatty liver disease. *Metabolism* **2011**, *60*, 404–413. [CrossRef]

103. Lin, X.; Zhang, Y.; Ye, G.; Li, X.; Yin, P.; Ruan, Q.; Xu, G. Classification and differential metabolite discovery of liver diseases based on plasma metabolic profiling and support vector machines. *J. Sep. Sci.* **2011**, *34*, 3029–3036. [CrossRef]

104. Krahenbuhl, S.; Reichen, J. Carnitine metabolism in patients with chronic liver disease. *Hepatology* **1997**, *25*, 148–153. [CrossRef]

105. Cheng, M.L.; Shiao, M.S.; Chiu, D.T.; Weng, S.F.; Tang, H.Y.; Ho, H.Y. Biochemical disorders associated with antiproliferative effect of dehydroepiandrosterone in hepatoma cells as revealed by LC-based metabolomics. *Biochem. Pharmacol.* **2011**, *82*, 1549–1561. [CrossRef]
106. Yaligar, J.; Teoh, W.W.; Othman, R.; Verma, S.K.; Phang, B.H.; Lee, S.S.; Wang, W.W.; Toh, H.C.; Gopalan, V.; Sabapathy, K.; et al. Longitudinal metabolic imaging of hepatocellular carcinoma in transgenic mouse models identifies acylcarnitine as a potential biomarker for early detection. *Sci. Rep.* **2016**, *6*, 20299. [CrossRef]
107. Ishikawa, H.; Takaki, A.; Tsuzaki, R.; Yasunaka, T.; Koike, K.; Shimomura, Y.; Seki, H.; Matsushita, H.; Miyake, Y.; Ikeda, F.; et al. L-carnitine prevents progression of non-alcoholic steatohepatitis in a mouse model with upregulation of mitochondrial pathway. *PLoS ONE* **2014**, *9*, e100627. [CrossRef]
108. Tan, Y.; Yin, P.; Tang, L.; Xing, W.; Huang, Q.; Cao, D.; Zhao, X.; Wang, W.; Lu, X.; Xu, Z.; et al. Metabolomics study of stepwise hepatocarcinogenesis from the model rats to patients: Potential biomarkers effective for small hepatocellular carcinoma diagnosis. *Mol. Cell. Proteom.* **2012**, *11*, M111 010694. [CrossRef]
109. Hu, J.; Lin, Y.Y.; Chen, P.J.; Watashi, K.; Wakita, T. Cell and animal models for studying hepatitis B virus infection and drug development. *Gastroenterology* **2019**, *156*, 338–354. [CrossRef]
110. Bukh, J. Animal models for the study of hepatitis C virus infection and related liver disease. *Gastroenterology* **2012**, *142*, 1279–1287. [CrossRef]
111. Lerat, H.; Higgs, M.; Pawlotsky, J.M. Animal models in the study of hepatitis C virus-associated liver pathologies. *Expert Rev. Gastroenterol. Hepatol.* **2011**, *5*, 341–352. [CrossRef]
112. Degos, F.; Christidis, C.; Ganne-Carrie, N.; Farmachidi, J.P.; Degott, C.; Guettier, C.; Trinchet, J.C.; Beaugrand, M.; Chevret, S. Hepatitis C virus related cirrhosis: Time to occurrence of hepatocellular carcinoma and death. *Gut* **2000**, *47*, 131–136. [CrossRef]
113. Ieluzzi, D.; Covolo, L.; Donato, F.; Fattovich, G. Progression to cirrhosis, hepatocellular carcinoma and liver-related mortality in chronic hepatitis B patients in Italy. *Dig. Liver Dis.* **2014**, *46*, 427–432. [CrossRef]
114. Zhou, T.C.; Lai, X.; Feng, M.H.; Tang, Y.; Zhang, L.; Wei, J. Systematic review and meta-analysis: Development of hepatocellular carcinoma in chronic hepatitis B patients with hepatitis e antigen seroconversion. *J. Viral Hepat.* **2018**, *25*, 1172–1179. [CrossRef]
115. Liu, S.Y.; Zhang, R.L.; Kang, H.; Fan, Z.J.; Du, Z. Human liver tissue metabolic profiling research on hepatitis B virus-related hepatocellular carcinoma. *World J. Gastroenterol.* **2013**, *19*, 3423–3432. [CrossRef]
116. Zhou, L.; Wang, Q.; Yin, P.; Xing, W.; Wu, Z.; Chen, S.; Lu, X.; Zhang, Y.; Lin, X.; Xu, G. Serum metabolomics reveals the deregulation of fatty acids metabolism in hepatocellular carcinoma and chronic liver diseases. *Anal. Bioanal. Chem.* **2012**, *403*, 203–213. [CrossRef]
117. Lin, X.; Yang, F.; Zhou, L.; Yin, P.; Kong, H.; Xing, W.; Lu, X.; Jia, L.; Wang, Q.; Xu, G. A support vector machine-recursive feature elimination feature selection method based on artificial contrast variables and mutual information. *J. Chromatogr. B Anal. Technol. Biomed. Life Sci.* **2012**, *910*, 149–155. [CrossRef]
118. Chen, S.L.; Kong, H.W.; Lu, X.; Li, Y.; Yin, P.Y.; Zeng, Z.D.; Xu, G.W. Pseudotargeted metabolomics method and its application in serum biomarker discovery for hepatocellular carcinoma based on ultra high-performance liquid chromatography/triple quadrupole mass spectrometry. *Anal. Chem.* **2013**, *85*, 8326–8333. [CrossRef]
119. Lu, Y.; Li, N.; Gao, L.; Xu, Y.J.; Huang, C.; Yu, K.; Ling, Q.; Cheng, Q.; Chen, S.; Zhu, M.; et al. Acetylcarnitine is a candidate diagnostic and prognostic biomarker of hepatocellular carcinoma. *Cancer Res.* **2016**, *76*, 2912–2920. [CrossRef]
120. Shariff, M.I.; Ladep, N.G.; Cox, I.J.; Williams, H.R.; Okeke, E.; Malu, A.; Thillainayagam, A.V.; Crossey, M.M.; Khan, S.A.; Thomas, H.C.; et al. Characterization of urinary biomarkers of hepatocellular carcinoma using magnetic resonance spectroscopy in a Nigerian population. *J. Proteome Res.* **2010**, *9*, 1096–1103. [CrossRef]
121. Shariff, M.I.; Gomaa, A.I.; Cox, I.J.; Patel, M.; Williams, H.R.; Crossey, M.M.; Thillainayagam, A.V.; Thomas, H.C.; Waked, I.; Khan, S.A.; et al. Urinary metabolic biomarkers of hepatocellular carcinoma in an Egyptian population: A validation study. *J. Proteome Res.* **2011**, *10*, 1828–1836. [CrossRef]
122. Cox, I.J.; Aliev, A.E.; Crossey, M.M.; Dawood, M.; Al-Mahtab, M.; Akbar, S.M.; Rahman, S.; Riva, A.; Williams, R.; Taylor-Robinson, S.D. Urinary nuclear magnetic resonance spectroscopy of a Bangladeshi cohort with hepatitis-B hepatocellular carcinoma: A biomarker corroboration study. *World J. Gastroenterol.* **2016**, *22*, 4191–4200. [CrossRef]

123. Xiao, J.F.; Varghese, R.S.; Zhou, B.; Nezami Ranjbar, M.R.; Zhao, Y.; Tsai, T.H.; Di Poto, C.; Wang, J.; Goerlitz, D.; Luo, Y.; et al. LC-MS based serum metabolomics for identification of hepatocellular carcinoma biomarkers in Egyptian cohort. *J. Proteome Res.* **2012**, *11*, 5914–5923. [CrossRef]

124. Liu, Y.; Hong, Z.; Tan, G.; Dong, X.; Yang, G.; Zhao, L.; Chen, X.; Zhu, Z.; Lou, Z.; Qian, B.; et al. NMR and LC/MS-based global metabolomics to identify serum biomarkers differentiating hepatocellular carcinoma from liver cirrhosis. *Int. J. Cancer* **2014**, *135*, 658–668. [CrossRef]

125. Gong, Z.G.; Zhao, W.; Zhang, J.; Wu, X.; Hu, J.; Yin, G.C.; Xu, Y.J. Metabolomics and eicosanoid analysis identified serum biomarkers for distinguishing hepatocellular carcinoma from hepatitis B virus-related cirrhosis. *Oncotarget* **2017**, *8*, 63890–63900. [CrossRef]

126. Chen, T.; Xie, G.; Wang, X.; Fan, J.; Qiu, Y.; Zheng, X.; Qi, X.; Cao, Y.; Su, M.; Wang, X.; et al. Serum and urine metabolite profiling reveals potential biomarkers of human hepatocellular carcinoma. *Mol. Cell. Proteom.* **2011**, *10*, M110 004945. [CrossRef]

127. Fitian, A.I.; Nelson, D.R.; Liu, C.; Xu, Y.L.; Ararat, M.; Cabrera, R. Integrated metabolomic profiling of hepatocellular carcinoma in hepatitis C cirrhosis through GC/MS and UPLC/MS-MS. *Liver Int.* **2014**, *34*, 1428–1444. [CrossRef]

128. Chen, J.; Wang, W.; Lv, S.; Yin, P.; Zhao, X.; Lu, X.; Zhang, F.; Xu, G. Metabonomics study of liver cancer based on ultra performance liquid chromatography coupled to mass spectrometry with HILIC and RPLC separations. *Anal. Chim. Acta* **2009**, *650*, 3–9. [CrossRef]

129. Yin, P.; Wan, D.; Zhao, C.; Chen, J.; Zhao, X.; Wang, W.; Lu, X.; Yang, S.; Gu, J.; Xu, G. A metabonomic study of hepatitis B-induced liver cirrhosis and hepatocellular carcinoma by using RP-LC and HILIC coupled with mass spectrometry. *Mol. Biosyst.* **2009**, *5*, 868–876. [CrossRef]

130. Zhang, Y.; Ding, N.; Cao, Y.F.; Zhu, Z.T.; Gao, P. Differential diagnosis between hepatocellular carcinoma and cirrhosis by serum amino acids and acylcarnitines. *Int. J. Clin. Exp. Pathol.* **2018**, *11*, 1763–1769.

131. Lu, X.; Nie, H.; Li, Y.Q.; Zhan, C.; Liu, X.; Shi, X.Y.; Shi, M.; Zhang, Y.B.; Li, Y. Comprehensive characterization and evaluation of hepatocellular carcinoma by LC-MS based serum metabolomics. *Metabolomics* **2015**, *11*, 1381–1393. [CrossRef]

132. Elsemman, I.E.; Mardinoglu, A.; Shoaie, S.; Soliman, T.H.; Nielsen, J. Systems biology analysis of hepatitis C virus infection reveals the role of copy number increases in regions of chromosome 1q in hepatocellular carcinoma metabolism. *Mol. Biosyst.* **2016**, *12*, 1496–1506. [CrossRef]

133. Zhou, L.; Ding, L.; Yin, P.; Lu, X.; Wang, X.; Niu, J.; Gao, P.; Xu, G. Serum metabolic profiling study of hepatocellular carcinoma infected with hepatitis B or hepatitis C virus by using liquid chromatography-mass spectrometry. *J. Proteome Res.* **2012**, *11*, 5433–5442. [CrossRef]

134. Lin, X.; Gao, J.; Zhou, L.; Yin, P.; Xu, G. A modified k-TSP algorithm and its application in LC-MS-based metabolomics study of hepatocellular carcinoma and chronic liver diseases. *J. Chromatogr. B Anal. Technol. Biomed. Life Sci.* **2014**, *966*, 100–108. [CrossRef]

135. Lu, X.; Zhang, X.; Zhang, Y.; Zhang, K.; Zhan, C.; Shi, X.; Li, Y.; Zhao, J.; Bai, Y.; Wang, Y.; et al. Metabolic profiling analysis upon acylcarnitines in tissues of hepatocellular carcinoma revealed the inhibited carnitine shuttle system caused by the downregulated carnitine palmitoyltransferase 2. *Mol. Carcinog.* **2019**. [CrossRef]

136. Lin, M.; Lv, D.; Zheng, Y.; Wu, M.; Xu, C.; Zhang, Q.; Wu, L. Downregulation of CPT2 promotes tumourigenesis and chemoresistance to cisplatin in hepatocellular carcinoma. *OncoTargets Ther.* **2018**, *11*, 3101–3110. [CrossRef]

137. Chen, S.; Wang, C.; Cui, A.; Yu, K.; Huang, C.; Zhu, M.; Chen, M. Development of a genetic and clinical data-based (gc) risk score for predicting survival of hepatocellular carcinoma patients after tumour resection. *Cell Physiol. Biochem.* **2018**, *48*, 491–502. [CrossRef]

 *metabolites* 

*Article*

# Assessment of L-Asparaginase Pharmacodynamics in Mouse Models of Cancer

Thomas D. Horvath [1], Wai Kin Chan [1], Michael A. Pontikos [1], Leona A. Martin [1], Di Du [1], Lin Tan [1], Marina Konopleva [2,3], John N. Weinstein [1] and Philip L. Lorenzi [1,*]

[1] Department of Bioinformatics and Computational Biology and The Proteomics and Metabolomics Core Facility, The University of Texas MD Anderson Cancer Center, Houston, TX 77030, USA; thomasdhorvath@gmail.com (T.D.H.); wkchan@mdanderson.org (W.K.C.); mapontikos@gmail.com (M.A.P.); lamartin1@mdanderson.org (L.A.M.); dudthu06@gmail.com (D.D.); ltan@mdanderson.org (L.T.); jweinste@mdanderson.org (J.N.W.)

[2] Department of Leukemia, The University of Texas MD Anderson Cancer Center, Houston, TX 77030, USA; mkonople@mdanderson.org

[3] Department of Stem Cell Transplantation, The University of Texas MD Anderson Cancer Center, Houston, TX 77030, USA

* Correspondence: PLLorenzi@mdanderson.org; Tel.: +1-713-792-9999

Received: 25 November 2018; Accepted: 4 January 2019; Published: 9 January 2019

**Abstract:** L-asparaginase (ASNase) is a metabolism-targeted anti-neoplastic agent used to treat acute lymphoblastic leukemia (ALL). ASNase's anticancer activity results from the enzymatic depletion of asparagine (Asn) and glutamine (Gln), which are converted to aspartic acid (Asp) and glutamic acid (Glu), respectively, in the blood. Unfortunately, accurate assessment of the in vivo pharmacodynamics (PD) of ASNase is challenging because of the following reasons: (i) ASNase is resilient to deactivation; (ii) ASNase catalytic efficiency is very high; and (iii) the PD markers Asn and Gln are depleted ex vivo in blood samples containing ASNase. To address those issues and facilitate longitudinal studies in individual mice for ASNase PD studies, we present here a new LC-MS/MS bioanalytical method that incorporates rapid quenching of ASNase for measurement of Asn, Asp, Gln, and Glu in just 10 µL of whole blood, with limits of detection (s:n $\geq$ 10:1) estimated to be 2.3, 3.5, 0.8, and 0.5 µM, respectively. We tested the suitability of the method in a 5-day, longitudinal PD study in mice and found the method to be simple to perform with sufficient accuracy and precision for whole blood measurements. Overall, the method increases the density of data that can be acquired from a single animal and will facilitate optimization of novel ASNase treatment regimens and/or the development of new ASNase variants with desired kinetic properties.

**Keywords:** Kidrolase; Erwinaze; asparaginase; glutaminase; pharmacodynamics; targeted metabolomics

---

## 1. Introduction

L-Asparaginase (ASNase; EC 3.5.1.1) is an amidohydrolase enzyme that catalytically deamidates L-asparagine (Asn) to L-aspartic acid (Asp) and ammonia, and, to a lesser degree, L-glutamine (Gln) to L-glutamic acid (Glu) and ammonia. After early reports of the anti-lymphoma activity possessed by guinea pig serum [1,2], ASNase was determined to be responsible for the activity [3]. Currently, only the variants from *Escherichia coli* (Medac® (Medac GmbH, Wedel, Germany), Kidrolase® (Jazz Pharmaceuticals, Dublin, Ireland), and Spectrila® (Medac GmbH, Wedel, Germany), and the pegylated enzyme, Oncaspar® (Takeda Pharmaceuticals, Osaka, Japan)) and *Erwinia chrysanthemi* (Erwinaze® (Jazz Pharmaceuticals, Dublin, Ireland)) have been approved for the treatment of cancer. Other forms have been tested but were found to be too toxic; for example, clinical trials with *Wolinella succinogenes* ASNase were terminated due to toxicity. It is generally thought that ASNase-mediated depletion of

Asn in the blood plasma is an effective therapy for cancer cells that express asparagine synthetase (ASNS; EC 6.3.5.4) at low levels and, hence, depend on systemic Asn to support their growth and proliferation. In fact, a causal association between ASNase anticancer activity and ASNS expression has been demonstrated [4–8]. Asn-starved leukemia cells exhibit a global decrease in protein biosynthesis that ultimately results in cell death [9–13].

Targeting metabolism is a prominent strategy in the treatment of cancer, and ASNase targets a key set of metabolic pathways centered on its targets Asn and Gln, which affect a wide range of downstream metabolites, as shown in Figure 1 and Table S1. Despite ongoing efforts to optimize the enzyme's ratio of asparaginase:glutaminase activity, numerous challenges persist with regard to optimizing clinical outcomes with ASNase therapy. One significant issue is that therapeutic drug monitoring of plasma ASNase activity must be conducted to ensure that Asn levels are effectively depleted [14].

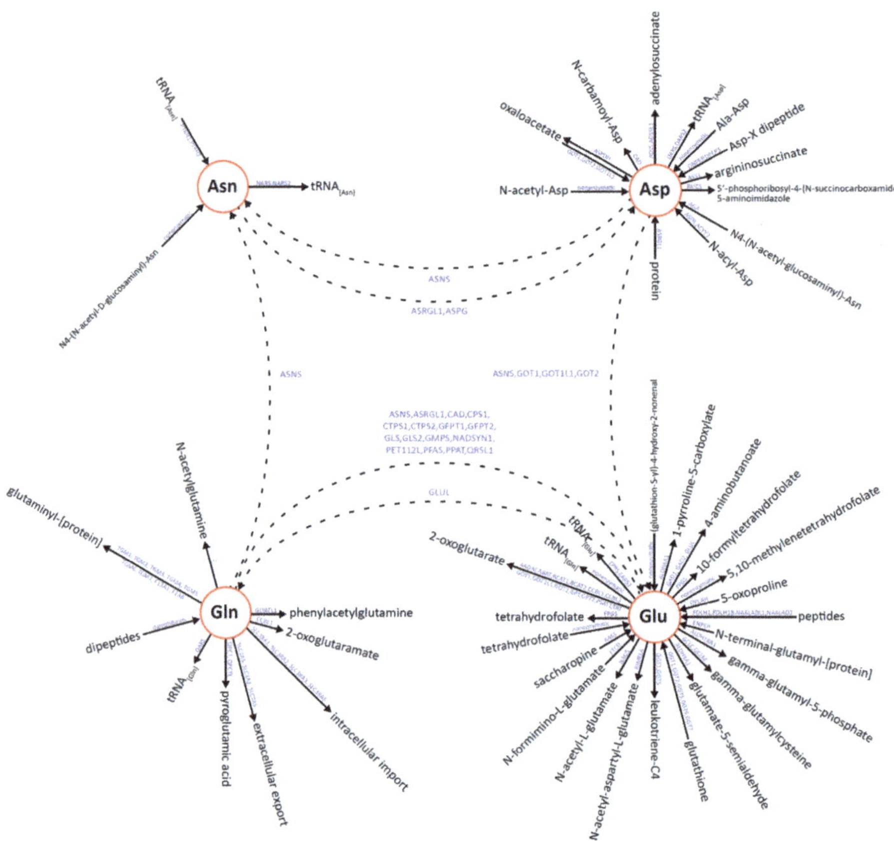

**Figure 1.** "Metaburst" of metabolic pathways modulated by ASNase, including biological reactions associated with the metabolites asparagine (Asn), aspartic acid (Asp), glutamine (Gln), and glutamic acid (Glu). All reactions are also listed in Table S1.

Unfortunately, technical challenges have hindered adoption of therapeutic drug monitoring methods. One challenge stems from the resilience of the enzyme to quenching [15,16]. A second challenge is its high catalytic efficiency ($k_{cat}/K_m$ approximately $1 \times 10^6$ M$^{-1}$s$^{-1}$) [17]. Consequently, even at the relatively low concentration of 0.1 IU/mL, ASNase fully depletes physiological concentrations of Asn within seconds [18]. Third, the pharmacodynamic (PD) markers Asn and Gln are depleted ex vivo in blood samples from patients treated with ASNase, thereby introducing analytical artifacts. A method that successfully quenches ASNase activity immediately upon blood collection by the addition of sulfosalicylic acid (SSA) has been reported [16] but requires large blood volumes (greater than 2 mL) and derivatization of the amino acids prior to chromatographic separation and fluorescence detection. Herein, we describe a liquid chromatography-tandem mass spectrometry (LC-MS/MS)-based bioanalytical method that rapidly quenches ASNase activity, demonstrates acceptable precision and accuracy across the normal range (NR) of Asn, Asp, Gln, and Glu that are typical in mouse whole blood, and has sufficient sensitivity to limit the sample volume to 10 µL, facilitating longitudinal studies in individual mice that have been treated with ASNase.

## 2. Results

### 2.1. Optimization of Amino Acid Acquisition Parameters and ASNase Activity Quenching

We first optimized the acquisition parameters on an Agilent 6460 triple quadrupole mass spectrometer using Agilent Optimizer Software (Version B.06.00) and post-column infusion; molecule-specific acquisition parameters for the analytes and internal standards are described in Table 1.

**Table 1.** Molecule-specific MS/MS parameters.

| Compound | SRM[a] (*m/z*) | Fragmentor Voltage (V) | Collision Energy (V) |
|---|---|---|---|
| $[^{13}C_0]$-Asn | 133.1 → 74.1 | 45 | 17 |
| $[^{13}C_4,^{15}N_2]$-Asn | 133.1 → 74.1 | 45 | 17 |
| $[^{13}C_0]$-Asp | 134.0 → 74.1 | 45 | 13 |
| $[^{13}C_4,^{15}N_1]$-Asp | 139.1 → 77.1 | 45 | 13 |
| $[^{13}C_0]$-Gln | 147.1 → 84.1 | 45 | 5 |
| $[^{13}C_4,^{15}N_2]$-Gln | 154.1 → 89.1 | 45 | 5 |
| $[^{13}C_0]$-Glu | 148.1 → 84.1 | 45 | 17 |
| $[^{13}C_4,^{15}N_2]$-Glu | 154.1 → 89.1 | 45 | 17 |

[a] Selected reaction monitoring.

Since quenching of ASNase is a key prerequisite for the accurate measurement of Asn, Asp, Gln, and Glu in the presence of ASNase, we first screened a range of organic solvents and organic acids for the ability to neutralize ASNase enzyme activity. The results illustrated in Figure 2 clearly show that methanol was found to be superior to acetonitrile in terms of ASNase quenching; Asn was almost completely converted to Asp in 20:80 water:acetonitrile. Given that 80% acetonitrile is widely used for the precipitation of protein from biological samples, our observations underscore the resilience of the ASNase enzyme to quenching. Another unexpected result was our observation of a chromatography issue for Asn by the presence of SSA in neat samples as shown in Supplementary Figure S1, which was shown previously to be an effective quencher of ASNase for a published LC-fluorescence based bioanalytical method [16]. Additional method development for an alternate extraction method (e.g., solid phase extraction) that removes SSA from the sample extract may eliminate the chromatography issue, but since we found alternative, effective ASNase quenching conditions that are compatible with our chromatographic system, we have not explored the use of SSA further. ASNase was successfully quenched by: (i) water containing 10% formic acid (FA), (ii) methanol containing 1% FA, and iii) acetonitrile containing 1% FA. Hence, we have chosen to incorporate those solvents into the method as the quencher, the protein precipitation solvent, and a component of the reconstitution solvent, respectively.

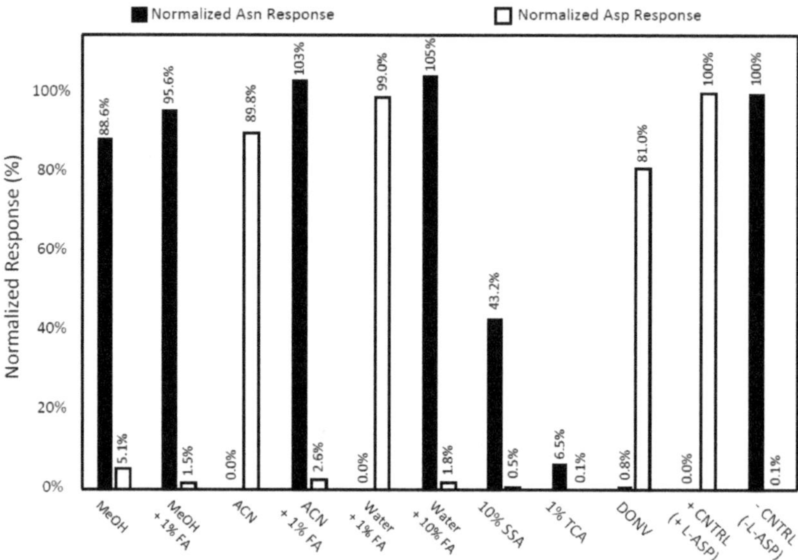


**Figure 2.** Screen for ASNase Activity Quenchers. The following conditions were tested for the ability to quench the conversion of Asn to Asp by ASNase: (1) 80:20 methanol (MeOH):water; (2) 80:20 MeOH containing 1% formic acid (FA):water; (3) 80:20 acetonitrile (ACN):water; (4) 80:20 ACN containing 1% FA:water; (5) 80:20 water containing 1% FA:water; (6) 80:20 water containing 10% FA:water; (7) 80:20 water containing 10% ($w/v$) sulfosalicylic acid (SSA):water; (8) 80:20 water containing 1% ($w/v$) trichloroacetic acid (TCA):water; (9) 80:20 water containing 40 mM 5-diazo-4-oxo-L-norvaline (DONV):water; (10) positive control sample containing ASNase in water; and (11) negative control sample containing water without ASNase. Data are presented as normalized response of Asn and Asp. The final activity of ASNase and solution concentration of Asn in each sample tested was approximately 20 IU/mL and 100 μM, respectively.

## 2.2. Accuracy, Precision, Recovery, Normalized Matrix Factor, and FTS Assessments

Analytical figures of merit were assessed through the preparation, extraction, and analysis of five analytical batches, each containing six replicates at each quality control (QC) level ($n$ = 30 replicates overall for each level) over five non-sequential days. Inter-day precision and accuracy at each QC level were defined as the coefficient of variation (%CV; standard deviation divided by the mean multiplied by 100) and percent relative error (%RE = (([AA]$_{mean}$/[AA]$_{nominal}$)-1) * 100), respectively. The resulting precision and accuracy data for the three QC levels studied are provided in Table 2. The accuracy of the mean concentrations for Gln and Asn were within 15% for all QC levels studied. The accuracy of the QC-Mid and QC-High levels for Glu and Asp were within 15%, but the accuracy of the QC-Low level in both instances was equal to or greater than 20%, which indicates that the method for these two analytes may lack the precision at the low end to discriminate between the dialyzed whole blood (DWB) matrix background and the exogenous levels of Asp and Glu contained in the QC-Low sample. Ultimately, because those two analytes are products of the ASNase reaction, the observed decrease in accuracy (increase in %RE) at QC-Low should not pose significant problems for the assay, since the in vivo whole blood concentration of Glu and Asp in the presence of ASNase are expected to increase over their empirically determined NR (79–122 μM for Glu and 30–47 μM for Asp in mouse whole blood). The background concentrations of Asn, Asp, Gln, and Glu remaining in the DWB matrix after the dialysis procedure (described in the Materials and Methods Section) were (fold-change below NR indicated in parentheses) 0.94 μM (46-fold), 1.49 μM (19-fold), 1.65 μM (348-fold), and 3.62 μM

(24-fold), respectively. The limit of detection for Asn, Asp, Gln, and Glu in just 10 μL of whole blood were estimated to be 2.3, 3.5, 0.8, and 0.5 μM, respectively.

**Table 2.** Inter-day mean concentration, accuracy, and precision for QC standards prepared in dialyzed whole blood matrix.

| [Asn]$_{nominal}$ (μM) | [Asn]$_{mean}$ (μM) [a] | Accuracy (%RE) [b] | Precision (%CV) [c] |
|---|---|---|---|
| 12.0 | 10.6 | −11.8% | 4.78% |
| 200 | 182 | −9.00% | 2.31% |
| 3200 | 3140 | −1.89% | 2.38% |
| [Asp]$_{nominal}$ (μM) | [Asp]$_{mean}$ (μM) [a] | Accuracy (%RE) [b] | Precision (%CV) [c] |
| 12.0 | 14.6 | 21.9% | 7.05% |
| 200 | 214 | 7.16% | 3.75% |
| 3200 | 3155 | 1.41% | 2.57% |
| [Gln]$_{nominal}$ (μM) | [Gln]$_{mean}$ (μM) [a] | Accuracy (%RE) [b] | Precision (%CV) [c] |
| 12.0 | 11.1 | −7.87% | 5.77% |
| 200 | 215 | 7.60% | 2.32% |
| 3200 | 3675 | 14.8% | 2.50% |
| [Glu]$_{nominal}$ (μM) | [Glu]$_{mean}$ (μM) [a] | Accuracy (%RE) [b] | Precision (%CV) [c] |
| 12.0 | 14.4 | 20.0% | 6.13% |
| 200 | 209 | 4.36% | 2.16% |
| 3200 | 3122 | −2.42% | 1.56% |

[a] The mean concentration was calculated from $1/x$ weighted linear least-squares regressions from the individual calibration curves in each batch ($n$ = 5 over five non-consecutive days) after correcting for the endogenous amino acid content contained in the dialyzed whole blood matrix; [b] percent relative error: %RE = (([AA]$_{mean}$/[AA]$_{nominal}$)-1) * 100; [c] coefficient of variation.

The normalized matrix factor (NMF) was calculated using Equation (1):

$$NMF = \left( \frac{\left( \frac{Area_{analyte,\ post}}{Area_{Analyte,\ neat}} \right)}{\left( \frac{Area_{IS,\ post}}{Area_{IS,\ neat}} \right)} \right) \tag{1}$$

where $Area_{analyte,\ post}$ and $Area_{IS,\ post}$, and $Area_{analyte,\ neat}$ and $Area_{IS,\ neat}$ are analyte and internal standard (IS) peak areas from the post-extraction DWB sample matrix and neat samples (a water blank is extracted and dried, and analyte and IS are added during the sample reconstitution step), respectively. The results for NMF and recovery assessments for each analyte and IS are provided in Table 3. Gln and Gln-IS exhibited the lowest mean recoveries at around 90% for the three levels studied, but all of the other analytes and IS compounds had recoveries near 100%. The NMF was approximately 1.0 for all analyte and IS compounds at all concentration levels tested, indicating that the degree of ion enhancement or suppression effects between each analyte/IS pair in the DWB matrix was equivalent.

**Table 3.** Mean recovery and mean normalized matrix factors (NMF) for three different quality control levels.

| [Analyte]/[IS] [a] | Mean Recovery Asn/Asn-IS [b] | Mean Recovery Asp/Asp-IS [c] | Mean Recovery Gln/Gln-IS [d] | Mean Recovery Glu/Glu-IS [e] |
|---|---|---|---|---|
| 8.00 μM/100 μM | 94%/98% | 97%/97% | 87%/89% | 96%/99% |
| 1000 μM/100 μM | 96%/96% | 101%/99% | 91%/92% | 100%/100% |
| 4000 μM/100 μM | 98%/98% | 100%/98% | 92%/92% | 99%/99% |
| [Analyte]/[IS] [a] | Mean NMF Asn/Asn-IS [b] | Mean NMF Asp/Asp-IS [c] | Mean NMF Gln/Gln-IS [d] | Mean NMF Glu/Glu-IS [e] |
| 8.00 μM/100 μM | 0.971 | 0.966 | 0.956 | 0.967 |
| 1000 μM/100 μM | 1.00 | 0.972 | 0.996 | 0.999 |
| 4000 μM/100 μM | 0.998 | 1.01 | 1.00 | 0.993 |

[a] IS: internal standard; [b] Asn-IS: [$^{13}$C$_4$,$^{15}$N$_2$]-asparagine; [c] Asp-IS: [$^{13}$C$_4$,$^{15}$N$_1$]-aspartic acid; [d] Gln-IS: [$^{13}$C$_5$,$^{15}$N$_2$]-glutamine; [e] Glu-IS: [$^{13}$C$_5$,$^{15}$N$_1$]-glutamic acid.

### 2.3. Pharmacodynamics of ASNase in NOD.Cg-PRKDC(scid) IL2RG(tm1Wjl) (NSG) Mice

We conducted a pilot study in three treatment groups of NSG mice (control, low-dose ASNase, and high-dose ASNase) to determine the suitability of the bioanalytical method for assessing ASNase PD in the mouse model of leukemia. The NR was defined as the largest absolute concentration range measured for each analyte, and the NR for each metabolite was: Asn: 40–50 µM; Asp: 30–47 µM; Gln: 528–623 µM; Glu: 79–122 µM as shown in Figure 3. Note: The connecting lines for each data series are used to visually connect the data points within each individual mouse; analyte concentrations should not be inferred from the lines between adjacent data points.

The ASNase field has historically used plasma as a biological matrix for ASNase PD. Our use of whole blood offers the notable advantage of rapid quenching of ASNase, but a weakness is that red blood cell (RBC) volume can be variable. However, since individual subjects are expected to have low variability in RBC volume, our combined use of whole blood and longitudinal methodology minimizes the variability associated with hematocrit. Although the pilot study was originally designed as a simple test case to assess the method performance, several interesting biological observations were made that warrant further studies. First, we found that Asn blood levels were detectable following treatment with ASNase, whereas previously published methods invariably yielded post-ASNase Asn levels "below the limit of detection/quantitation" [19–22]. The improved detection of Asn with our new method is partially due to the fact that we used whole blood, which captures target analytes in both the red blood cell (RBC) and plasma compartments, whereas other methods typically use serum or plasma. From the perspective of ASNase neutralization/quenching, whole blood sampling may yield more accurate results, since whole blood can be quickly quenched after collection to eliminate artifactual ex vivo depletion of Asn, whereas plasma and serum require additional time (between 3 and 10 min of centrifugation time for whole blood to plasma processing) during which even low levels of ASNase activity are able to deplete large quantities of Asn. Second, although previously published methods [23–25] report lower Limit of Detection (LOD) and Limit of Quantitation (LOQ) levels, those methods typically require blood volumes that range from 200 µL to 2 mL—volumes that are not compatible with longitudinal studies in individual mice (total blood volume in an individual mouse is ~1 mL). The precision and accuracy obtained from just 10 µL of whole blood now make it possible to conduct longitudinal studies of individual mice. Third, the results suggest that the method is suitable for measuring repletion of Asn after ASNase cessation as shown in Figure 3A; mean Asn concentrations at 96 h (48 h after the final L-ASP dose) for the 1,000 IU/(kg·day) and the 5,000 IU/(kg·day) dosages were 6.66 µM (*n* = 2; blue lines with blue squares and circles) and 4.72 µM (*n* = 2; red lines with red squares and circles), respectively. Overall, these biological observations suggest that new investigations should be undertaken to interrogate features of the biological response to ASNase treatment, including efforts to identify the true whole blood concentration of Asn that is thought to trigger cancer cell death.

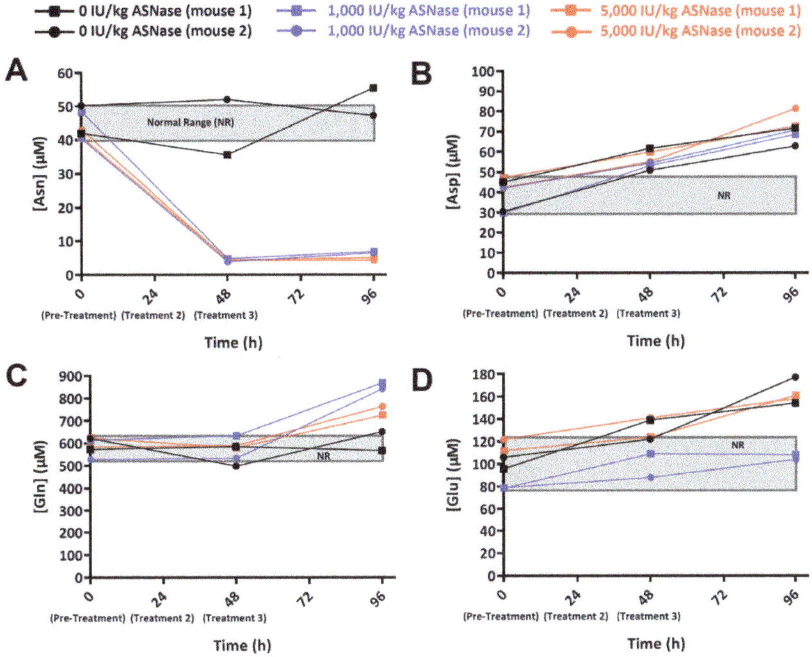

**Figure 3.** Amino acid concentration (μM) versus time (h) in whole blood of mice treated with ASNase. (**A**) asparagine (Asn), (**B**) aspartic acid (ASP), (**C**) glutamine (Gln), and (**D**) glutamic acid (Glu). Each data series represents an individual mouse, and the mice were arranged into the following three cohorts that were given intraperitoneal injections of either vehicle (1x PBS) or ASNase at 0, 24, and 48 h: (1) Control mice ($n = 2$; blackline with black squares and circles); (2) mice treated with 1000 IU/(kg·day) ASNase ($n = 2$; blue lines with blue squares and circles); (3) mice treated with 5000 IU/(kg·day) ASNase ($n = 2$ red lines with red squares and circles). ASNase was administered at 0, 24, and 48 h, and whole blood (10 μL) was collected from each mouse at 0, 48, and 96 h, and was processed as described. The gray box represents the NR levels of Asn, Asp, Gln, and Glu defined by the 0 h (pre-treatment) samples.

Blood levels of Asp tended to increase above the NR for all animals (including the control mice) by the 48 h timepoint as shown in Figure 3B. Since the disappearance of Asn was not matched by a commensurate (stoichiometric) appearance of Asp within the ASNase-treated groups, the results prompt the hypothesis that Asp is tightly regulated with excess amounts being shunted to other metabolite pools/pathways. Candidate pathways include: (1) ASNS-mediated conversion of Asp to Asn with a corresponding conversion of Gln to Glu; and (2) aspartate aminotransferase (AST/GOT1/GOT2; EC 2.6.1.1)-mediated conversion of Asp to Glu with a corresponding conversion of alpha-ketoglutaric acid to oxaloacetic acid as show in Figure 1. Further work to discern those possibilities is warranted.

The blood level of Gln at the 48 h time point (24 h after the second dose of ASNase) appeared to be near the NR defined by the 0 h time points, but at 96 h Gln levels exhibited dose-dependent up-regulation, with the 1,000 IU/kg ASNase dose yielding a larger up-regulation of Gln than the 5,000 IU/kg ASNase dose as shown in Figure 3C. The corresponding levels of Glu, by contrast, did not differ commensurately with Gln, and that was also the case for Asp, which was not modulated commensurately with Asn. One potential explanation for those results is that glutamine synthetase (GLUL; EC 6.3.1.2), which catalyzes the condensation of Glu and ammonia to produce Gln, is up-regulated at the tissue-level. Indeed GLUL has been reported to be up-regulated by ASNase treatment [26]. A number of additional enzymes that may rapidly metabolize ASNase-generated Glu to maintain steady-state levels are listed in Figure 1.

## 3. Discussion

Here we present a bioanalytical method that enables the direct measurement of the ASNase PD markers Asn, Asp, Gln, and Glu by LC-MS/MS without sample derivatization, building upon previously reported methods [27]. The method is sufficiently sensitive to measure the PD markers in as little as 10 µL of whole blood, thus facilitating longitudinal studies in individual small animals such as mice. Moreover, the method uses experimental conditions (e.g., water containing 10% FA for ASNase quenching, methanol containing 1% FA for protein precipitation, and acetonitrile containing 1% FA as a component of the reconstitution solution) that we have shown to be effective at quenching ASNase activity, thereby eliminating ex vivo turnover of Asn and Gln in the processed sample extracts prior to analysis. Application of the method to a small pilot study that included three cohorts of mice treated with increasing doses of ASNase yielded interesting biological observations that warrant further study: (i) ASNase treatment did not appear to modulate blood levels of Asp, contrary to our expectation of a stoichiometric increase in Asp concentration commensurate with ASNase-mediated depletion of Asn; (ii) unexpectedly large increases of Gln after cessation of ASNase treatment prompt the hypothesis that glutamine synthetase (GLUL) is up-regulated at a systems-level (perhaps in specific organs) in response to ASNase treatment. Overall, the method promises to improve our understanding of the mechanisms that mediate sensitivity and resistance to ASNase.

## 4. Materials and Methods

### 4.1. Reagents and Chemicals

Optima™ LC-MS-grade acetonitrile, methanol, and water, and a 1M hydrochloric acid solution were purchased from Thermo Fisher Scientific (Waltham, MA, USA). Formic acid (98%), ammonium formate, SSA, and trichloroacetic acid (TCA) were purchased from Sigma-Aldrich (St. Louis, MO, USA). Authentic reference materials and stable-isotope labeled internal standards including Asn, $[^{13}C_4,^{15}N_2]$-Asn, Asp, $[^{13}C_4,^{15}N_1]$-Asp, Gln, $[^{13}C_5,^{15}N_2]$-Gln, Glu, and $[^{13}C_5,^{15}N_1]$-Glu were purchased from Sigma-Aldrich. Kidrolase® (ASNase) was purchased from Jazz Pharmaceuticals. The L-Asn analog and irreversible ASNase inhibitor, 5-diazo-4-oxo-L-norvaline (DONV), was synthesized by Acme Bioscience (Palo Alto, CA, USA).

### 4.2. Equipment and Consumables

In this study, 1.4 mL polypropylene Matrix tubes and SepraSeal caps were purchased from Thermo Fisher Scientific and were used for frozen storage of dialyzed whole blood (DWB) matrix, DWB-derived quality control (QC) standards, and for the extraction of all study samples. Micrewtube® tubes (2.0 mL) and caps were purchased from Simport (Montreal, QC, Canada) and were used for frozen storage of the calibration (calibrator) standard spiking solutions. Slide-A-Lyzer Dialysis Cassette G2 (10,000 MWCO; 15 mL capacity) from Thermo Fisher Scientific was used to dialyze lysed whole blood matrix. A Kinetex HILIC (100 × 2.1 mm, 1.7 µm particle size) analytical column from Phenomenex (Torrance, CA, USA) was used for chromatographic separation of ASNase PD markers. Chromatographic data were acquired using a 1290 Infinity Binary UHPLC system coupled to a 6460 tandem-mass spectrometer produced by Agilent Technologies (Santa Clara, CA, USA). MassHunter LC/MS Data Acquisition Software (Version B.06.00) was used for control and operation of the LC-MS/MS system, and MassHunter Quantitative Analysis Software (Version B.06.00) was used for chromatographic peak integration. Weighted (1/x) linear regression analysis for each calibration curve was performed using GraphPad Prism (Version 6.05) from GraphPad Software (La Jolla, CA, USA).

### 4.3. Optimization of Experimental Conditions for the Effective Quenching of ASNase Activity

Overall, nine different experimental conditions were tested for ASNase quenching efficacy in freshly prepared neat solutions (i.e., fresh solvent mixtures, or solvents containing acids or inhibitors in the absence of biological material). We tested the following solutions: (1) four organic solvent solutions with and without formic acid (methanol, methanol with 1% formic acid, acetonitrile, and acetonitrile with 1% formic acid); (2) two aqueous solutions that consisted of water with 1% formic acid, and water with 10% formic acid; (3) two aqueous solutions of protein precipitation reagents that consisted of water with 10% (*w/v*) SSA and water with 1% (*w/v*) TCA; and (4) one aqueous solution that consisted of water with 40 mM DONV. A non-quenched positive control sample (+ASNase) and a negative control sample lacking ASNase (-ASNase) were also prepared with a consistent Asn concentration for normalization purposes. All of the solutions, samples, and controls were prepared fresh on the day of the experiment and tested as a single replicate in a single batch.

The quenching study samples were prepared by adding 100 µL of an ASNase spiking solution (100 IU/mL) prepared in water, and 400 µL of the appropriate quenching test solution to a labeled sample tube, and all of the samples were briefly vortex-mixed. Next, 10 µL of an Asn spiking solution (5000 µM) solution was added to each sample, and each sample was vortex-mixed briefly. The samples were allowed to incubate at room temperature for 10 min, then they were centrifuged at 17,000× $g$ for 5 min at 4 °C. The supernatant was transferred to polypropylene autosampler vials and analyzed on the LC-MS/MS system. The final activity of ASNase and solution concentration of Asn in each sample tested was approximately 20 IU/mL and 100 µM, respectively.

### 4.4. Dialysis of Mouse Whole Blood

A mixed sex, pooled lot of K2-EDTA CD-1 mouse whole blood was purchased from BioreclamationIVT (New York, NY, USA). Upon receipt, the whole blood matrix was frozen at −20 °C for a minimum of 24-h to ensure complete RBC lysis. Prior to dialysis, the whole blood matrix was thawed, then centrifuged at 4500× $g$ for 10 min to pellet the RBC debris, and finally the whole blood supernatant was dialyzed (10,000 MWCO membrane) against 5 L of 1× PBS over five 24-h passages at +4 °C. The DWB matrix was then transferred to polypropylene tubes and stored at −20 °C until needed. The mouse DWB matrix was used in the preparation of calibrators and QCs. The endogenous background levels of Asn, Asp, Gln, and Glu that remained in the DWB matrix after dialysis were determined by triplicate analysis of blank samples in each analytical run.

### 4.5. Preparation of Amino Acid Stock, Combined Intermediate, and Calibrator Spiking Solutions

Individual Asn and Gln stock solutions were prepared in a solution of water that contained 1% formic acid. Individual Asp and Glu stock solutions were prepared in a solution of water that contained 1 M hydrochloric acid to ensure solubility. The nominal concentration for each stock solution was corrected to account for salt form, purity, and water content reported on the reference material product literature. Individual stock solutions were used to prepare a combined intermediate solution that contained each analyte component at a concentration of 10,000 µM each. The combined intermediate solution was serially diluted further to prepare calibrator spiking solutions at the following concentrations: 4.00, 8.00, 100, 400, 1000, 2000, 3600, and 4000 µM. All dilutions were made using a solution of water that contained 1% formic acid as the diluent. The individual stock solutions, combined intermediate, and the individual calibrator spiking solutions were stored in polypropylene tubes and at −80 °C when not in use.

### 4.6. Preparation of Stable-Isotope Labeled Internal Standard Stock and Working Internal Standard Solutions

Individual [$^{13}C_4$,$^{15}N_2$]-Asn and [$^{13}C_5$,$^{15}N_2$]-Gln stock solutions were prepared in a solution of water that contained 1% formic acid. Individual [$^{13}C_4$,$^{15}N_1$]-Asp and [$^{13}C_5$,$^{15}N_1$]-Glu stock solutions were prepared in a solution of water that contained 1 M hydrochloric acid to ensure solubility.

The nominal concentration for each internal standard stock solution was corrected to account for salt form, purity, and water content reported on the reference material product literature. The individual internal standard stock solutions were used to prepare a working internal standard (WIS) solution that contained each internal standard component at a concentration of 1,000 μM each. All dilutions were made using a solution of water that contained 1% formic acid as the diluent. The individual internal standard stock and WIS solutions were stored in polypropylene tubes and at −80 °C when not in use.

### 4.7. Sample Extraction Procedure

The following sample extraction procedures were performed for all analytical batch preparations described in this study. All samples immediately received a 30 μL aliquot of water that contained 20% formic acid, which was added to quench ASNase activity in the study samples. All samples were capped and vortex-mixed for 30 s. All samples received a 240 μL aliquot of a solution of methanol that contained 1% formic acid to precipitate protein. All samples were vortex-mixed for two min and centrifuged for ten min at $17,000 \times g$ at ambient temperature. Supernatant was then transferred to a new sample tube and evaporated to dryness in the Savant vacuum concentrator with sample heating at 45 °C. The collection and extraction procedures used for the ASNase PD study samples is described below.

### 4.8. Sample Reconstitution

Samples were reconstituted according to the following two-step process: (1) Each sample received a 20 μL aliquot of a solution of water that contained 1% formic acid, was capped, and was vortex-mixed for 30 s to solubilize glutamic acid and aspartic acid prior to the addition of organic solvent; (2) Each sample then received a 180 μL aliquot of a solution of acetonitrile that contained 1% formic acid, was capped, and was vortex-mixed for an additional 2 min. All samples were again centrifuged at $17,000 \times g$ for ten min, and the supernatant was transferred to polypropylene injection vials (Thermo Fisher Scientific). The sample extracts were stored in a refrigerator until analysis, and a 5 μL sample volume was injected onto the instrument for analysis.

### 4.9. Liquid Chromatography/Mass Spectrometry Conditions

Hydrophilic Interaction Chromatography (HILIC) mobile phase A (MPA; weak) and mobile phase B (MPB; strong) solutions used for this study were acetonitrile/200 mM ammonium formate/formic acid (950:50:20; *v:v:v*), and acetonitrile/water/200 mM ammonium formate/formic acid (500:450:50:20; *v:v:v:v*), respectively. The chromatographic column was a Phenomenex Kinetex HILIC analytical column (2.1 × 100 mm, 1.7 μm particle size). The chromatographic method included column heating at 30 °C, autosampler tray chilling at +4 °C, a mobile phase flowrate of 0.200 mL/min, and a gradient elution program specified as follows: 0–2.5 min, 5% MPB; 2.5–10.5 min, 5–90% MPB; 10.5–12 min, 90–5% MPB; 12–14 min, 5% MPB. The overall cycle-time for a single injection was approximately 14.5 min. Representative extracted-ion chromatograms for the analytes and internal standards can be found in Supplementary Figure S2.

The Agilent Jet Stream-electrospray ionization (AJS-ESI) source was installed on the mass spectrometer and operated in unit/unit resolution and positive ionization mode with the following acquisition parameters: gas temperature: 325 °C; gas flow: 6 L/min; nebulizer: 40 psi; sheath gas temperature: 350 °C; sheath gas flow: 9 L/min; capillary voltage: +1250 V; nozzle voltage: +500 V. All reference and internal standard compounds were optimized using the Agilent Optimizer Software (Version B.06.00) and post-column infusion; molecule-specific acquisition parameters for the analytes and internal standards are described in Table 1.

### 4.10. Study Design

This study was performed in a pathogen-free vivarium at The University of Texas MD Anderson Cancer Center under an approved Institutional Animal Care and Use Committee (IACUC) study protocol (ACUF #00001658-RN00). The study included five NOD.Cg-PRKDC(scid) IL2RG(tm1Wjl) (NSG; stock #005557) mice purchased from The Jackson Laboratory (Bar Harbor, ME). The five mice were arranged into three treatment groups: (1) control ($n$ = 2; 100 μL PBS administered by intraperitoneal (IP) injection every day for three days); (2) low-dose ($n$ = 2; 100 μL PBS containing 1000 IU Kidrolase per kg body weight administered IP every day for three days); (3) high-dose ($n$ = 2; 100 μL PBS containing 5000 IU Kidrolase per kg body weight administered IP every day for three days). Each treatment was administered at 0, 24, and 48 h after study initiation. Whole blood (10 μL) was collected from each mouse before study initiation (pre-dose) and again at 48 and 96 h (48 h after cessation of treatment). When whole blood was collected on treatment days, it was collected immediately before the administration of the treatment.

### 4.11. Preparation of Sample Extraction Tubes

Prior to blood collection, the following solutions were added to individual extraction tubes corresponding to each sample: (1) 10 μL of WIS solution; (2) 10 μL of a solution of water that contained 1% formic acid to make up for the spiking solution volume in the preparation of the Calibrators; and (3) 30 μL of a solution of water that contained 20% formic acid to quench the enzymatic activity of ASNase present in the whole blood samples. Finally, all sample tubes were vortex-mixed for approximately 30 s, centrifuged at 3,000× $g$ for approximately 1 min, appropriately labeled for a specific sample, and stored on ice for transport to the vivarium.

### 4.12. Collection, ASNase Activity Quenching, and Extraction of Mouse Whole Blood Study Samples

Using a sterile taiL-snip method, a 10 μL whole blood sample was collected from the tail of each mouse and was immediately transferred to the appropriately labeled extraction tube containing the stable isotope-labeled IS compounds. Each sample was briefly vortexed to thoroughly mix the whole blood with the contents of the extraction tube. Complete RBC lysis and protein precipitation was ensured by the addition of 240 μL of an ice-cold solution of methanol that contained 1% formic acid. Samples were stored on ice, transported back to the laboratory, vortex-mixed for two min, and centrifuged for ten min at 17,000× $g$ at ambient temperature. Supernatant was then transferred to a new sample tube and evaporated to dryness in a Savant vacuum concentrator with sample heating at 45 °C. Dried sample extracts were capped and stored at −80 °C until reconstitution.

### 4.13. Quantitative Analysis Workflow

A custom quantitative analysis workflow had been devised to correct for the detectable levels of Asn, Asp, Gln, and Glu contained in the DWB matrix that was used to prepare the calibrator samples. Chromatographic peak integrations were performed using the MassHunter Quantitative Analysis software package. When peak integrations were completed for each individual analyte, the results table was exported into a custom Excel spreadsheet that was designed to automate the following computational tasks: (1) compute the mean area response for the analyte contained in the blank whole blood matrix; (2) subtract the mean analyte response in the whole blood matrix from the analyte response measured for each calibrator level; and (3) compute a corrected instrument response (IR ≡ [Corrected Area]$_{analyte}$/[Area]$_{IS}$) factor for each calibrator level. Individual calibration curves were generated in GraphPad Prism by performing a least-squares linear regression with 1/x weighting on plots of IR factors versus nominal analyte concentration for each calibrator. Individual slope and intercept outputs from the linear regression analysis for each calibration curve were input into the spreadsheet in order to compute the analyte concentration of each sample present in that batch.

**Supplementary Materials:** The following are available online at http://www.mdpi.com/2218-1989/9/1/10/s1, Figure S1: Sulfosalicylic acid exhibits an effect on Asn chromatography; Figure S2: Extracted-ion chromatograms (XIC) for the transitions monitored; Table S1: Compiled biological reactions involving asparagine, aspartic acid, glutamine, and glutamic acid.

**Author Contributions:** Conceptualization, T.D.H., J.N.W., and P.L.L.; Methodology, T.D.H., W.K.C., L.A.M., and P.L.L.; Formal Analysis, T.D.H. and L.A.M.; Investigation, T.D.H., W.K.C., M.A.P., D.D., L.T., M.K., and J.N.W.; Resources, M.K., J.N.W., and P.L.L.; Data Curation, T.D.H.; Writing-Original Draft Preparation, T.D.H. and P.L.L.; Writing-Review & Editing, M.K., J.N.W., and P.L.L.; Funding Acquisition, J.N.W. and P.L.L.

**Funding:** This research was funded by Cancer Prevention and Research Institute of Texas (CPRIT) grant number RP130397, NIH grant number 1S10OD012304-01, and NIH/NCI grant number P30CA016672.

**Conflicts of Interest:** The authors declare no conflict of interest. The funders had no role in the design of the study; in the collection, analyses, or interpretation of data; in the writing of the manuscript, or in the decision to publish the results.

## References


1. Clementi, A. La désamidation enzymatique de l'asparagine chez les différentes espéces animales et la signification physio logique de sa presence dans L'organisme. *Arch. Int. Physiol.* **1922**, *19*, 369–398. [CrossRef]

2. Kidd, J.G. Regression of transplated lymphomas induced in vivo by means of normal guinea pig serum. *J. Exp. Med.* **1953**, *98*, 565–583. [CrossRef] [PubMed]

3. Broome, J.D. Evidence that the L-asparaginase of guinea pig serum is responsible for its antilymphoma effects. *J. Exp. Med.* **1963**, *118*, 99–120. [CrossRef] [PubMed]

4. Avramis, V.I. Asparaginases: Biochemical pharmacology and modes of drug resistance. *Anticancer Res.* **2012**, *32*, 2423–2438.

5. Aslanian, A.M.; Fletcher, B.S.; Kilberg, M.S. Asparagine synthetase expression alone is sufficient to induce L-asparaginase resistance in molt-4 human leukaemia cells. *Biochem. J.* **2001**, *357*, 321–328. [CrossRef]

6. Lorenzi, P.L.; Reinhold, W.C.; Rudelius, M.; Gunsior, M.; Shankavaram, U.; Bussey, K.J.; Scherf, U.; Eichler, G.S.; Martin, S.E.; Chin, K.; et al. Asparagine synthetase as a causal, predictive biomarker for L-asparaginase activity in ovarian cancer cells. *Mol. Cancer Ther.* **2006**, *5*, 2613–2623. [CrossRef] [PubMed]

7. Lorenzi, P.L.; Llamas, J.; Gunsior, M.; Ozbun, L.; Reinhold, W.C.; Varma, S.; Ji, H.; Kim, H.; Hutchinson, A.A.; Kohn, E.C.; et al. Asparagine synthetase is a predictive biomarker of L-asparaginase activity in ovarian cancer cell lines. *Mol. Cancer Ther.* **2008**, *7*, 3123–3128. [CrossRef]

8. Chan, W.K.; Lorenzi, P.L.; Anishkin, A.; Purwaha, P.; Rogers, D.M.; Sukharev, S.; Rempe, S.B.; Weinstein, J.N. The glutaminase activity of L-asparaginase is not required for anticancer activity against asns-negative cells. *Blood* **2014**, *123*, 3596–3606. [CrossRef] [PubMed]

9. Story, M.D.; Voehringer, D.W.; Stephens, L.C.; Meyn, R.E. L-asparaginase kills lymphoma cells by apoptosis. *Cancer Chemother. Pharmacol.* **1993**, *32*, 129–133. [CrossRef] [PubMed]

10. Bussolati, O.; Belletti, S.; Uggeri, J.; Gatti, R.; Orlandini, G.; Dall'Asta, V.; Gazzola, G.C. Characterization of apoptotic phenomena induced by treatment with L-asparaginase in NIH3T3 cells. *Exp. Cell Res.* **1995**, *220*, 283–291. [CrossRef]

11. Ueno, T.; Ohtawa, K.; Mitsui, K.; Kodera, Y.; Hiroto, M.; Matsushima, A.; Inada, Y.; Nishimura, H. Cell cycle arrest and apoptosis of leukemia cells induced by L-asparaginase. *Leukemia* **1997**, *11*, 1858–1861. [CrossRef] [PubMed]

12. Fumarola, C.; Zerbini, A.; Guidotti, G.G. Glutamine deprivation-mediated cell shrinkage induces ligand-independent cd95 receptor signaling and apoptosis. *Cell Death Differ.* **2001**, *8*, 1004–1013. [CrossRef] [PubMed]

13. Holleman, A.; den Boer, M.L.; Kazemier, K.M.; Janka-Schaub, G.E.; Pieters, R. Resistance to different classes of drugs is associated with impaired apoptosis in childhood acute lymphoblastic leukemia. *Blood* **2003**, *102*, 4541–4546. [CrossRef] [PubMed]

14. Marini, B.L.; Perissinotti, A.J.; Bixby, D.L.; Brown, J.; Burke, P.W. Catalyzing improvements in all therapy with asparaginase. *Blood Rev.* **2017**, *31*, 328–338. [CrossRef] [PubMed]

15. Asselin, B.L.; Lorenson, M.Y.; Whitin, J.C.; Coppola, D.J.; Kende, A.S.; Blakley, R.L.; Cohen, H.J. Measurement of serum L-asparagine in the presence of L-asparaginase requires the presense of and L-asparaginase inhibitor. *Cancer Res.* **1991**, *51*, 6568–6573. [PubMed]

16. Gentili, D.; Zucchetti, M.; Conter, V.; Masera, G.; D'Incalci, M. Determination of L-asparagine in biological samples in the presence of L-asparaginase. *J. Chromatogr. B* **1994**, *657*, 47–52. [CrossRef]

17. Anishkin, A.; Vanegas, J.M.; Rogers, D.M.; Lorenzi, P.L.; Chan, W.K.; Purwaha, P.; Weinstein, J.N.; Sukharev, S.; Rempe, S.B. Catalytic role of the substrate defines specificity of therapeutic L-asparaginase. *J. Mol. Biol.* **2015**, *427*, 2867–2885. [CrossRef] [PubMed]

18. Purwaha, P.; Lorenzi, P.L.; Silva, L.P.; Hawke, D.H.; Weinstein, J.N. Targeted metabolomic analysis of amino acid response to L-asparaginase in adherent cells. *Metabolomics* **2014**, *10*, 909–919. [CrossRef] [PubMed]

19. Hawkins, D.S.; Park, J.R.; Thomson, B.G.; Felgenhauer, J.L.; Holcenberg, J.S.; Panosyan, E.H.; Avramis, V.I. Asparaginase pharmacokinetics after intensive polyethylene glycoL-conjugated L-asparaginase therapy for children with relapsed acute lymphoblastic leukemia. *Clin. Cancer Res.* **2004**, *10*, 5335–5341. [CrossRef]

20. Pieters, R.; Hunger, S.P.; Boos, J.; Rizzari, C.; Silverman, L.; Baruchel, A.; Goekbuget, N.; Schrappe, M.; Pui, C.-H. L-asparaginase treatment in acute lymphoblastic leukemia: A focus on erwinia asparaginase. *Cancer* **2011**, *117*, 238–249. [CrossRef]

21. Appel, I.M.; Kazemier, K.M.; Boos, J.; Lanvers, C.; Huijmans, J.; Veerman, A.J.P.; van Wering, E.; den Boer, M.L.; Pieters, R. Pharmacokinetic, pharmacodynamic and intracellular effects of peg-asparaginase in newly diagnosed childhood acute lymphoblastic leukemia: Results from a single agent window study. *Leukemia* **2008**, *22*, 1665–1679. [CrossRef] [PubMed]

22. Panetta, J.C.; Gajjar, A.; Hijiya, N.; Hak, L.J.; Cheng, C.; Liu, W.; Pui, C.-H.; Relling, M.V. Comparison of native *E. coli* and peg asparaginase pharmacokinetics and pharmacodynamics in pediatric acute lymphoblastic leukemia. *Clin. Pharmacol. Ther.* **2009**, *86*, 651–658. [CrossRef] [PubMed]

23. Jarrar, M.; Gaynon, P.S.; Periclou, A.P.; Fu, C.; Harris, R.E.; Stram, D.; Altman, A.; Bostrom, B.; Breneman, J.; Steele, D.; et al. Asparagine depletion after pegylated *E. coli* asparaginase treatment and induction outcome in children with acute lymphoblastic leukemia in first bone marrow relapse: A children's oncology group study (CCG-1941). *Pediatr. Blood Cancer* **2006**, *47*, 141–146. [CrossRef] [PubMed]

24. Tong, W.H.; Pieters, R.; Kaspers, G.J.; te Loo, D.M.; Bierings, M.B.; van den Bos, C.; Kollen, W.J.; Hop, W.C.; Lanvers-Kaminsky, C.; Relling, M.V.; et al. A prospective study on drug monitoring of pegasparaginase and erwinia asparaginase and asparaginase antibodies in pediatric acute lymphoblastic leukemia. *Blood* **2014**, *123*, 2026–2033. [CrossRef] [PubMed]

25. Gentili, D.; Conter, V.; Rizzari, C.; Tschuemperlin, B.; Zucchetti, M.; Orlandoni, D.; D'Incalci, M.; Masera, G. L-asparagine depletion in plasma and cerebro-spinal fluid of children with acute lymphoblastic leukemia during subsequent exposures to erwinia L-asparaginase. *Ann. Oncol.* **1996**, *7*, 725–730. [CrossRef]

26. Rotoli, B.M.; Uggeri, J.; Dall'Asta, V.; Visigalli, R.; Barilli, A.; Gatti, R.; Orlandini, G.; Gazzola, G.C.; Bussolati, O. Inhibition of glutamine synthetase triggers apoptosis in asparaginase-resistant cells. *Cell Physiol. Biochem.* **2005**, *15*, 281–292. [CrossRef]

27. Prinsen, H.C.; Schiebergen-Bronkhorst, B.G.; Roeleveld, M.W.; Jans, J.J.; de Sain-van der Velden, M.G.; Visser, G.; van Hasselt, P.M.; Verhoeven-Duif, N.M. Rapid quantification of underivatized amino acids in plasma by hydrophilic interaction liquid chromatography (HILIC) coupled with tandem mass-spectrometry. *J. Inherit. Metab. Dis.* **2016**, *39*, 651–660. [CrossRef] [PubMed]

 *metabolites*

*Review*

# Critical Review of Volatile Organic Compound Analysis in Breath and In Vitro Cell Culture for Detection of Lung Cancer

Zhunan Jia [1,2], Abhijeet Patra [1], Viknish Krishnan Kutty [1] and Thirumalai Venkatesan [1,2,3,4,5,*]

1   NUSNNI-Nanocore, National University of Singapore, Singapore 117411, Singapore; nnijz@nus.edu.sg (Z.J.); abhijeet@nus.edu.sg (A.P.); viknish@nus.edu.sg (V.K.K.)
2   NUS Graduate School for Integrative Sciences and Engineering, National University of Singapore, Singapore 117456, Singapore
3   Department of Electrical Engineering, National University of Singapore, Singapore 117583, Singapore
4   Department of Materials Science and Engineering, National University of Singapore, Singapore 117574, Singapore
5   Department of Physics, National University of Singapore, Singapore 117581, Singapore
*   Correspondence: venky@nus.edu.sg; Tel.: +(65)-65165187

Received: 10 January 2019; Accepted: 13 March 2019; Published: 18 March 2019

**Abstract:** Breath analysis is a promising technique for lung cancer screening. Despite the rapid development of breathomics in the last four decades, no consistent, robust, and validated volatile organic compound (VOC) signature for lung cancer has been identified. This review summarizes the identified VOC biomarkers from both exhaled breath analysis and in vitro cultured lung cell lines. Both clinical and in vitro studies have produced inconsistent, and even contradictory, results. Methodological issues that lead to these inconsistencies are reviewed and discussed in detail. Recommendations on addressing specific issues for more accurate biomarker studies have also been made.

**Keywords:** volatile organic compound; lung cancer; breath analysis; in vitro study; biomarker

## 1. Introduction

Cancer is the second leading cause of death by disease worldwide, exceeded only by heart disease [1]. Among all types of cancer, lung cancer accounts for 1.6 million deaths each year, exceeding those of the next three most common cancers combined (prostate, breast, and colon cancer) [2]. Lung cancer is typically silent in its early stages; symptoms such as coughing, chest pain, weight loss, etc. are often ignored by patients as typical signs of the onset of old age. Histologically, lung cancer is divided into non-small cell lung cancer (NSCLC) and small cell lung cancer (SCLC), with the former accounting for about 85% of cases and the latter, the remaining 15%. NSCLC can further be classified into adenocarcinoma, squamous cell carcinoma, and large cell carcinoma [3]. Treatment options and prognosis are critically dependent on the stage and histology of the disease. Using current diagnostic techniques, such as computer tomography (CT), sputum cytology, and biopsy, 85% of lung cancer cases are diagnosed at a stage when treatment is ineffective at curing the disease [4]. Overall, the 5-year survival is about 10–15% due to late diagnosis. However, if the disease is diagnosed at stage 1, the 5-year survival increases dramatically to 80% [5]. With lung cancer incidences rising around the world, the need for an early detection tool is both critical and urgent. Breath volatile organic compound (VOC) analysis is one such promising technique.

In the last few decades, extensive effort has been focused on searching for VOC biomarkers for lung cancer, either from the headspace of lung cancer cells or from the exhaled breath of patients. However, both clinical and in vitro studies have failed to produce a consistent and validated list.

This review aims to summarize the volatile markers produced by these studies in the last 30 years and discuss the methodological issues that have led to the inconsistencies between different studies.

The most commonly used techniques for VOC analysis include mass spectrometry and sensor technologies. Mass spectrometry-based studies usually produce a list of molecules that could be used as biomarkers, while studies using sensor arrays only produce a pattern without individual compound identification. Though detecting lung cancer using various sensor technologies has produced meaningful and promising results [6–14], it is beyond the scope of the current review. Comprehensive reviews on lung cancer VOC studies using sensors could be found elsewhere [15,16].

In this review, we focus on mass spectrometry-based clinical, as well as in vitro, studies that provided individual compound identification. We compare the results from these studies and discuss in detail the methodological issues that have led to the inconsistencies across different studies. A short and concise review on breath analysis of lung cancer can be found in [17]. Saalberg et al. and Hua et al. did systematic reviews on breath analysis as a screening technique for lung cancer [18,19]. Zhou et al. discussed the recent developments in the analytical techniques of breath analysis for lung cancer detection [20]. None of these reviews discussed in vitro studies or evaluated the methodological issues of these studies.

## 2. VOC Biomarkers of Lung Cancer in Exhaled Breath

The pioneering study on VOC in exhaled breath from lung cancer patients was done by Gordon et al. in 1985 using gas chromatography mass spectrometry (GC-MS) [21]. Since then, interest in the clinical diagnostic potential of breath analysis in lung cancer detection has risen, evidenced by a rapidly increasing number of publications in the last 30 years. In Table 1, we summarize 25 clinical studies on the breath analysis of lung cancer patients who have identified biomarkers. A majority of these studies adopted a case control approach. Lung cancer patients were recruited as a case group, subjects not clinically diagnosed with lung cancer were recruited as a control group, and the breath VOC profile was compared between them. An identified VOC is considered as a biomarker if its concentration is statistically different between these two groups. Almost all studies used GC-MS as the analytical platform, with the exception of two studies that used proton transfer reaction mass spectrometry (PTR-MS) [22] and ion mobility mass spectrometry (IMS) [23]. Bajtarevic et al. reported results from both PTR-MS and GC-MS [24].

The lung cancer biomarkers identified by these studies are largely inconsistent. To better illustrate the biomarker results of Table 1, we filtered the biomarkers that have been identified by at least four studies and ranked them based on the occurrence (Figure 1). The most frequently emerging biomarkers of lung cancer include propanol, isoprene, acetone, pentane, hexanal, toluene, benzene, and ethylbenzene. Michael Philips, one of the pioneers in the breath research field, conducted three independent biomarker discovery studies for lung cancer using GC-MS [25–27]. In view of the different lists produced from these studies, he commented that although the exact identities of markers derived from these three studies are not the same, the major biomarkers were mainly alkane derivatives, which are consistent in all three of his studies. The relative abundance of most of these VOCs was found to have decreased in the participants with lung cancer, as compared to the healthy control; this difference could be attributed to the increased catabolism of lipid peroxidation products due to the activated CYP450 genotypes in lung cancer [26]. However, there are many other studies in which alkanes were not found to be associated with lung cancer [28–30]. None of these studies evaluated the origin of the detected VOCs. In fact, the mechanism of most VOCs in exhaled breath remains unknown. Hakim et al. reviewed the possible biochemical pathways of lung cancer related VOCs [31].

Generally, it is accepted that until now there has been no consistent and validated list of VOC biomarkers for lung cancer in the literature [31–33]. Reasons for these inconsistencies are manifold. There is a large variation in different studies in terms of breath sampling procedures, study designs (selection of control group, selection of patients, etc.), and data analysis protocols. Insightful accounts on the advantages and drawbacks of various data analysis techniques can be found in [34,35].

**Table 1.** Identified volatile biomarkers of lung cancer through breath (Chronological order).

| Year | First Author | Sample Size | | Biomarker |
|------|--------------|-------------|---------|-----------|
| | | Lung Cancer | Control | |
| **1985** | Gordon [21] | 12 | 17 | acetone, 2-butanone, n-propanol |
| **1988** | O'Neill [36] | 8 | 0 | hexane, 2-methylpentane, trimethyl heptane, isoprene, benzene, toluene, ethylbenzene, cumene, trimethyl benzene, alkylbenzene, styrene, naphthalene, 1-methylnaphthalene, propanal, acetone, 2-butanone, phenol, benzaldehyde, acetophenone, nonanal, ethyl propanoate, methyl isobutanoate, dichloromethane, dichlorobenzene, trichloroethane, trichlorofluoromethane, tetrachloroethylene |
| **1999** | Philips [25] | 60 | 48 | styrene, 2,2,4,6,6-pentamethylheptane, 2-methylheptane, decane, n-propylbenzene undecane, methyl cyclopentane, 1-methyl-2-pentylcyclopropane, trichlorofluoromethane, benzene, 1,2,4-trimethylbenzene, isoprene, 3-methyloctane, 1-hexene, 3-methylnonane, 1-heptene, 1,4-dimethylbenzene, 2,4-dimethylheptane, hexanal, cyclohexane, 1-methylethenylbenzene, heptanal |
| **2003** | Philips [26] | 178 | 102 | butane, 3-methyltridecane, 7-methyltridecane, 4-methylctane,3-methylhexane, heptane, 2-methylhexane, pentane, 5-methyldecane |
| **2005** | Poli [37] | 36 | 85 | 2-methylpentane, pentane, ethylbenzene, xylenes, trimethylbenzene, toluene, benzene, heptane, decane, styrene, octane, pentamethyl heptane |
| **2007** | Philips [27] | 193 | 211 | 1,5,9-trimethyl-1,5,9-cyclododecatriene, 2,2,4-trimethyl-1,3-pentanediol tributyrate, ethyl 4-ethoxybenzoate, 2-methyl- propanoic acid, (1,1-dimethylethyl)-2-methyl-1,3-propanediyl ester, 10,11-dihydro-5H-dibenz-(b,f)-azepine, 2,5-2,6-bis(1,1-dimethylethyl)-cyclohexadiene-1,4-dione, 1,1-oxybi-benzene, 2,5-dimethyl-furan, 2,2-diethyl-1,1-biphenyl, 2,4-dimethyl-3-pentanone, trans-caryophyllene, 2,3-dihydro-1,1,3-trimethyl-3-phenyl-1H-indene, 1-propanol, 4-methyl-decane, 1,2-benzenedicarboxylic acid, diethyl ester, 2,5-dimethyl-2,4-hexadiene |
| **2007** | Wehinger [38] | 17 | 170 | formaldehyde, isopropanol |
| **2009** | Bajtarevic [24] | 220 | 441 | isoprene, acetone, methanol, 2-butanone, benzaldehyde, 2,3-butanedione, 1-propanol |
| **2010** | Fuchs [28] | 12 | 12 | pentanal, hexanal, octanal, nonanal |
| **2010** | Peng [39] | 30 | 22 | p-cymene, toluene, dodecane, 3,3-dimethylpentane, 2,3,4-trimethylhexane, (1-phenyl-1-butenyl)benzene 1,3-dimethylbenzene, 1-iodononane, [(1,1-dimethylethyl) thiol]acetic acid, 4-(4-propylcyclohexyl)-4′-cyano[1,1′-biphenyl]4-yl ester benzoic acid, 2-amino-5-isopropyl-8-methyl-1-azulenecarbonitrile, 5-(2-methylpropyl)nonane, 2,3,4-trimethyldecane, 6-ethyl-3-octanyl 2-(trifluoromethyl)benzoate, p-xylene, and 2,2-dimethyldecane |
| **2010** | Song [40] | 43 | 41 | 1-butanol, 3-hydroxy-2-butanone |

**Table 1.** *Cont.*

| Year | First Author | Sample Size | | Biomarker |
|---|---|---|---|---|
| | | Lung Cancer | Control | |
| 2010 | Kischkel [41] | 31 | 31 | isoprene, acetone, 2-butanone, cyclohexanone, dimethyl sulfide, acetonitrile, ethanol, isopropanol, acetaldehyde, propanal, butanal, pentanal, hexanal, heptanal, octanal, 2-propenal, 2-butenal, propane, butane, pentane, hexane, heptane, 2-methylbutane, 2-methylpropanal, 2,2-dimethylbutane, 2,3-dimethylbutane, 2-methylpentane, 3-methylpentane, 2,2-dimethylpentane, 2,4-dimethylpentane, 3,3-dimethylpentane, 2-methylhexane, cyclohexane, benzene, toluene, chlorobenzene, 1,2-dimethylbenzene, 1,2-dichlorobenzene, carbon disulfide, dimethyl formamide, 2,5-dimethylfuran, 1-propanol |
| 2011 | Rudnicka [42] | 23 | 30 | propane, carbon disulfide, 2-propenal, ethylbenzene, isopropyl alcohol |
| 2011 | Ulanowska [43] | 134 | 143 | ethanol, acetone, butane, dimethyl sulfide, isoprene, propanal, 1-propanol, 2-pentanone, furan, o-xylene, ethylbenzene, pentanal, hexanal, nonane |
| 2012 | Peled [44] | 53 | 19 | 1-octene |
| 2012 | Wang [45] | 88 | 85 | 2,4,6-trimethyloctane, 2-methyldodecane, 2-tridecanone, 2-pentadecanone, 8-methylheptadecane, 2-heptadecanone, nonadecane, eicosane |
| 2012 | Buszewski [46] | 29 | 44 | butanal, ethyl acetate, 2-pentanone, ethylbenzene, 1-propanol, 2-propanol |
| 2014 | Handa [23] | 50 | 39 | 3-methyldodecane, 1-butanol, 2-methylbutylacetate/2-hexanol/nonanal cyclohexanone, isopropylamine, ethylbenzene, hexanal, cyclohexanone, heptanal, 3-methyl-1-butanol |
| 2014 | Wang [29] | 18 | 0 | caprolactam, propanoic acid |
| 2014 | Zou [47] | 79 | 38 | 2-methyl-5-propylnonane, butylated hydroxytoluene, 2,6,11-trimethyl-dodecane, hexadecanal, 8-hexylpentadecane |
| 2015 | Capuano [48] | 20 | 10 | ethanol, 2-butanone, thiophene, 4-heptanone, butanoic acid, , acetic acid, cyclohexanone, 2,2,-dimethyl-hexanal, 1,1-diethoxy-3-methylbutane; 1-(1-ethoxyethoxy)-pentane, 2,2,6-trimethyloctane, 2-ehtyl-1-hexanol, undecane, thymol, 2-methyl-1-decanol, 3,7-dimethyl- decane, |
| 2015 | Corradi [49] | 71 | 67 | pentane, 2-methylpentane, hexane, benzene, ethylbenzene, trimethylbenzene, heptane, pentamethyl heptane, toluene, total xylenes, styrene, propanal, butanal, pentanal, hexanal, heptanal, octanal, nonanal, trans-2-hexenal, trans-2-heptenal, trans-2-nonenal |
| 2016 | Monila [30] | 68 | 60 | p-cresol, eicosenamide, 1-hexadecylindane and cumyl alcohol |
| 2016 | Schallschmidt [50] | 37 | 23 | propanal, butanal, decanal, butanal, 2-butanone, ethylbenzene |
| 2017 | Sakumura [51] | 107 | 29 | hydrogen cyanide, methanol, acetonitrile, isoprene, 1-propanol |

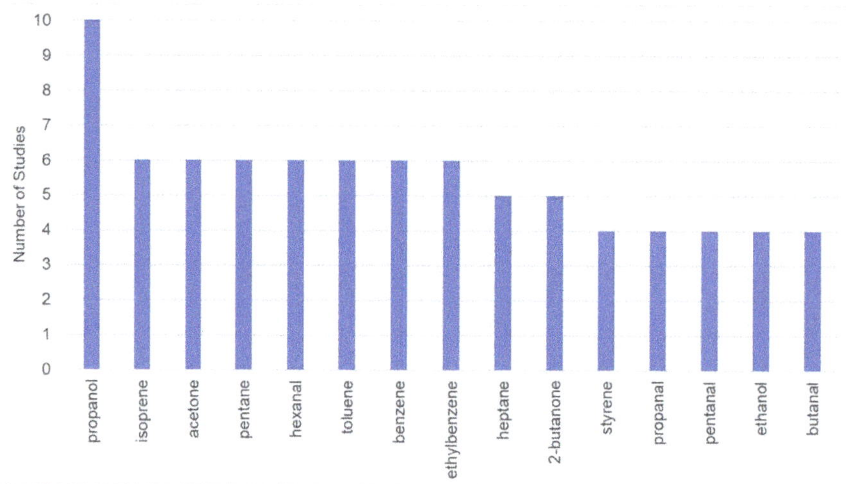


**Figure 1.** Volatile organic compounds (VOCs) identified as lung cancer biomarker in four or more studies.

## 2.1. Methodological Issues of Clinical Studies

In Table 2, we have summarized and listed the various methodological issues in breath sampling and study design. In this section, we will discuss in detail the effect of various methodological issues, what researchers have done in the included 25 studies to account for these issues, and, consequently, what the current best practices are, based on these studies.

**Table 2.** Methodological Issues of Clinical Studies.

| Breath Sampling | Study Design |
|---|---|
| | 6. Age/gender |
| | 7. Diet |
| 1. Environmental VOCs | 8. Exercise |
| 2. Phase of breath (alveolar vs. whole breath) | 9. Smoking |
| 3. Expiratory flow rate and Hyperventilation | 10. Medication |
| 4. Temperature and humidity of environment | 11. Comorbidities |
| 5. Contaminations from collection system | 12. Disease Stage |
| | 13. Histology |

## 2.2. Environmental VOCs

More than 1000 VOCs have been detected in human breath and the majority of these VOCs have exogenous origins [52]. The effect of environmental VOCs on breath analysis was first recognized by Philips, and he has proposed that this problem can be solved by determination of the "alveolar gradient" of a VOC [42,43]. The alveolar gradient is defined as the concentration of the VOC in breath minus the concentration in the room air. A positive alveolar gradient means more of the VOC was exhaled than inhaled and vice versa. Philips measured the alveolar gradients of various VOCs and concluded that VOCs with negative alveolar gradients are metabolized by the body and those with positive alveolar gradients are manufactured in the body [53,54]. However, later studies proved this to be an incorrect assertion. VOCs with positive alveolar gradients may result from VOCs absorbed from food [55], drugs [56,57], or even bacteria in the GI tract, airways, or mouth cavity [58]. On the other hand, VOCs with a negative alveolar gradient may, in fact, have metabolic origins. The journey of environmental VOCs in the human body is a complex process of mixing, diffusion, distribution in blood and fat tissues, and metabolism, as shown in the report by Philips et al. [59]. The rate and degree to which environmental VOCs are removed from the body depend on the concentration of the VOC, the duration of exposure [60], the solubility in blood and lipid tissues [61], and individual physiology.

Early theoretical modelling experiments, aimed at evaluating the health effects of industrial VOC exposure, have shown that the partition coefficient of a VOC in lungs, blood, and tissue is specific to its physical and chemical properties, and varies immensely [62,63]. Schubert et al. measured inspired, expired, and blood concentrations for four VOCs (pentane, acetone, isoprene, and isoflurane) and found that only when the inspired concentration was less than 5% of the expired concentration did the disappearance rate of VOC from the blood correlate significantly with the rate of exhalation [64]. Another study by Spanel et al. found that all seven studied VOCs (pentane, isoprene, acetone, ammonia, methanol, formaldehyde, and deuterated water) are partially retained by the body, and there are close linear relationships between the exhaled and inhaled concentrations [65]. They also introduced a useful parameter called a retention coefficient, which is the ratio of the increase of the exhaled concentration to the increase of the inhaled concentration. The retention coefficients measured for these seven VOCs vary numerically from 0.06 for formaldehyde to 0.76 for pentane [65].

With these initial efforts, it is evident, unfortunately, that there is no general rule that can be applied to all VOCs when accounting for the effects of environmental VOCs. Apart from using the concept of alveolar gradient, researchers have addressed this issue either by using an inspiratory filter [12] or by letting the patients stay in a ventilated room for a predetermined amount of time before collection [45].

For discovery-type studies, it seems that an inspiratory filter is the best solution, for now. However, the time it takes to clear out environmental VOCs from the body is compound specific. Much more effort is, therefore, needed to understand the origins and dynamics of various VOCs observed in human breath.

### 2.3. Phase of Breath Sample Collected

Each exhalation can be divided into three phases based on the $CO_2$ pressure in the breath. Phase 1 and phase 2 are air from the dead space in oral cavity and upper airways. Phase 3 is alveolar air from deep inside the lungs [66]. End-tidal breath refers to the portion of alveolar air nearer to the end of one exhalation. For the purpose of disease diagnostics, the alveolar phase is desired because VOCs in this portion are from the blood-gas exchange in the alveoli and thus more closely reflect metabolic conditions. Concentrations of certain VOCs differ in whole breath versus end-tidal breath. For example, VOCs such as carbonic acid, dimethyl ester, and methyl format were found to be significantly higher in end-tidal breath, while methylene chloride and 3-ethyl pentane were lower in end-tidal breath than in whole breath [67].

Most studies on lung cancer breath analysis either collected whole breath [26,38] or collected alveolar or end-tidal breath based on a crude estimation. The end-tidal breath was collected either by discarding the front portion of the breath [37,45] or by filling the dead space air into separate bags [39]. Kischkel et al. collected alveolar breath based on a fast responding $CO_2$ sensor [41].

To selectively collect alveolar phase accurately, monitoring $CO_2$ pressure is a good idea. $CO_2$ level is a reliable indicator of the phase, and $CO_2$ sensors are readily available. Birken et al. integrated a capnography setup into the breath collection procedure to visually monitor the phase of breath and manually draw alveolar air using a syringe [68]. Later, the same group developed an automatic $CO_2$ controlled sampling device and demonstrated the performance of the automatic sampler to be comparable to manual sampling [69]. In 2016, Owlstone Medical developed a breath collection device named RECIVA™ [70]. This is the first and only commercially available breath collection device that allows an accurate selection of phase. Such controlled and standardized methods in selecting phase of breath could significantly improve the consistency of breath biomarker research.

### 2.4. Expiratory Flow Rate, Breath-Holding, and Hyperventilation

Studies have shown that expiratory flow rate and hyperventilation affect the levels of various VOCs. Contradictory results were reported on the effect of flow rates on common breath VOCs. Doran et al. found that at a higher flow rate, lower levels of acetone and phenols were observed [67].

However, in another study, acetone level was not affected by the expiratory flow rate [71], and in yet another study, higher levels of acetone at a higher flow rate were reported [72]. Reports on trends of isoprene at different expiratory flow rates are also contradictory [73,74].

When asked to provide a breath sample, some people tend to hold their breath before exhalation. Therefore, the effect of breath-holding on breath VOC was investigated, and results from different studies are in agreement. Pentane and isoprene levels increased significantly after 20 s of breath-bolding [73]. A similar trend was observed in another study for isoprene after a 40 s breath holding [74]. Other VOCs found to be increased after breath holding in this study include 2-propanol and acetaldehyde [74]. In a third study, acetone, methanol, isoprene, and dimethyl sulphide increased significantly after 30 s of breath holding [72]. The unanimous increase of various VOCs with breath holding may be attributed to the prolonged time for them to diffuse from the alveoli to the airway.

Hyperventilation was found to have a negative effect on levels of methanol, dimethyl sulfide, acetaldehyde, and ethanol [72], as well as isoprene [75]. On the other hand, acetone level was not notably affected by hyperventilation [75]. This is attributed to the fact that acetone has much lower solubility in blood than the rest of the VOCs, and, therefore, can be quickly released from blood during hyperventilation.

The diameter of the mouthpiece used during sampling affects airway resistance and, subsequently, affects levels of certain VOCs. It was found that a smaller mouthpiece diameter caused a 19% increase in isoprene levels. Furan hydrogen sulfide also increased significantly [76]. A mouthpiece with a diameter larger than 1 cm was recommended for future studies [76].

In all the studies included in Table 1, subjects were asked to breath "normally", flow rate was not measured, and no information on the diameter of mouthpiece was given.

The learning outcome from these studies is that sampling parameters, such as exhalation rate, breath-holding, and airway resistance, must be controlled and recorded in a standardized manner across different subjects for consistent and reliable results. Hyperventilation should be avoided.

### 2.5. Temperature and Humidity of Environmental Air

This is a seldom noticed, but important, confounder for longitudinal studies, in which breath collection spans over a long period of time, during which temperature and humidity change drastically (e.g., winter to summer) [44]. This factor is also critical for multi-center clinical studies, when collection is done in different regions of the world with significantly different climates [26]. Thekedar et al. [74] assessed the changes in exhaled VOC concentrations sampled after a 5 min stay under 3 °C, 47% relative humidity and 27 °C, 19% relative humidity using PTR-MS. Acetonitrile, ethanol, methanol, and propanol showed higher concentrations in the samples collected under the warm air compared to those collected in the cold air. On the other hand, VOCs with a proton transfer reaction product of ion $m/z$ s of 85, 86, 99 and 169 showed higher concentrations in samples collected under cold air. The fact that studies on lung cancer biomarkers were conducted under vastly different temperature and humidity conditions is another significant reason for the inconsistent results.

### 2.6. Contamination from Collection Systems

Unfortunately, most commercially available breath collection apparatuses involve materials that could, themselves, be sources of contamination. For example, the widely accepted Tedlar bags for breath collection are made of polyvinyl fluoride and are known to emit many VOCs from the bag material, which, therefore, poses a threat of contamination [77,78]. Several studies used tedlar bags for lung cancer biomarker discovery [38,39,42,47]. Inert materials such as Teflon, stainless steel, or glass should be used for breath collection systems and other types of materials should be avoided as much as possible.

### 2.7. Age/Gender

Isoprene, alkanes and methylated alkanes were found to be related to age [79–81]. With an increase in oxidative stress level with advancing age, levels of these VOCs in breath increase gradually.

Spanel et al. also found that breath ammonia increases with increasing age, but acetone and hydrogen cyanide do not vary greatly with age [82]. Gender also influences the breath VOC profile. isoprene and several other VOCs were found to be gender specific [41,80,83].

In case control studies, results could be biased due to unmatched age and gender. In the reviewed studies, lung cancer patients are often significantly older than control subjects [23,24,38,41,43–45,47], and usually there are more male subjects in the case group than in the control group [24,26,41,43].

## 2.8. Diet

The effect of diet on breath VOCs is also a complex one. Certain types of food, such as yoghurt [84] and seafood [85] contains a number of VOCs that rapidly and directly appear after ingestion. Food also affects breath VOC by changing metabolism, inflammation, or redox status, or by interacting with gut flora. Some studies required subjects to fast for 12 h, or overnight, before breath collection [25,42,45,47], while other studies had no restrictions; It is not understood how long it takes for the VOCs from diet to be eliminated from breath. Whether overnight fasting helps in eliminating these effects needs to be studied further. On the other hand, dietary style may have a prolonged effect that could not be eliminated by fasting.

## 2.9. Smoking

Smoking was identified as one of the key risk factors for lung cancer, and smoke contain many VOCs. It was found smokers have higher levels of benzene and acetonitrile in their breath. Although the level of benzene in a smoker rapidly decreases to a similar level as a non-smoker within an hour, the level of acetonitrile, since the last smoking, takes about a week to become that of a non-smoker [86]. Alcohol consumption leads to increased levels of acetaldehyde in the breath, and it was found long-term smoking elevates the production of acetaldehyde from alcohol [56]. This finding was confirmed by another study [87]. These results show that smoking can affect other metabolic pathways that are not directly related to VOCs from cigarettes. Smoking cigarettes is also known to increase oxidative stress. As a result, levels of isoprene and pentane were found to be increased after smoking [88]. Other smoking-related VOCs include 2,5-dimethyl furan and 1,3-butadiene. Smoking related VOCs need to be clearly distinguished from endogenous compounds that are related to disease conditions.

All studies reported the smoking history of recruited subjects but adopted vastly different strategies for data analysis. Some studies did not discuss the possible effects of smoking on their results, even though the case and control group had highly uneven smoking histories [40,47]. Coraddi et al. found the value "pack-year" alone had a fair diagnostic power [49]. Combining this value with VOC markers could help in developing a more robust biomarker panel for lung cancer detection. Wang et al. identified smoking related VOCs using ROC and excluded these molecules from the lung cancer biomarker list [45].

Currently, there are two ways to minimize the influence of smoking. One is to design the study carefully, so that case and control groups have matched smoking histories. In two studies where the smoking histories of the case and control groups were closely matched, no effect of tobacco smoke was found on the diagnostic power of the identified biomarker panel [26,27]. The other strategy is to exclude smoking-related VOCs from the biomarkers for lung cancer detection. However, it is not fully known yet what other metabolic pathways may be affected by smoking.

## 2.10. Comorbidity

Many target subjects for disease diagnostic studies often have more than one medical condition. These diseases will also change the VOC profile and confound the biomarker discovery for the targeting disease. Most studies recruited healthy subjects as a control group [29,41,50], while a few studies recruited subjects with similar comorbidity as a control group [44]. The variances in the choice of control groups contribute to the inconsistencies across the different studies.

### 2.11. Disease Staging

One key advantage of breath analysis is its potential for early detection. It is of keen interest to know if stage influences VOC profile. Philips et al. identified 22 breath VOCs that could differentiate control and lung cancer subjects, regardless of stages. For stage 1 patients, the 22 VOCs had 100% sensitivity and 81.3% specificity [25]. Other studies also showed no discrimination between early stage and advanced stage [26,38,40,45,47]. However, Corradi et al. showed that although lung cancer patients have higher levels of ethylbenzene in their breath, the difference is less pronounced between early stage lung cancer and control subjects [49]. Peled et al. analyzed breath samples using a combination of GC-MS and chemical nanoarray, GC-MS analysis did not show any discrimination between early stage and late stage, and also for sub-histological types of lung cancer; however, chemical nanoarray-based techniques could discriminate between early and late stage, and between adenocarcinoma and squamous cell carcinoma, with an accuracy of 88% [44]. Due to the limitations of the chemical sensor array, the identity of the VOCs that contribute to these discriminations is unknown.

### 2.12. Histology

Lung cancer is a complex disease with different histologies. Very few studies compared the VOC profile between different histological types. Song et al. found that patients with adenocarcinoma showed higher concentrations of 1-butanol and 3-hydroxy-2-butanone [40]. In the study by Corradi et al., adenocarcinoma showed higher levels of hexane and ethyl benzene compared to squamous cell carcinoma [49]. Other studies showed that histology has no significant impact on breath VOCs [26,27,47].

Most of the methodological issues discussed above are not limited to lung cancer. Rather, they are shared by breath VOC studies with various objectives. Establishing a standardized practice for these methodological issues is a challenging task and requires a collective effort from all researchers in the field. Jens et al. suggested a framework for standardizing breath analysis at different technical levels [89]. In 2017, the European Respiratory Society published a technical standard on exhaled biomarkers in lung disease [90] and highlighted a few key areas for future research. These are important first steps towards standardized protocols in breath analysis. For highly complex and heterogenous diseases, such as lung cancer, implementing standardized practice is especially critical in developing biomarkers with a clinical value. Though much more needs to be done to establish a standardized methodological procedure, this area of study is well worth pursuing due to the huge potential of breath analysis for non-invasive and early disease detection.

## 3. In Vitro Studies

In vitro cell culture provides a convenient alternative for studying volatile signatures of lung cancer while bypassing many confounding factors associated with breath sampling. Many studies have identified the VOC biomarker of cultured lung cells, and the results show that different types of lung cell lines can generate different panels of VOCs (Table 3). Studies from the same cell line using different techniques produced inconsistent results. For example, a study of the NSCLC cell line Calu-1 using selected ion flow tube-mass spectrometry (SIFT-MS) [91–93] consistently showed higher levels of acetaldehyde, while a study by GC-MS [94] showed that acetaldehyde was decreased in this cell line. Sporning et al. also found decreased level of acetaldehyde in another type of lung cancer cell line [95]. Most studies used only one or two cell lines. Two studies included more than six cell lines [96,97]. These studies brought in vitro studies one step further to investigating whether VOCs from cells in vitro could discriminate between different histologies. Barash et al. showed that VOCs could discriminate between (1) lung cancer and normal lung epithelial cells; (2) NSCLC and SCLC cells; and (3) two subtype of NSCLC: adenocarcinoma and squamous cell carcinoma [97]. Jia et al. demonstrated that although NSCLC and SCLC showed distinct VOC profiles, adenocarcinoma and squamous cell carcinomas could not be differentiated among NSCLCs. On the other hand, large cell carcinomas show different VOC profile with the rest of the NSCLCs [97].

**Table 3.** Identified volatile biomarker of lung cancer from in vitro studies.

| First Author | Cancer Cell | Normal Cell | Analytical Technique | VOC-Increased Concentration | VOC-Decreased Concentration |
|---|---|---|---|---|---|
| David [91] | SK-MES and CALU-1 | | SIFT-MS | acetaldehyde | |
| Chen [98] | primary tissues | | SPME-GC-MS | styrene, decane, isoprene and benzene | |
| Filipiak [94] | CALU-1 | | GC-MS | 2,3,3-trimethylpentane, 2,3,5-trimethylhexane, 2,4-dimethylheptane, 4-methyloctane | acetaldehyde, 3-methylbutanal, n-butyl acetate, acetonitrile, acrolein, methacrolein, 2-methylpropanal, 2-butanone, 2-methoxy-2-methylpropane, 2-ethoxy-2-methylpropane, hexanal |
| Sule-Suso [92] | CALU-1 | NL20 and 35FL121 Tel+ | SIFT-MS | acetaldehyde | |
| Sponring [95] | NCI-H2087 | | GC-MS | 2-ethyl-1-hexanol and 2-methylpenthane | acetaldehyde, 2-methylpropanal, 3-methylbutanal, 2-methylbutanal, hexanal, n-butyl acetate |
| Brunner [99] | A549 | | PTR-MS | 2-pentanone, 2-methyl-1-pentene, 2,4-dimethyl-1-heptene, acetone, ethanol, isobutene, n-octane, tert-butyl methyl ether, tert-butyl ethyl ether | n-butyl acetate, 3-methylbutanal, 2-methylpropanal, methacrolein, 2-methyl-2-butenal, 2-ethylacrolein, pyrrole |
| Hanai [100] | A549 | | SPME-GC-MS | dimethyl succinate, 2-pentanone, phenol, 2-methylpyrazine, 2-hexanone and acetophenone | benzophenone, maltol, dimethyl disulfide, methanethiol, 1-butanol, acetonitrile, cyclohexanone, tributyl phosphate, 2-methyl-1-propanal, benzyl alcohol, styrene |
| Rutter [93] | CALU-1 | NL20 | SIFT-MS | acetaldehyde | |
| Barash [96] | H1650, H820, A549, H1975, H4006, H1435, CALU-3, H2009, HCC95, HCC15, H226, NE18, H774, H69, H187, H526 | Minna 3KT | GC-MS | | decanal |
| Wang [45] | A549, NCI-H446, SK-MES-1 | BEAS-2B | GC-MS | 2-pentadecanone, nonadecane and eicosane | |
| Jia [97] | A549, HCC827, H226, H520, H460, H526 | SAEC | SPME-GC-MS | benzaldehyde, 2-ehtyl-1-hexanol, 2,4-decadien-1-ol | |
| Thriumani [101] | A549, Calu-3 | WI38VA13 | SPME-GC-MS | decane, ethylbenzene, n-propyl benzene, 1-ethyl-2-methylbenzene, styrene, dodecane, cyclohexanol, decanal, nonanal, 1,3-Di-tert-butylbenzene, tetradecane, 2-ethyl-1-dodecanol, 2-ethylhexanol, benzaldehyde, acetophenone, 2-Ethyl-m-xylene, 1-methyl-2-pyrrolidinone, heneicosane | ethanedioic acid |

*Limitations of In Vitro Studies*

Though analyzing VOCs from cultured cells faces fewer problems compared to analyzing human breath samples, there are many methodological issues in the current literature.

The first issue is that almost all studies have used standard cell culture flasks made of polystyrene. Polymer materials like polystyrene emit VOCs themselves. The background from the vessel should be measured and corrected in cell experiments. Alternatively, if cells can survive, glass vessels should be used instead, as adopted by some studies [91,102]. Schallschmidt et al. measured the background from plastic culture vessels and identified several alkanes and aromatics [103]. These molecules are often also found in cultures with living cells. As a result, the background from plastic cell culture vessels may easily lead to misinterpretations.

The second issue is that different cell growth media other than those recommended by the supplier were used to get a uniform VOC background. A culture medium contains nutrients, such as glucose, amino acids, and vitamins, that are essential to cell growth. Certain cell lines require special formulations for optimum growth. The most commonly used basic medium is called DMEM (Dulbecco's modified Eagle's medium). A culture medium has a considerable VOC background and differs from one type to another. In some studies, a cell line with a special medium requirement was cultured in a basic medium, in order to get the same VOC background across different cell lines. Filipiak el al. studied three cell lines: lung cancer cell line A549, primary human bronchial epithelial cells (HBEpC), and human fibroblasts (hFB) [104]. Although the authors cultivated HBEpC in an airway epithelial cell growth medium with special supplements, as recommended by the ATCC (American Type Culture Collection) for initial propagation, for the VOC experiment they cultured all three types of cells in DMEM for 21 h. It is questionable whether the HBEpC cells remained in a healthy condition in DMEM, as no cell viability data or pictures of the cells were shown. It is beyond doubt that different VOC backgrounds from different types of cell growth media should be taken into consideration. Instead of compromising on the growth condition of cells, we believe the method adopted by Barash et al. is more acceptable [96], where each cell line was grown in its recommended medium and the VOC effect of the medium was corrected during data analysis before comparing across different cell lines.

Another limitation is that cells in an in vitro culture live in a drastically different environment than tumor cells in the human body. As a result, none of the identified biomarkers achieved clinical relevance [105,106]. Kalluri et al. showed that hypoxia influences the VOCs that the cancer cells produce and suggested future in vitro studies to culture cells in hypoxic conditions [106]. Lung cancer cell grown in a 3D environment was found to emit higher levels of VOCs than in 2D cultures [93]. These studies indicate that cell culture experiments could be more relevant when the conditions better mimic the real situation.

Despite these limitations, in vitro cell culture provides a convenient way to directly assess the effect of certain stimuli on the VOC profile produced by cancer cells. Lawal et al. used cultured lung cells to study the effect of a bacterial infection on the VOC profile [107]. They co-cultured lung epithelial cells with *Pseudomonas aeruginosa*, a bacterium commonly found in pneumonia, and measured the VOC profile with and without the bacteria. Acetone, ethanol, 3-methyl-1butanol, and three other VOCs were found to be elevated in bacteria infected cells, indicating the bacterial origin of these VOCs. They also simulated the effect of oxidative stress induced by bacterial infection by adding hydrogen peroxide to the cell culture and identified several alkanes as potential markers for oxidative stress. Feinberg et al. blocked glycolysis in cultured lung cancer cells and identified unique signatures in all cells studied [108]. A recent study identified the unique VOC profiles of lung cancer cells with a different p53 mutation status at a single cell level [102]. These studies demonstrated the usefulness of in vitro cell cultures in identifying the possible biochemical origins of VOCs.

## 4. Conclusions

Breath analysis for lung cancer screening is a rapidly developing field. Accelerating the pace of the development of a robust panel of markers that can be translated for clinical use will require progress in three key areas: (1) development of standardized and flexible breath sampling protocols, (2) longitudinal multi-centre clinical trials with careful study design and external validation, and (3) understanding of the biochemical pathways involved in lung cancer development and progression. Measuring VOCs in vitro, after blocking specific pathways or knocking out specific genes, provides direct evidence of the biochemical origins of the VOCs. We believe that these discoveries will ultimately contribute to the development of breath analysis as a technique for the early detection of lung cancer, allowing breath analysis to realize its long-held potential and to become a critical tool in personalized medicine.

**Funding:** This research was supported by NUSNNI-Nanocore core fund.

**Conflicts of Interest:** The authors declare no conflict of interest.

## References


1. WHO. Cancer. Available online: http://www.who.int/news-room/fact-sheets/detail/cancer (accessed on 15 May 2018).
2. World Health Organization. Cancer. Available online: http://www.who.int/mediacentre/factsheets/fs297/en/ (accessed on 18 April 2017).
3. Weinberg, R.A. *The Biology of Cancer*, 2nd ed.; Garland Science: New York, NY, USA, 2013.
4. Jemal, A.; Siegel, R.L.; Miller, K.D. Cancer statistics, 2017. *CA Cancer J. Clin.* **2017**, *67*, 7–30. [CrossRef]
5. The National Lung Screening Trial Research Team. Reduced Lung-Cancer Mortality with Low-Dose Computed Tomographic Screening. *N. Engl. J. Med.* **2011**, *365*, 395–409. [CrossRef] [PubMed]
6. Gasparri, R.; Santonico, M.; Valentini, C.; Sedda, G.; Borri, A.; Petrella, F.; Maisonneuve, P.; Pennazza, G.; D'Amico, A.; Di Natale, C.; et al. Volatile signature for the early diagnosis of lung cancer. *J. Breath Res.* **2016**, *10*, 16007. [CrossRef]
7. Barash, O.; Peled, N.; Hirsch, F.R.; Haick, H. Sniffing the Unique Odor Print of Non-Small-Cell Lung Cancer with Gold Nanoparticles. *Small* **2009**, *5*, 2618–2624. [CrossRef] [PubMed]
8. Macagnano, A.; D'Arcangelo, G.; Roscioni, C.; Finazzi-Agrò, A.; D'Amico, A.; Di Natale, C.; Martinelli, E.; Paolesse, R. Lung cancer identification by the analysis of breath by means of an array of non-selective gas sensors. *Biosens. Bioelectron.* **2003**, *18*, 1209–1218. [CrossRef]
9. Mazzone, P.J.; Wang, X.-F.; Xu, Y.; Mekhail, T.; Beukemann, M.C.; Na, J.; Kemling, J.W.; Suslick, K.S.; Sasidhar, M. Exhaled Breath Analysis with a Colorimetric Sensor Array for the Identification and Characterization of Lung Cancer. *J. Thorac. Oncol.* **2012**, *7*, 137–142. [CrossRef] [PubMed]
10. Mazzone, P.J.; Hammel, J.; Dweik, R.; Na, J.; Czich, C.; Laskowski, D.; Mekhail, T. Diagnosis of lung cancer by the analysis of exhaled breath with a colorimetric sensor array. *Thorax* **2007**, *62*, 565–568. [CrossRef] [PubMed]
11. D'Amico, A.; Pennazza, G.; Santonico, M.; Martinelli, E.; Roscioni, C.; Galluccio, G.; Paolesse, R.; Di Natale, C. An investigation on electronic nose diagnosis of lung cancer. *Lung Cancer* **2010**, *68*, 170–176. [CrossRef] [PubMed]
12. Machado, R.F.; Laskowski, D.; Deffenderfer, O.; Burch, T.; Zheng, S.; Mazzone, P.J.; Mekhail, T.; Jennings, C.; Stoller, J.K.; Pyle, J.; et al. Detection of Lung Cancer by Sensor Array Analyses of Exhaled Breath. *Am. J. Respir. Crit. Care Med.* **2005**, *171*, 1286–1291. [CrossRef]
13. Dragonieri, S.; Annema, J.T.; Schot, R.; Van Der Schee, M.P.; Spanevello, A.; Carratu, P.; Resta, O.; Rabe, K.F.; Sterk, P.J. An electronic nose in the discrimination of patients with non-small cell lung cancer and COPD. *Lung Cancer* **2009**, *64*, 166–170. [CrossRef]
14. Peng, G.; Tisch, U.; Adams, O.; Hakim, M.; Shehada, N.; Broza, Y.Y.; Billan, S.; Abdah-Bortnyak, R.; Kuten, A.; Haick, H. Diagnosing lung cancer in exhaled breath using gold nanoparticles. *Nat. Nanotechnol.* **2009**, *4*, 669–673. [CrossRef]

15. Adiguzel, Y.; Külah, H. Breath sensors for lung cancer diagnosis. *Biosens. Bioelectron.* **2015**, *65*, 121–138. [CrossRef]

16. Queralto, N.; Berliner, A.N.; Goldsmith, B.; Martino, R.; Rhodes, P.; Lim, S.H. Detecting cancer by breath volatile organic compound analysis: A review of array-based sensors. *J. Breath Res.* **2014**, *8*, 27112. [CrossRef] [PubMed]

17. Iniesta, C.B.; Carpeño, J.D.C.; Carrasco, J.A.; Moreno, V.; Saenz, E.C.; Feliú, J.; Sereno, M.; Río, F.G.; Barriuso, J.; Barón, M.G.; et al. New screening method for lung cancer by detecting volatile organic compounds in breath. *Clin. Transl. Oncol.* **2007**, *9*, 364–368. [CrossRef]

18. Saalberg, Y.; Wolff, M. VOC breath biomarkers in lung cancer. *Clin. Chim. Acta* **2016**, *459*, 5–9. [CrossRef]

19. Hua, Q.; Zhu, Y.; Liu, H. Detection of volatile organic compounds in exhaled breath to screen lung cancer: A systematic review. *Future Oncol.* **2018**, *14*, 1647–1662. [CrossRef]

20. Zhou, J.; Huang, Z.-A.; Kumar, U.; Chen, D.D. Review of recent developments in determining volatile organic compounds in exhaled breath as biomarkers for lung cancer diagnosis. *Anal. Chim. Acta* **2017**, *996*. [CrossRef]

21. Davies, T.; Hayward, N. Volatile products from acetylcholine as markers in the rapid urine test using head-space gas—liquid chromatography. *J. Chromatogr. B Biomed. Sci. Appl.* **1984**, *307*, 11–21. [CrossRef]

22. Feinberg, T.; Alkoby-Meshulam, L.; Herbig, J.; Cancilla, J.C.; Torrecilla, J.S.; Mor, N.G.; Bar, J.; Ilouze, M.; Haick, H.; Peled, N. Cancerous glucose metabolism in lung cancer—Evidence from exhaled breath analysis. *J. Breath Res.* **2016**, *10*, 26012. [CrossRef] [PubMed]

23. Handa, H.; Usuba, A.; Maddula, S.; Baumbach, J.I.; Mineshita, M.; Miyazawa, T. Exhaled Breath Analysis for Lung Cancer Detection Using Ion Mobility Spectrometry. *PLoS ONE* **2014**, *9*, e114555. [CrossRef]

24. Bajtarevic, A.; Ager, C.; Pienz, M.; Klieber, M.; Schwarz, K.; Ligor, T.; Filipiak, W.; Denz, H.; Fiegl, M.; Jamnig, H.; et al. Noninvasive detection of lung cancer by analysis of exhaled breath. *BMC Cancer* **2009**, *9*, 348. [CrossRef] [PubMed]

25. Phillips, M.; Gleeson, K.; Hughes, J.M.B.; Greenberg, J.; Cataneo, R.N.; Baker, L.; McVay, W.P. Volatile organic compounds in breath as markers of lung cancer: A cross-sectional study. *Lancet* **1999**, *353*, 1930–1933. [CrossRef]

26. Phillips, M.; Cataneo, R.N.; Cummin, A.R.; Gagliardi, A.J.; Gleeson, K.; Greenberg, J.; Maxfield, R.A.; Rom, W.N. Detection of Lung Cancer With Volatile Markers in the Breatha. *Chest* **2003**, *123*, 2115–2123. [CrossRef] [PubMed]

27. Phillips, M.; Altorki, N.; Austin, J.H.; Cameron, R.B.; Cataneo, R.N.; Greenberg, J.; Kloss, R.; Maxfield, R.A.; Munawar, M.I.; Pass, H.I.; et al. Prediction of lung cancer using volatile biomarkers in breath. *Cancer Biomark.* **2007**, *3*, 95–109. [CrossRef] [PubMed]

28. Fuchs, P.; Loeseken, C.; Schubert, J.K.; Miekisch, W. Breath gas aldehydes as biomarkers of lung cancer. *Int. J. Cancer* **2009**, *126*, 2663–2670. [CrossRef] [PubMed]

29. Wang, C.; Dong, R.; Wang, X.; Lian, A.; Chi, C.; Ke, C.; Guo, L.; Liu, S.; Zhao, W.; Xu, G.; et al. Exhaled volatile organic compounds as lung cancer biomarkers during one-lung ventilation. *Sci. Rep.* **2014**, *4*, 7312. [CrossRef] [PubMed]

30. Peralbo-Molina, A.; Calderón-Santiago, M.; Priego-Capote, F.; Gamez, B.J.; De Castro, M.D.L. Identification of metabolomics panels for potential lung cancer screening by analysis of exhaled breath condensate. *J. Breath Res.* **2016**, *10*, 26002. [CrossRef] [PubMed]

31. Hakim, M.; Broza, Y.Y.; Barash, O.; Peled, N.; Phillips, M.; Amann, A.; Haick, H. Volatile Organic Compounds of Lung Cancer and Possible Biochemical Pathways. *Chem. Rev.* **2012**, *112*, 5949–5966. [CrossRef]

32. Krilaviciute, A.; Heiss, J.A.; Leja, M.; Kupcinskas, J.; Haick, H.; Brenner, H. Detection of cancer through exhaled breath: A systematic review. *Oncotarget* **2015**, *6*, 38643–38657. [CrossRef]

33. Amann, A.; Corradi, M.; Mazzone, P.; Mutti, A. Lung cancer biomarkers in exhaled breath. *Expert Rev. Mol. Diagn.* **2011**, *11*, 207–217. [CrossRef]

34. Miekisch, W.; Herbig, J.; Schubert, J.K. Data interpretation in breath biomarker research: Pitfalls and directions. *J. Breath Res.* **2012**, *6*, 36007. [CrossRef] [PubMed]

35. Smolinska, A.; Hauschild, A.-C.; Fijten, R.; Dallinga, J.W.; Baumbach, J.; Van Schooten, F.J. Current breathomics—A review on data pre-processing techniques and machine learning in metabolomics breath analysis. *J. Breath Res.* **2014**, *8*, 27105. [CrossRef]

36. O'Neill, H.J.; Gordon, S.M.; O'Neill, M.H.; Gibbons, R.D.; Szidon, J.P. A computerized classification technique for screening for the presence of breath biomarkers in lung cancer. *Clin. Chem.* **1988**, *34*, 1613–1618. [PubMed]

37. Poli, D.; Carbognani, P.; Corradi, M.; Goldoni, M.; Acampa, O.; Balbi, B.; Bianchi, L.; Rusca, M.; Mutti, A. Exhaled volatile organic compounds in patients with non-small cell lung cancer: Cross sectional and nested short-term follow-up study. *Respir. Res.* **2005**, *6*, 71. [CrossRef]

38. Wehinger, A.; Schmid, A.; Mechtcheriakov, S.; Ledochowski, M.; Grabmer, C.; Gastl, G.A.; Amann, A. Lung cancer detection by proton transfer reaction mass-spectrometric analysis of human breath gas. *Int. J. Mass Spectrom.* **2007**, *265*, 49–59. [CrossRef]

39. Hakim, M.; Broza, Y.Y.; Billan, S.; Abdah-Bortnyak, R.; Kuten, A.; Tisch, U.; Haick, H.; Peng, G. Detection of lung, breast, colorectal and prostate cancers from exhaled breath using a single array of nanosensors. *Br. J. Cancer* **2010**, *103*, 542–551. [CrossRef]

40. Song, G.; Qin, T.; Liu, H.; Xu, G.-B.; Pan, Y.-Y.; Xiong, F.-X.; Gu, K.-S.; Sun, G.-P.; Chen, Z.-D. Quantitative breath analysis of volatile organic compounds of lung cancer patients. *Lung Cancer* **2010**, *67*, 227–231. [CrossRef] [PubMed]

41. Kischkel, S.; Miekisch, W.; Sawacki, A.; Straker, E.M.; Trefz, P.; Amann, A.; Schubert, J.K. Breath biomarkers for lung cancer detection and assessment of smoking related effects—Confounding variables, influence of normalization and statistical algorithms. *Clin. Chim. Acta* **2010**, *411*, 1637–1644. [CrossRef] [PubMed]

42. Rudnicka, J.; Kowalkowski, T.; Ligor, T.; Buszewski, B. Determination of volatile organic compounds as biomarkers of lung cancer by SPME–GC–TOF/MS and chemometrics. *J. Chromatogr. B Biomed. Sci. Appl.* **2011**, *879*, 3360–3366. [CrossRef]

43. Ulanowska, A.; Kowalkowski, T.; Trawińska, E.; Buszewski, B. The application of statistical methods using VOCs to identify patients with lung cancer. *J. Breath Res.* **2011**, *5*, 46008. [CrossRef]

44. Peled, N.; Hakim, M.; Bunn, P.A.; Miller, Y.E.; Kennedy, T.C.; Mattei, J.; Mitchell, J.D.; Hirsch, F.R.; Haick, H. Non-Invasive Breath Analysis of Pulmonary Nodules. *J. Thorac. Oncol.* **2012**, *7*, 1528–1533. [CrossRef] [PubMed]

45. Wang, Y.; Hu, Y.; Wang, D.; Yu, K.; Wang, L.; Zou, Y.; Zhao, C.; Zhang, X.; Wang, P.; Ying, K. The analysis of volatile organic compounds biomarkers for lung cancer in exhaled breath, tissues and cell lines. *CBM* **2012**, *11*, 129–137. [CrossRef] [PubMed]

46. Buszewski, B.; Ligor, T.; Jezierski, T.; Wenda-Piesik, A.; Walczak, M.; Rudnicka, J. Identification of volatile lung cancer markers by gas chromatography–mass spectrometry: Comparison with discrimination by canines. *Anal. Bioanal. Chem.* **2012**, *404*, 141–146. [CrossRef] [PubMed]

47. Hu, Y.; Zou, Y.; Zhang, X.; Chen, X.; Ying, K.; Wang, P. Optimization of volatile markers of lung cancer to exclude interferences of non-malignant disease. *Cancer Biomark.* **2014**, *14*, 371–379. [CrossRef]

48. Capuano, R.; Santonico, M.; Pennazza, G.; Ghezzi, S.; Martinelli, E.; Roscioni, C.; Lucantoni, G.; Galluccio, G.; Paolesse, R.; Di Natale, C.; et al. The lung cancer breath signature: A comparative analysis of exhaled breath and air sampled from inside the lungs. *Sci. Rep.* **2015**, *5*, 16491. [CrossRef] [PubMed]

49. Corradi, M.; Poli, D.; Banda, I.; Bonini, S.; Mozzoni, P.; Pinelli, S.; Alinovi, R.; Andreoli, R.; Ampollini, L.; Casalini, A.; et al. Exhaled breath analysis in suspected cases of non-small-cell lung cancer: A cross-sectional study. *J. Breath Res.* **2015**, *9*, 27101. [CrossRef] [PubMed]

50. Schallschmidt, K.; Becker, R.; Jung, C.; Bremser, W.; Walles, T.; Neudecker, J.; Leschber, G.; Frese, S.; Nehls, I. Comparison of volatile organic compounds from lung cancer patients and healthy controls—Challenges and limitations of an observational study. *J. Breath Res.* **2016**, *10*, 46007. [CrossRef] [PubMed]

51. Sakumura, Y.; Koyama, Y.; Tokutake, H.; Hida, T.; Sato, K.; Itoh, T.; Akamatsu, T.; Shin, W.; Seitz, W.R. Diagnosis by Volatile Organic Compounds in Exhaled Breath from Lung Cancer Patients Using Support Vector Machine Algorithm. *Sensors* **2017**, *17*, 287. [CrossRef] [PubMed]

52. Costello, B.D.L.; Amann, A.; Alkateb, H.; Flynn, C.; Filipiak, W.; Khalid, T.; Osborne, D.; Ratcliffe, N.M. A review of the volatiles from the healthy human body. *J. Breath Res.* **2014**, *8*, 14001. [CrossRef] [PubMed]

53. Phillips, M.; Greenberg, J.; Sabas, M. Alveolar gradient of pentane in normal human breath. *Free Radic. Res.* **1994**, *20*, 333–337. [CrossRef] [PubMed]

54. Phillips, M.; Herrera, J.; Krishnan, S.; Zain, M.; Greenberg, J.; Cataneo, R.N. Variation in volatile organic compounds in the breath of normal humans. *J. Chromatogr. B Biomed. Sci. Appl.* **1999**, *729*, 75–88. [CrossRef]

55. Ajibola, O.A.; Smith, D.; Španěl, P.; Ferns, G.A.A. Effects of dietary nutrients on volatile breath metabolites. *J. Nutr. Sci.* **2013**, *2*, e34. [CrossRef]

56. Kamysek, S.; Fuchs, P.; Sukul, P.; Schubert, J.; Miekisch, W.; Trefz, P. Drug detection in breath: Non-invasive assessment of illicit or pharmaceutical drugs. *J. Breath Res.* **2017**, *11*, 24001. [CrossRef]

57. Beauchamp, J.; Kirsch, F.; Buettner, A. Real-time breath gas analysis for pharmacokinetics: Monitoring exhaled breath by on-line proton-transfer-reaction mass spectrometry after ingestion of eucalyptol-containing capsules. *J. Breath Res.* **2010**, *4*, 26006. [CrossRef] [PubMed]

58. Schulz, S.; Dickschat, J.S. Bacterial volatiles: The smell of small organisms. *Nat. Prod. Rep.* **2007**, *24*, 814–842. [CrossRef]

59. Phillips, M.; Greenberg, J.; Awad, J. Metabolic and environmental origins of volatile organic compounds in breath. *J. Clin. Pathol.* **1994**, *47*, 1052–1053. [CrossRef] [PubMed]

60. Beauchamp, J. Inhaled today, not gone tomorrow: Pharmacokinetics and environmental exposure of volatiles in exhaled breath. *J. Breath Res.* **2011**, *5*, 37103. [CrossRef] [PubMed]

61. Haick, H.; Broza, Y.Y.; Mochalski, P.; Ruzsanyi, V.; Amann, A. Assessment, origin, and implementation of breath volatile cancer markers. *Chem. Soc. Rev.* **2013**, *43*, 1423–1449. [CrossRef] [PubMed]

62. Jakubowski, M.; Czerczak, S. Calculating the retention of volatile organic compounds in the lung on the basis of their physicochemical properties. *Environ. Toxicol. Pharmacol.* **2009**, *28*, 311–315. [CrossRef] [PubMed]

63. Buist, H.E.; De Wit-Bos, L.; Bouwman, T.; Vaes, W.H. Predicting blood:air partition coefficients using basic physicochemical properties. *Regul. Toxicol. Pharmacol.* **2012**, *62*, 23–28. [CrossRef] [PubMed]

64. Schubert, J.K.; Miekisch, W.; Birken, T.; Geiger, K.; Nöldge-Schomburg, G.F.E.; Schubert, J. Impact of inspired substance concentrations on the results of breath analysis in mechanically ventilated patients. *Biomarkers* **2005**, *10*, 138–152. [CrossRef] [PubMed]

65. Španěl, P.; Dryahina, K.; Smith, D. A quantitative study of the influence of inhaled compounds on their concentrations in exhaled breath. *J. Breath Res.* **2013**, *7*, 17106. [CrossRef] [PubMed]

66. D'mello, J.; Butani, M. Capnography. *Indian J. Anaesth.* **2002**, *46*, 269–278.

67. Doran, S.L.F.; Romano, A.; Hanna, G.B. Optimisation of sampling parameters for standardised exhaled breath sampling. *J. Breath Res.* **2017**, *12*, 016007. [CrossRef] [PubMed]

68. Birken, T.; Schubert, J.; Miekisch, W.; Nöldge-Schomburg, G. A novel visually CO2 controlled alveolar breath sampling technique. *THC* **2006**, *14*, 499–506.

69. Miekisch, W.; Hengstenberg, A.; Kischkel, S.; Beckmann, U.; Mieth, M.; Schubert, J.K. Construction and Evaluation of a Versatile CO2 Controlled Breath Collection Device. *IEEE Sens. J.* **2010**, *10*, 211–215. [CrossRef]

70. ReCIVA® Breath Sampler. Available online: https://www.owlstonemedical.com/products/reciva/ (accessed on 19 July 2018).

71. Bikov, A.; Paschalaki, K.; Logan-Sinclair, R.; Horváth, I.; A Kharitonov, S.; Barnes, P.J.; Usmani, O.S.; Paredi, P. Standardised exhaled breath collection for the measurement of exhaled volatile organic compounds by proton transfer reaction mass spectrometry. *BMC Pulm. Med.* **2013**, *13*, 43. [CrossRef] [PubMed]

72. Boshier, P.R.; Priest, O.H.; Hanna, G.B.; Marczin, N. Influence of respiratory variables on the on-line detection of exhaled trace gases by PTR-MS. *Thorax* **2011**, *66*, 919–920. [CrossRef]

73. Lärstad, M.A.E.; Torén, K.; Bake, B.; Olin, A.-C. Determination of ethane, pentane and isoprene in exhaled air ? effects of breath-holding, flow rate and purified air. *Acta Physiol.* **2007**, *189*, 87–98. [CrossRef] [PubMed]

74. Španěl, P.; Smith, D. Comment on 'Influences of mixed expiratory sampling parameters on exhaled volatile organic compound concentrations'. *J. Breath Res.* **2011**, *5*, 48001. [CrossRef] [PubMed]

75. Herbig, J.; Titzmann, T.; Beauchamp, J.; Kohl, I.; Hansel, A. Buffered end-tidal (BET) sampling—A novel method for real-time breath-gas analysis. *J. Breath Res.* **2008**, *2*, 37008. [CrossRef] [PubMed]

76. Sukul, P.; Schubert, J.K.; Kamysek, S.; Trefz, P.; Miekisch, W. Applied upper-airway resistance instantly affects breath components: A unique insight into pulmonary medicine. *J. Breath Res.* **2017**, *11*, 47108. [CrossRef] [PubMed]

77. Beauchamp, J.; Herbig, J.; Gutmann, R.; Hansel, A. On the use of Tedlar® bags for breath-gas sampling and analysis. *J. Breath Res.* **2008**, *2*, 46001. [CrossRef]

78. Pet'Ka, J.; Etiévant, P.; Callement, G. Suitability of different plastic materials for head or nose spaces short term storage. *Analusis* **2000**, *28*, 330–335. [CrossRef]

79. Phillips, M.; Cataneo, R.N.; Greenberg, J.; Gunawardena, R.; Naidu, A.; Rahbari-Oskoui, F. Effect of age on the breath methylated alkane contour, a display of apparent new markers of oxidative stress. *J. Lab. Clin. Med.* **2000**, *136*, 243–249. [CrossRef] [PubMed]

80. Lechner, M.; Moser, B.; Niederseer, D.; Karlseder, A.; Holzknecht, B.; Fuchs, M.; Colvin, S.; Tilg, H.; Rieder, J. Gender and age specific differences in exhaled isoprene levels. *Respir. Physiol. Neurobiol.* **2006**, *154*, 478–483. [CrossRef] [PubMed]

81. Phillips, M.; Greenberg, J.; Cataneo, R.N. Effect of age on the profile of alkanes in normal human breath. *Free Radic. Res.* **2000**, *33*, 57–63. [CrossRef] [PubMed]

82. Španěl, P.; Dryahina, K.; Smith, D. Acetone, ammonia and hydrogen cyanide in exhaled breath of several volunteers aged 4–83 years. *J. Breath Res.* **2007**, *1*, 11001. [CrossRef] [PubMed]

83. Das, M.K.; Bishwal, S.C.; Das, A.; Dabral, D.; Varshney, A.; Badireddy, V.K.; Nanda, R. Investigation of Gender-Specific Exhaled Breath Volatome in Humans by GCxGC-TOF-MS. *Anal. Chem.* **2013**, *86*, 1229–1237. [CrossRef]

84. Cheng, H. Volatile Flavor Compounds in Yogurt: A Review. *Crit. Rev. Food Sci. Nutr.* **2010**, *50*, 938–950. [CrossRef]

85. Varlet, V.; Fernandez, X. Review. Sulfur-containing Volatile Compounds in Seafood: Occurrence, Odorant Properties and Mechanisms of Formation. *Food Sci. Technol. Int.* **2010**, *16*, 463–503. [CrossRef] [PubMed]

86. Jordan, A.; Hansel, A.; Lindinger, W.; Holzinger, R. Acetonitrile and benzene in the breath of smokers and non-smokers investigated by proton transfer reaction mass spectrometry (PTR-MS). *Int. J. Mass Spectrom. Ion Process.* **1995**, *148*, 68–70. [CrossRef]

87. Hesselbrock, V.M.; Shaskan, E.G. Endogenous breath acetaldehyde levels among alcoholic and non-alcoholic probands: Effect of alcohol use and smoking. *Prog. Neuro-Psychopharmacol. Biol. Psychiatry* **1985**, *9*, 259–265. [CrossRef]

88. Euler, D.E.; Davé, S.J.; Guo, H. Effect of cigarette smoking on pentane excretion in alveolar breath. *Clin. Chem.* **1996**, *42*, 303–308. [PubMed]

89. Herbig, J.; Beauchamp, J. Towards standardization in the analysis of breath gas volatiles. *J. Breath Res.* **2014**, *8*, 37101. [CrossRef]

90. Horváth, I.; Barnes, P.J.; Loukides, S.; Sterk, P.J.; Högman, M.; Olin, A.-C.; Amann, A.; Antus, B.; Baraldi, E.; Bikov, A.; et al. A European Respiratory Society technical standard: Exhaled biomarkers in lung disease. *Eur. Respir. J.* **2017**, *49*, 1600965. [CrossRef]

91. Smith, D.; Wang, T.; Sulé-Suso, J.; Španěl, P.; El Haj, A. Quantification of acetaldehyde released by lung cancer cellsin vitrousing selected ion flow tube mass spectrometry. *Rapid Commun. Mass Spectrom.* **2003**, *17*, 845–850. [CrossRef]

92. Sulé-Suso, J.; Pysanenko, A.; Španěl, P.; Smith, D. Quantification of acetaldehyde and carbon dioxide in the headspace of malignant and non-malignant lung cells in vitro by SIFT-MS. *Analyst* **2009**, *134*, 2419. [CrossRef]

93. Rutter, A.V.; Chippendale, T.W.E.; Yang, Y.; Smith, D.; Španěl, P.; Sulé-Suso, J. Quantification by SIFT-MS of acetaldehyde released by lung cells in a 3D model. *Analyst* **2013**, *138*, 91–95. [CrossRef]

94. Filipiak, W.; Sponring, A.; Mikoviny, T.; Ager, C.; Troppmair, J.; Schubert, J.; Miekisch, W.; Amann, A. Release of volatile organic compounds (VOCs) from the lung cancer cell line CALU-1 in vitro. *Cancer Cell Int.* **2008**, *8*, 17. [CrossRef]

95. Sponring, A.; Filipiak, W.; Mikoviny, T. Release of Volatile Organic Compounds from the lung cancer cell line NCI-H2087 in vitro. *Anticancer Res.* **2009**, *29*, 419–426. [PubMed]

96. Barash, O.; Peled, N.; Tisch, U.; Bunn, P.A.; Hirsch, F.R.; Haick, H. Classification of lung cancer histology by gold nanoparticle sensors. *Nanomed. Nanotechnol. Biol. Med.* **2012**, *8*, 580–589. [CrossRef] [PubMed]

97. Jia, Z.; Zhang, H.; Ong, C.N.; Patra, A.; Lu, Y.; Lim, C.T.; Venkatesan, T. Detection of Lung Cancer: Concomitant Volatile Organic Compounds and Metabolomic Profiling of Six Cancer Cell Lines of Different Histological Origins. *ACS Omega* **2018**, *3*, 5131–5140. [CrossRef] [PubMed]

98. Chen, X.; Xu, F.; Wang, Y.; Pan, Y.; Lu, D.; Wang, P.; Ying, K.; Chen, E.; Zhang, W. A study of the volatile organic compounds exhaled by lung cancer cells in vitro for breath diagnosis. *Cancer* **2007**, *110*, 835–844. [CrossRef] [PubMed]

99. Brunner, C.; Szymczak, W.; Höllriegl, V.; Mörtl, S.; Oelmez, H.; Bergner, A.; Huber, R.M.; Hoeschen, C.; Oeh, U. Discrimination of cancerous and non-cancerous cell lines by headspace-analysis with PTR-MS. *Anal. Bioanal. Chem.* **2010**, *397*, 2315–2324. [CrossRef] [PubMed]

100. Hanai, Y.; Shimono, K.; Oka, H.; Baba, Y.; Yamazaki, K.; Beauchamp, G.K. Analysis of volatile organic compounds released from human lung cancer cells and from the urine of tumor-bearing mice. *Cancer Cell Int.* **2012**, *12*, 7. [CrossRef] [PubMed]
101. Thriumani, R.; Zakaria, A.; Hashim, Y.Z.H.-Y.; Jeffree, A.I.; Helmy, K.M.; Kamarudin, L.M.; Omar, M.I.; Shakaff, A.Y.M.; Adom, A.H.; Persaud, K.C. A study on volatile organic compounds emitted by in-vitro lung cancer cultured cells using gas sensor array and SPME-GCMS. *BMC Cancer* **2018**, *18*, 362. [CrossRef] [PubMed]
102. Serasanambati, M.; Broza, Y.Y.; Marmur, A.; Haick, H. Profiling Single Cancer Cells with Volatolomics Approach. *IScience* **2018**, *11*, 178–188. [CrossRef]
103. Schallschmidt, K.; Becker, R.; Jung, C.; Rolff, J.; Fichtner, I.; Nehls, I. Investigation of cell culture volatilomes using solid phase micro extraction: Options and pitfalls exemplified with adenocarcinoma cell lines. *J. Chromatogr. B Biomed. Sci. Appl.* **2015**, *1006*, 158–166. [CrossRef]
104. Filipiak, W.; Sponring, A.; Filipiak, A.; Ager, C.; Schubert, J.; Miekisch, W.; Amann, A.; Troppmair, J. TD-GC-MS Analysis of Volatile Metabolites of Human Lung Cancer and Normal Cells In vitro. *Cancer Epidemiol. Biomark. Prev.* **2010**, *19*, 182–195. [CrossRef]
105. Schallschmidt, K.; Becker, R.; Zwaka, H.; Menzel, R.; Johnen, D.; Fischer-Tenhagen, C.; Rolff, J.; Nehls, I. In vitrocultured lung cancer cells are not suitable for animal-based breath biomarker detection. *J. Breath Res.* **2015**, *9*, 27103. [CrossRef] [PubMed]
106. Kalluri, U.; Naiker, M.; A Myers, M. Cell culture metabolomics in the diagnosis of lung cancer-the influence of cell culture conditions. *J. Breath Res.* **2014**, *8*, 27109. [CrossRef] [PubMed]
107. Lawal, O.; Knobel, H.; Weda, H.; Bos, L.D.; Nijsen, T.M.E.; Goodacre, R.; Fowler, S.J. Volatile organic compound signature from co-culture of lung epithelial cell line with Pseudomonas aeruginosa. *Analyst* **2018**, *143*, 3148–3155. [CrossRef] [PubMed]
108. Feinberg, T.; Herbig, J.; Kohl, I.; Las, G.; Cancilla, J.C.; Torrecilla, J.S.; Ilouze, M.; Haick, H.; Peled, N. Cancer metabolism: The volatile signature of glycolysis—In vitro model in lung cancer cells. *J. Breath Res.* **2017**, *11*, 16008. [CrossRef] [PubMed]

 *metabolites*

*Article*

# Delta-Tocotrienol Modulates Glutamine Dependence by Inhibiting ASCT2 and LAT1 Transporters in Non-Small Cell Lung Cancer (NSCLC) Cells: A Metabolomic Approach

**Lichchavi Dhananjaya Rajasinghe, Melanie Hutchings and Smiti Vaid Gupta ***

Department of Nutrition and Food Science, Wayne State University, Detroit, MI 48202, USA;
lichchavi.rajasinghe@wayne.edu (L.D.R.); dv2329@wayne.edu (M.H.)
* Correspondence: sgupta@wayne.edu; Tel.: +1-313-577-5565

Received: 30 November 2018; Accepted: 4 March 2019; Published: 13 March 2019

**Abstract:** The growth and development of non-small cell lung cancer (NSCLC) primarily depends on glutamine. Both glutamine and essential amino acids (EAAs) have been reported to upregulate mTOR in NSCLC, which is a bioenergetics sensor involved in the regulation of cell growth, cell survival, and protein synthesis. Seen as novel concepts in cancer development, ASCT2 and LAT transporters allow glutamine and EAAs to enter proliferating tumors as well as send a regulatory signal to mTOR. Blocking or downregulating these glutamine transporters in order to inhibit glutamine uptake would be an excellent therapeutic target for treatment of NSCLC. This study aimed to validate the metabolic dysregulation of glutamine and its derivatives in NSCLC using cellular 1H-NMR metabolomic approach while exploring the mechanism of delta-tocotrienol ($\delta$T) on glutamine transporters, and mTOR pathway. Cellular metabolomics analysis showed significant inhibition in the uptake of glutamine, its derivatives glutamate and glutathione, and some EAAs in both cell lines with $\delta$T treatment. Inhibition of glutamine transporters (ASCT2 and LAT1) and mTOR pathway proteins (P-mTOR and p-4EBP1) was evident in Western blot analysis in a dose-dependent manner. Our findings suggest that $\delta$T inhibits glutamine transporters, thus inhibiting glutamine uptake into proliferating cells, which results in the inhibition of cell proliferation and induction of apoptosis via downregulation of the mTOR pathway.

**Keywords:** cancer; mTOR; vitamin E; SLC1A5; tocotrienols; apoptosis; cell growth; cell transporters; essential amino acids; ASCT2; glutaminolysis; alanine; glutathione; glutamate; lung; bio actives; nutraceuticals

---

## 1. Introduction

Non-small cell lung cancer (NSCLC) presents itself as aggressive tumors arise from the airway epithelial cells (majority) and interior parts of the lungs [1]. It remains one of the leading causes of disease-related mortalities in the world. The current therapeutic options for NSCLC, which include surgery, radiotherapy, and chemotherapy [1], have slightly improved NSCLC survival rate at some developmental stages in both men and women. However, there has been a plateauing of the overall five-year survival rate, hovering ~12–18% between the years 1975 and 2011 [2]. Also, several studies report that there is a high probability of reoccurrence and development of resistance to drug therapies in NSCLC after treatment with chemotherapeutic agents, surgical resection, and radiation therapy [3]. This warrants efforts to identify novel therapeutic agents and targets for preventing and treating NSCLC.

Research in nutrition-based modulation against diseases has opened up new horizons in cancer prevention, contributing to drug discovery and development processes for numerous chronic diseases, including cancer [4,5]. Most bioactive agents extracted from plants show minimum cell cytotoxicity while simultaneously targeting multiple signaling pathways involved in cell growth, apoptosis, invasion, angiogenesis, and metastasis in cancer cells [6,7]. Tocotrienols ($\alpha$, $\beta$, $\gamma$, and $\delta$), isomers of vitamin E, are found in vegetable oils, including rice bran oil and palm oil, wheat germ, barley, annatto, and certain other types of seeds, nuts, and grains [8]. They exert biological effects including antiangiogenesis, antioxidant activities, and anticancer activities [9,10]. Our previous studies clearly demonstrated that delta-tocotrienol ($\delta$T) inhibits the proliferation and metastatic/invasion potential while concurrently inducing apoptosis in NSCLC cells, in a dose-dependent manner [11]. We also identified some of the probable molecular targets of $\delta$T treatments on NSCLC [11–13]. Therefore, $\delta$T is multitargeted and can be considered a valuable potential approach to further investigate for treatment of NSCLC.

Metabolomics, a novel, versatile, and comprehensive approach, can provide unbiased information about metabolite concentrations, altered signaling pathways, and their interactions. Most current cancer metabolomics studies focus on finding diagnostic biomarkers and understanding fundamental mechanisms in cancer [14]. Nonetheless, this approach could also be used effectively for identifying the efficacy of treatments [15]. The NSCLC metabolome is a potentially informative reflection of the impact of the disease and its dynamics which could lead to promising developments in cancer research, strongly geared toward the discovery of new biomarkers of disease onset, progression, and effects of treatment regimens. Given that cancer cells, including NSCLC, show aberrant energy metabolism [16,17], it is of interest to investigate the changes in energy metabolism in NSCLC cells upon $\delta$T treatment, utilizing the global advantage of the metabolomic approach [18].

Glutamine plays a role as an indirect energy source in NSCLC, which produces ATP through glutamine-driven oxidative phosphorylation [19]. Extra consumption of glutamine in tumors is used for generating metabolic precursors for uncontrolled cell proliferation. These precursors include elevated levels of nucleic acids, lipids, and proteins for cell proliferation [20], as well as increased GSH production for cell death resistance [21]. Current literature provides further evidence that glutamine in cancer facilitates exchange of EAAs (essential amino acids) with glutamine into proliferating cells via glutamine transporters, which induces mTOR (mammalian target of rapamycin) activation in NSCLC and other types of cancer [22,23]. Activated mTOR then promotes protein translation and cell growth via activation of its downstream genes such as S6k1 and 4EBP1 [24]. Alanine, serine, cysteine-preferring transporter 2 (ASCT2), also known as (SLC1A5), and bidirectional L-type amino acid transporter 1 (LAT1) are the two primary transporters for glutamine uptake [25,26]. LAT1 enables transport of the EAAs to improve cancer cell growth via mTOR-induced translations, and ASCT2 sustains the cytoplasmic amino acid pool to drive LAT1 function [27]. This collaboration of ASCT2 and LAT1 reduce apoptosis and enhance the energy production and cell growth via net delivery of glutamine inside the cell [27].

A recent study reported that A549 and H1229 lung cancer cells show glutamine dependency, and that deprivation of glutamine inhibits cell growth [28]. Decreases in glutamine uptake, cell cycle progression, and mTORC1 pathway after inhibition of ASCT2 functionality by chemicals or shRNA in vitro was observed in prostate and pancreatic cancer cell lines [29]. Also, inhibition of LAT1 using BCH (2-aminobicyclo-(2,2,1)-heptane-2-carboxylic acid) in H1395 lung cancer cell line reduced the cellular leucine uptake and consequently inhibited mTOR pathway activity, which finally reduced cell proliferation and viability [30]. Induction of apoptosis was also reported in hepatoma, hybridoma, leukemia, myeloma, and fibroblast cells after glutamine deprivation [31,32]. Our preliminary metabolomics studies showed that $\delta$T treatments inhibited glutamine levels in A549 and H1299 cells. Also, in our previous studies, induction of apoptosis and inhibition of cell growth was observed in A549 and H1299 cells in a dose-dependent manner after $\delta$T treatments [11,33–36]. Therefore, the aim of this study was to verify the metabolic dysregulation of glutamine and its derivatives upon $\delta$T treatment while investigating the effect of $\delta$T on the expression of glutamine transporters (ASCT2 and LAT1) and the mTOR pathway.

## 2. Results

### 2.1. δT Changes Metabolite Profiles in A549 and H1299 Cells

To investigate the changes in metabolism and metabolites with δT intervention, supervised OPLS-DA analysis was performed using NMR spectral data acquired from intracellular cell lysate. The OPLS-DA score plot of cellular NMR metabolic profile resulting from 30 μM δT treated and control cells lines are shown in Figure 1A. The OPLS-DA score plot exhibited clear separation between control and treatment groups in A549 cells and H1299 cells with δT treatment; the high Q2 and R2 values indicate a considerable difference in the cellular metabolic profile of treated cells compared to control cells while validating the model that we used for OPLS-DA analysis.

To identify the metabolites represented in the NMR spectral regions (bins) that varied significantly between control and treatment groups, the corresponding loading S-Line plot from the OPLS-DA model was generated. Figure 1B shows a representative S-Line plot corresponding to the score plot of Figure 1. These bin numbers were further analyzed to identify the significant metabolites (using Chenomx) that contributed to the separation of the control and treatment groups seen in the OPLS-DA model. Based on the analysis of S-Line plot bin numbers, the key bin numbers responsible for the differences could be attributed to glutamine, glutamate and glutathione, and some amino acids in both cell lines.

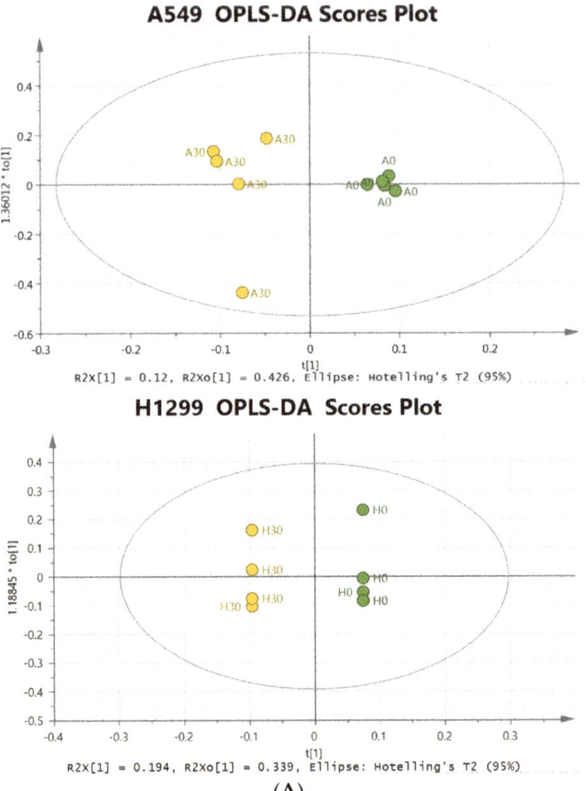



**Figure 1.** *Cont.*

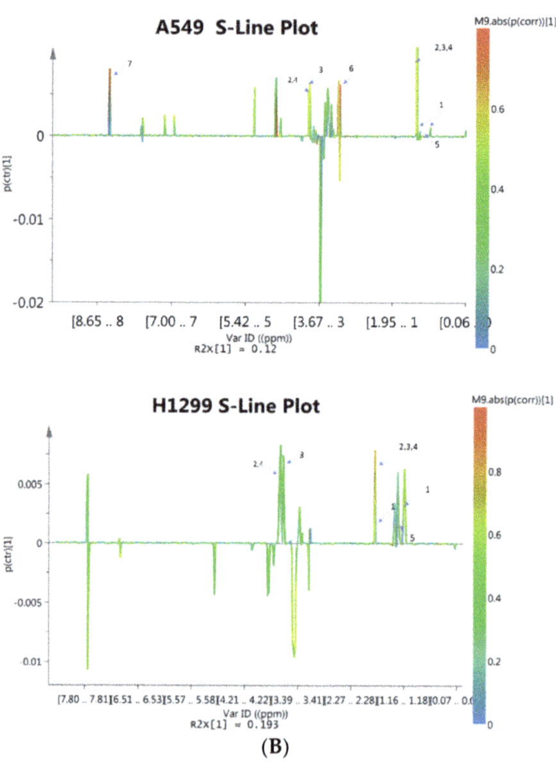



**Figure 1.** OPLS-DA analysis of metabolome of lung cancer cell lines after treating with/without δT for 72 h. (**A**) OPLS-DA Scores plot based on the cellular metabolic profiling of lung cancer cell lines, namely A549 (Top) and H1299 (Bottom); the 30 μM treatment (Yellow) and control (green) were generated using SIMCA+ software; the results indicated that cellular metabolic profiling of lung cancer cell lines was significantly changed after δT treatment for 72 h. (**B**) The S-Line plots of OPLS-DA analysis of A549 (top) and H1299 (Bottom) from treatment (30 μM) and control (0 μM) cells. The key metabolites that changed significantly are marked on the S-Line plot and include (1) leucine, (2) glutamine, (3) glutamate, (4) glutathione, (5) lactate, (6) taurine, and (7) formate.

## 2.2. Quantification of Metabolites Reveals That δT Alters Glutamine Metabolism

Chenomx 7.6 Suite NMR software was used to probe the metabolome profiles in the treatment and control groups. 1H-NMR spectra provided information on over 45 metabolites (both cell lines), including amino acids, intermediates of the tricarboxylic acid cycle (TCA), energy molecules, and nucleic acid associated molecules (Table 1).

The table shows the detailed results including p-values, mean and standard deviation from the *t*-test for the groups (with or without 30 μM of δT treatment) tested. Among the metabolites that were significantly different in concentration in the δT treated vs. control cells, we identified several metabolites from the glutamine metabolism and related pathways that were significantly decreased ($p < 0.05$) in the treatment group as compared to controls. In addition, we found that metabolites such as leucine and some essential amino acids had significantly lower concentrations in both cell lines after δT treatment. These essential amino acids include isoleucine, leucine, lysine, methionine, and tryptophan. Moreover, the metabolites related to cell proliferation such as 2-oxoglutarate, citrate, succinate, malate, aspartame, ATP, ADP, NADPH, and uracil significantly decreased ($p < 0.05$) in the treatment group as compared to controls (Table 1).

**Table 1.** List of metabolite concentrations determined using NMR in A549 (A) H1299 (B) cells. *p*-values less than 0.05 were considered statistically significant for univariate analysis. Treatment column indicates the samples with the 30 μM treatment of δT. All the concentrations are reported in μM.

(A)

| Metabolite Name | Mean ± SD (Control) | Mean ± SD (Treatment) | *p*-Value | Fold Changes Control/Trt |
|---|---|---|---|---|
| *Amino Acids* | | | | |
| Aspartate | 102.3 ± 11.9 | 55.9 ± 4.7 | 0.0016 | 1.8 |
| Glutamate | 80.8 ± 7.9 | 48.7 ± 4.7 | 0.0019 | 1.7 |
| Leucine | 33.7 ± 4.1 | 17 ± 3.7 | 0.0030 | 2.0 |
| Glycine | 33.1 ± 1.2 | 20.4 ± 4.2 | 0.0035 | 1.6 |
| Alanine | 31 ± 1.4 | 19.8 ± 3.9 | 0.0045 | 1.6 |
| Glutamine | 99.9 ± 6.7 | 64.7 ± 13.3 | 0.0073 | 1.5 |
| Histidine | 54 ± 8.4 | 85.9 ± 31.3 | 0.0815 | 0.6 |
| Asparagine | 116.9 ± 16.2 | 54.5 ± 13.1 | 0.0033 | 2.1 |
| Taurine | 90.3 ± 19.9 | 78.2 ± 26.8 | 0.2822 | 1.2 |
| Valine | 23.8 ± 1.4 | 21.6 ± 6.3 | 0.2878 | 1.1 |
| Tryptophan | 81.3 ± 15 | 72.7 ± 28.7 | 0.3340 | 1.1 |
| Proline | 51.9 ± 49.3 | 63.7 ± 25.7 | 0.3659 | 0.8 |
| Lysine | 41.6 ± 22.8 | 37.2 ± 6.1 | 0.4075 | 1.1 |
| Isoleucine | 31.5 ± 9.9 | 30.6 ± 7 | 0.4499 | 1.0 |
| Methionine | 5.8 ± 5.3 | 5.5 ± 3.4 | 0.4653 | 1.1 |
| Arginine | nd | nd | | |
| *Intermediate of TCA Cycle and Energy Metabolism* | | | | |
| Lactate | 138.5 ± 5.6 | 99.9 ± 3.6 | 0.0003 | 1.4 |
| 2-Oxoglutarate | 43.6 ± 3.3 | 29.3 ± 4.7 | 0.0061 | 1.5 |
| AMP | 32.1 ± 5 | 45 ± 1.7 | 0.0063 | 0.7 |
| Glutaric acid monomethyl ester | 17.8 ± 6.4 | 34 ± 2.8 | 0.0077 | 0.5 |
| Malate | 90.2 ± 10.7 | 48.7 ± 10.3 | 0.0111 | 1.9 |
| Succinate | 9.3 ± 2.6 | 5.2 ± 2.8 | 0.0645 | 1.8 |
| Glucose | 119.1 ± 53.4 | 187.3 ± 63.7 | 0.1139 | 0.6 |
| ADP | 47.8 ± 8.3 | 40.8 ± 4.8 | 0.1370 | 1.2 |
| Citrate | 42.4 ± 3.8 | 35.6 ± 11.6 | 0.1959 | 1.2 |
| NADH | 38.4 ± 3.5 | 43.4 ± 16 | 0.3040 | 0.9 |
| NADPH | 47 ± 6.3 | 51.3 ± 12.5 | 0.3118 | 0.9 |
| ATP | 42.2 ± 5.4 | 42.9 ± 11.3 | 0.4653 | 1.0 |
| *Nucleic acid Associataed Metabolites* | | | | |
| Uracil | 98 ± 14.1 | 60.1 ± 24 | 0.0387 | 1.6 |
| UDP-N-Acetylglucosamine | 6.9 ± 2.1 | 3.9 ± 3.4 | 0.1266 | 1.8 |
| *Other* | | | | |
| Glutathione | 69.6 ± 2.1 | 41.7 ± 6.7 | 0.0011 | 1.7 |
| Citrulline | 81.9 ± 5.1 | 63.9 ± 13 | 0.0438 | 1.3 |
| Cystine | 81.4 ± 6.3 | 58.4 ± 19 | 0.0582 | 1.4 |
| N-Acetylglucosamine | 21.9 ± 9.3 | 12.8 ± 5.2 | 0.1065 | 1.7 |
| Formate | 294.3 ± 68.5 | 312.8 ± 8.9 | 0.3334 | 0.9 |
| Fumarate | 25 ± 3.2 | 27.7 ± 5 | 0.2363 | 0.9 |

**Table 1.** *Cont.*

(B)

| Metabolite Name | Mean ± SD (Control) | Mean ± SD (Treatment) | *p*-Value | Fold Changes Control/Trt |
|---|---|---|---|---|
| *Amino Acids* | | | | |
| Aspartate | 105.5 ± 3.5 | 77.4 ± 4.3 | 0.0010 | 1.4 |
| Glutamate | 80.1 ± 5.7 | 49.3 ± 6.2 | 0.0033 | 1.6 |
| Leucine | 31.8 ± 1.3 | 18.3 ± 0.8 | <0.0001 | 1.7 |
| Glycine | 28.2 ± 4.7 | 18.1 ± 3.2 | 0.0561 | 1.6 |
| Alanine | 28.8 ± 2.2 | 18.2 ± 2.3 | 0.0044 | 1.6 |
| Glutamine | 75.3 ± 5.1 | 53.7 ± 8.4 | 0.0177 | 1.4 |
| Histidine | ND | ND | | |
| Asparagine | 105 ± 21 | 84 ± 23.3 | 0.1986 | 1.3 |
| Taurine | ND | ND | | |
| Valine | 28.8 ± 4.9 | 21.7 ± 5.6 | 0.1706 | 1.3 |
| Tryptophan | 36.8 ± 2 | 17.8 ± 11.4 | 0.0401 | 2.1 |
| Proline | 90.2 ± 39.3 | 74.3 ± 34.9 | 0.3453 | 1.2 |
| Lysine | 38.8 ± 11.3 | 19.4 ± 7.1 | 0.0547 | 2 |
| Isoleucine | 37.2 ± 4.9 | 23.8 ± 2.7 | 0.0138 | 1.6 |
| Methionine | 8.7 ± 0.8 | 6.7 ± 1.9 | 0.1247 | 1.3 |
| Arginine | 43.8 ± 2.7 | 28.4 ± 6.6 | 0.0189 | 1.5 |
| *Intermediate of TCA Cycle and Energy Metabolism* | | | | |
| Lactate | 125.8 ± 7.3 | 122 ± 15.4 | 0.3857 | 1 |
| 2-Oxoglutarate | 32.5 ± 7.9 | 17.2 ± 1.5 | 0.0272 | 1.9 |
| AMP | 27.5 ± 0.2 | 13.7 ± 2 | 0.0003 | 2 |
| Glutaric acid monomethyl ester | 27.4 ± 0 | 20.6 ± 7.4 | | 1.3 |
| Malate | 130.9 ± 7.8 | 84.7 ± 9 | 0.0027 | 1.5 |
| Succinate | 13.9 ± 1.7 | 5.3 ± 3.8 | 0.0215 | 2.6 |
| Glucose | 196.4 ± 50.1 | 147.1 ± 19.4 | 0.1324 | 1.3 |
| ADP | 33.6 ± 5.1 | 14.9 ± 7.7 | 0.0227 | 2.3 |
| Citrate | 35.2 ± 0.8 | 25.6 ± 4.3 | 0.0183 | 1.4 |
| NADH | 65.3 ± 11.7 | 43.7 ± 30.7 | 0.2024 | 1.5 |
| NADPH | 48.6 ± 11.1 | 38.1 ± 23.5 | 0.2996 | 1.3 |
| ATP | 43.5 ± 7.8 | 22.2 ± 5.5 | 0.0171 | 2 |
| *Nucleic acid Associated Metabolites* | | | | |
| Uracil | 88.5 ± 11.9 | 40.2 ± 16.3 | 0.0139 | 2.2 |
| UDP-N-Acetylglucosamine | ND | | | |
| *Other* | | | | |
| Glutathione | 42.3 ± 4.5 | 28 ± 6.5 | 0.0319 | 1.5 |
| Citrulline | 65.4 ± 20.6 | 53.4 ± 25.4 | 0.3156 | 1.2 |
| Cystine | 61 ± 7.2 | 26.3 ± 14.1 | 0.0338 | 2.3 |
| N-Acetylglucosamine | | | | |
| Fumarate | | | | |
| Formate | 354.5 ± 90.9 | 346.7 ± 41 | 0.4585 | 1 |
| Tyrosine | 12.9 ± 0.6 | 67.8 ± 9.1 | 0.0134 | 0.2 |

Heatmap analysis from MetaboAnalyst 3.0 revealed that A549 and H1299 cell lysates had similar changing trends in metabolites of δT treated groups versus control (Figure 2A), which suggests that the supplement of δT impacts both cell lines in a similar manner. At the same time, our heatmap results also revealed that control and treatment groups supplemented with δT were clustered into two major groups (Green and Red groups at the top of the Heatmap) which suggest clear separation in two groups with their metabolites and also validates the separation in OPLS-DA analysis. The random forest importance plot identified 15 metabolites key in classifying the data with aspartame, alanine, leucine, glutamate glutathione, and glutamine having the most influence on classification (Figure 2B).

To further comprehend the biological relevance of the identified metabolites from Chenomx analysis, we performed pathway analysis using MetaboAnalyst 3.0 software [25]. Some of the key altered pathways identified from pathway analysis include lysine biosynthesis, purine metabolism, alanine, aspartate and glutamate metabolism, glutamine and glutamate metabolism, citrate cycle (TCA cycle), and pyruvate metabolism for both cell lines (Figure 3A).

## A549

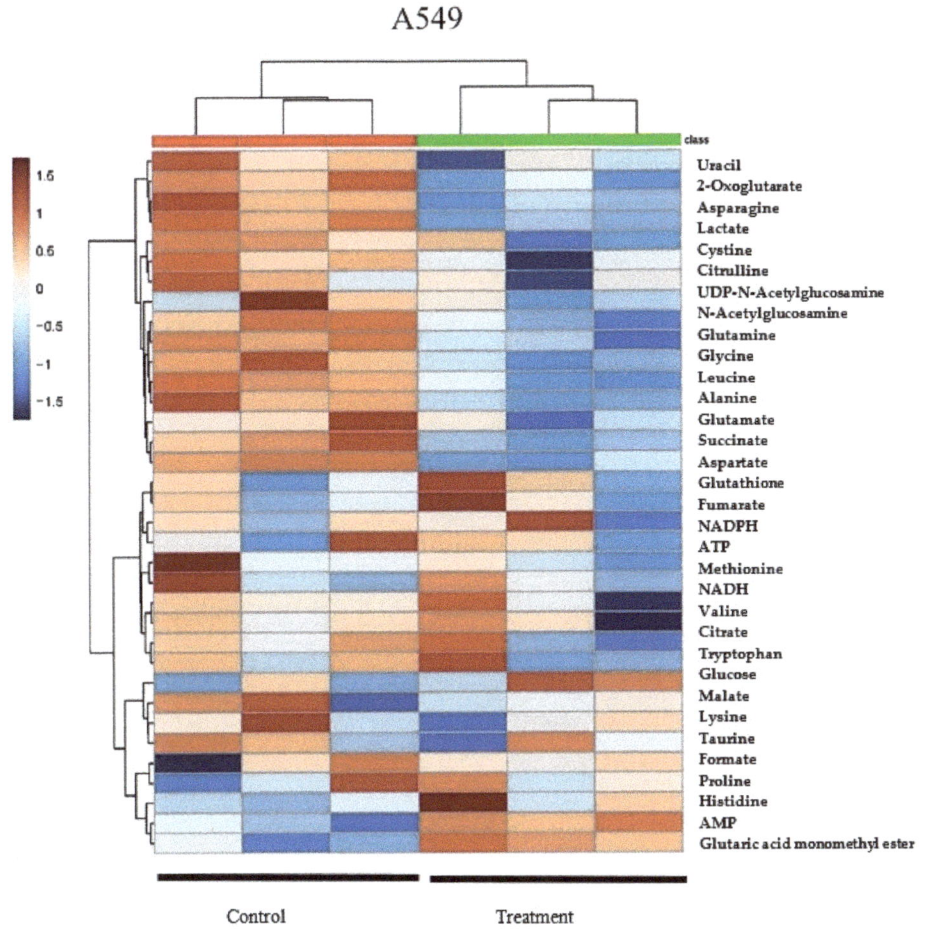


**Figure 2.** *Cont.*

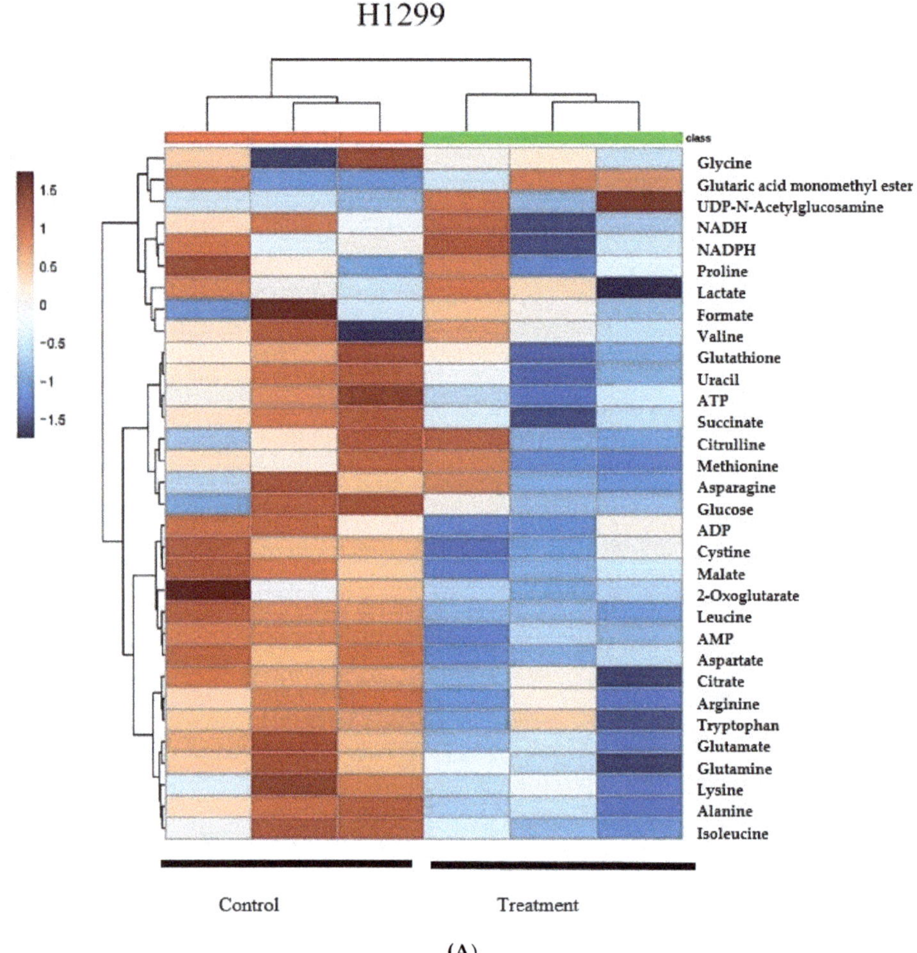


(A)

**Figure 2.** *Cont.*

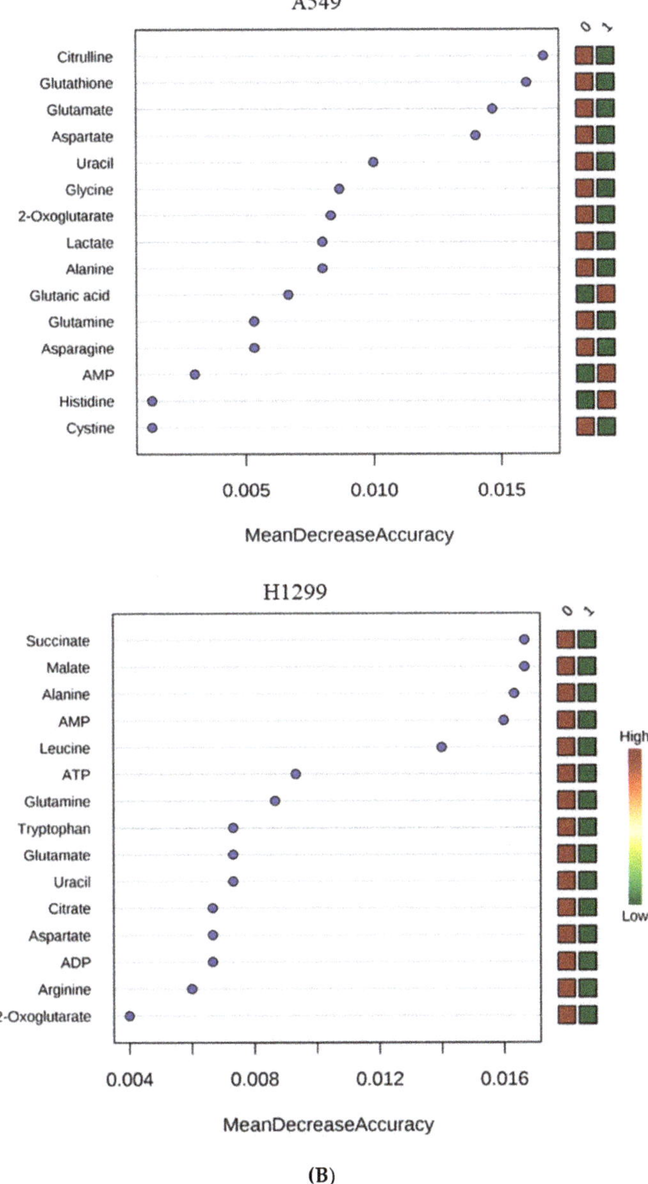



**Figure 2.** Hierarchical clustering analysis of δT-altered metabolites (Heatmap) and contribution of metabolites in A549 and H1299. The metabolites, quantified with Chenomx software analysis of NMR spectra of A549 and H1299 cells after incubating with or without δT for 72 h, were used to generate the heat map (**A**) using Metaboanalyst software. Each column represents a sample, and each row represents the expression profile of metabolites. Blue color represents a decrease, and red color an increase. The very top row with green color indicates the control samples and red color row indicates the samples with the 30 μM treatment of δT. Random Forest (**B**) showed in bottom graphs identifies the significant features. The features are ranked by the mean decrease in classification accuracy when they are permuted.

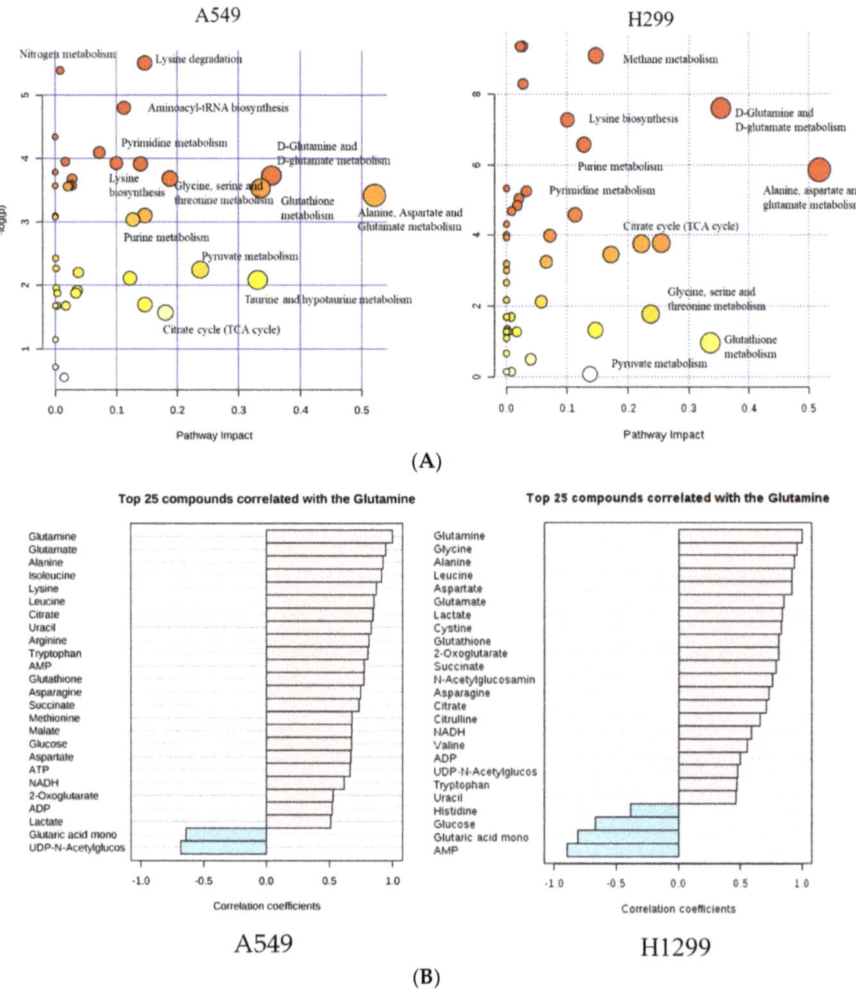

**Figure 3.** The most predominant altered metabolic pathways (**A**) and top 25 metabolites correlated with glutamine (**B**). Summary of the altered metabolism pathways (**A**) after treating with/without δT for 72 h, as analyzed using MetaboAnalyst 3.0. The size and color of each circle was based on pathway impact value and *p*-value, respectively. Circles, larger and higher along the *Y* axis, show higher impact of pathway on the organism. The top 25 metabolites, correlating with glutamine level (**B**) after treating with/without δT for 72 h. *X*-axis shows maximum correlation; pink color shows positive correlation whereas blue shows negative correlation.

As random forest importance plot and pathway analysis indicate that glutamine-based metabolites play a significant contribution to glutamine metabolism and related pathways, correlation between other metabolites were assessed using Pearson correlation analysis to validate the relationship between glutamine and metabolites in other pathways. Interestingly, nearly 20 metabolites showed more than (>0.7) correlation with glutamine and metabolites belonging to the key impaired pathways identified from pathway analysis using MetaboAnalyst 3.0 software. The metabolites in glutamine and glutamate metabolism include glutathione, glutamate, 2-oxoglutarate which show a 0.9, 0.7, and 0.6 correlation in A549 and 0.8, 0.8, and 0.8 correlation in H1299 (Figure 3B).

### 2.3. δT Inhibits Glutamine Transporters (LAT-1 and ASCT2) and the mTOR Pathway in A549 and H1299 Cells

Metabolomic analysis and subsequent quantification of metabolites using Chenomx NMR suite (Edmonton, AB, Canada) revealed the potent effect of δT on glutamine metabolism, downstream metabolites of glutamine and essential amino acids (Figures 1 and 2, Table 1). Current literature provides evidence that glutamine uptake and some essential amino acids, including leucine, are associated with the activation of the mTOR pathway [37]. Thus, Western blot analysis was performed to investigate the effect of δT on the mTOR pathway and glutamine transporters. Upon intervention with δT (30 μM), the glutamine transporters (LAT-1 and ASCT2) and key mTOR pathway proteins (P-mTOR and p-4EBP-1) were found to be inhibited, relative to the untreated controls (Figure 4).

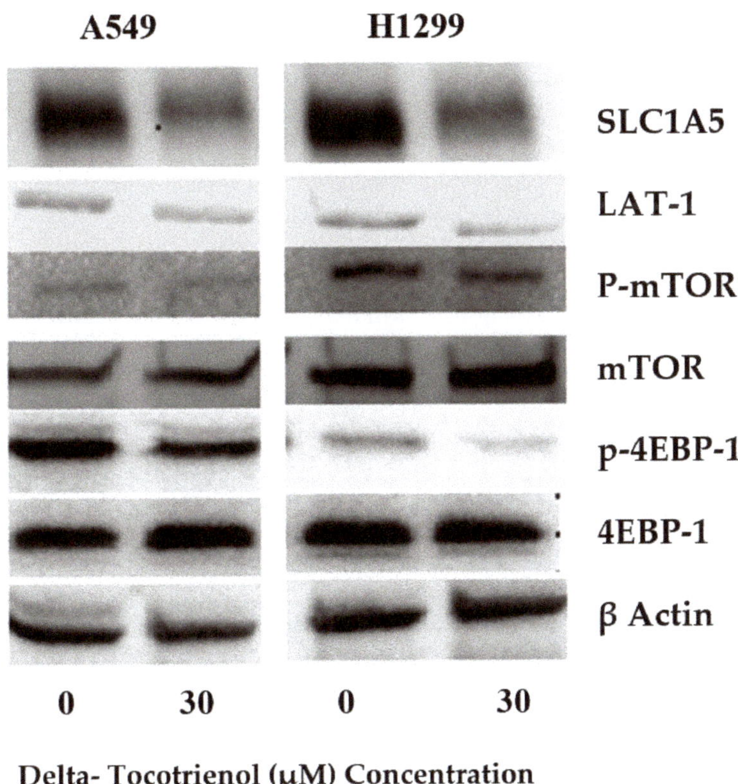


**Figure 4.** *Cont.*

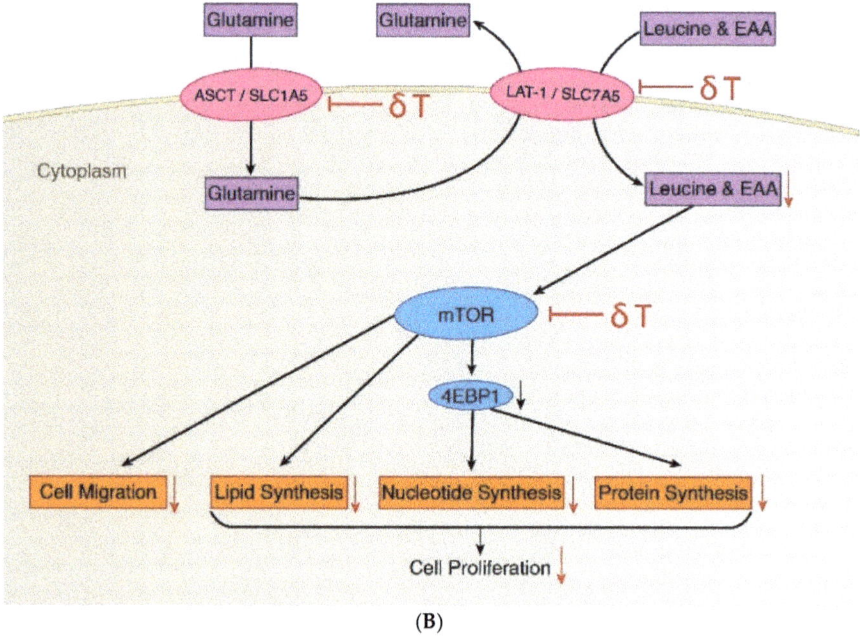



**(B)**

**Figure 4.** δT inhibits glutamine transporters (LAT-1 and ASCT2) and the mTOR pathway in A549 and H1299 cells. (**A**) The expressions of LAT-1, ASCT2, p-mTOR, mTOR, p-4EBP-1, 4EBP1, and β-actin proteins were detected by Western blot analysis in A549 and H1299 after treating with 0 μM and 30 μM concentrations of δT for 72 h. (**B**) The fate of glutamine uptake in A549 and H1299 involving metabolites (purple), associated key proteins (pink), and functions (orange). Glutamine in cancer facilitates exchanging of EAAs (essential amino acids) into proliferating cells via glutamine transporters (LAT1 and ASCT2), which induces mTOR activation in A549 and H1299. Activated mTOR then promotes protein translation and cell growth via activation of its downstream genes 4EBP1. The black arrows indicate pathway direction, while the red downward arrows indicate inhibition.

## 3. Discussion

In this study, we used multivariate analysis of NMR spectra and NMR quantification data to observe differences in the intracellular metabolomes. We discovered clear differences in the intracellular metabolomes, and subsequently the contributing metabolites, of the control and δT treated cells using OPLS-DA and Heat map analysis (Figures 1 and 2A). Also, we observed a minor difference in the results obtained through multivariate analysis of NMR spectra and NMR quantification variation in this analysis which is common in metabolomic data sets. This type of variation is well documented in several publications in the current literature [6]. Most variations arise from the metabolites present in very low concentrations. In addition, metabolites whose resonances yield a very high number of overlapping peaks also suffer from variations in quantitation [6]. The two different methods were therefore used in conjunction to verify the data.

Previously, using histone ELISA and ANNEXIN V stain-based flow cytometry analysis, we reported that the 10 to 30 μM range of δT was not necrotic to A549 and H1299 cells, and that it induced apoptosis in a dose-dependent manner [11,12]. Also, using MTS and clonogenic assays in the previous studies, we demonstrated that 30 μM of δT inhibited cell growth significantly in the A549 and H1299 cells lines [12]. Other metabolomics investigations have also reported changes in metabolism after inducing apoptosis in different cancer types, namely leukemia cell lines [38]. Our data suggests that metabolite changes in the control vs. δT treated lung cancer cell populations are a result of induction of apoptosis after δT treatment.

The role of natural dietary components in cancer growth and progression has become a very popular subject with minimum effect or no effect on normal cells. Several cell culture studies showed that δT was not causing apparent impairment towards the noncancerous cell lines, although it significantly effects different cancer cell types, including lung cancer. For instance, Human Fetal Lung Fibroblast Cells treated with 100 μm or higher of δT did not show any toxic effect including induction of apoptosis and DNA damage [18]. In another study, 10 μM DT3, a lower dose than our treatment, was determined to be nontoxic, and enhanced cell viability and proliferative potential in the human lung fibroblast cell lines MRC-5 and HFL1, as shown by WST-1 and clonogenic assays [39]. In addition, Immortal human pancreatic duct epithelial cell lines did not show any significant inhibitory effect on cell proliferation and cell cycle progression when they were incubated with δT [40]. Similarly, normal human melanocytes treated with δT (5–20 μg/mL) for 24 h or 48 h did not affect cell growth at both time intervals [41]. Preclinical and clinical evidence also supports the use of δT to reduce tumor growth with no effects on healthy humans or animals, making δT attractive compounds. No adverse effects were observed upon administration of 300 mg/kg dose of δT, in any tissues or organs of mice [42]. In humans, δT can be safely administered at doses up to 1600 mg twice daily [43]. In another study with osteopenic women, supplementation for 12 weeks did not affect body composition, physical activity, quality of life, or intake of macro- and micronutrients [44]. All of the aforementioned studies used δT concentrations above 30 μM that we used for this study, and it is obvious that δT does not affect healthy cells including human fetal lung fibroblast cells. Therefore, a control arm of normal lung cells with expressed or unexpressed LAT1 and/or ASCT2 were not included in our study design.

Further, LAT1 or ASCT2 transporters with cancer is nowadays well-assessed [9]. Overexpression of LAT1 is well described in many human cancers and it certainly relates to metabolic changes occurring in cancer development and progression [45]. LAT-1 is expressed in cancers of most human tissues according to GENT database [46], which suggests an important role of LAT-1 expression on cancer development. In contrast, it is poorly expressed or, in some cases, absent in most of the corresponding noncancer human tissues [46]. In the immunohistochemistry analysis of the normal lung, LAT1 protein was identified only on granular regions in the cytoplasm of chondrocytes of the bronchial cartilage, serous cells of the bronchial glands, and alveolar macrophages within the normal lung, whereas the expression was zero for nonciliated bronchiolar epithelial cells (Clara cells), goblet cells of the bronchus, mucinous cells of the bronchial glands, and alveolar type I or type II cells [47]. In the same study, expression of LAT1 protein appeared in the cytoplasm of bronchial surface epithelial cells as a single nodular spot, which was considered to represent an intracellularly localized nonfunctional protein [47]. ASCT2 transporters also are poorly expressed or, in some cases, absent in most of the corresponding noncancer human tissues according to GENT database [46]. Hassanein et al. identified ASCT2 transporters expressed in stage I NSCLC when compared to matched controls using shotgun proteomic analysis [48]. In addition, ASCT2 deficient mice showed regular functions such as normal B-cell development, proliferation, and antibody production [49]. Therefore, control arms of normal lung cells that are expressed or unexpressed (LAT1 and ASCT2) was also not included in our study design as there was a minimum expression and/or functionality observed for LAT1 and ASCT2 in other tissues and noncancerous tissues.

A significant reduction of glutamine, glutamate, GSH and 2-oxoglutarate after treating with 30 μM of δT on NSCLC cell lines was observed (Table 1). The key aberrant pathways identified using the pathway analysis tool include glutamate and glutamine, alanine, aspartate, glutathione metabolism, and the TCA cycle (Figure 3). In addition, the metabolites identified from these pathways show a strong correlation with glutamine levels (Figure 3B). Further, glutamine and its related metabolites were identified in the S-plot of OPLS-DA analysis and the Random Forest importance plot as the key players causing the separation, reflecting the differences in their metabolomic profiles (Figures 1 and 2B). Glutamine deprivation has been shown to induce apoptosis in hepatoma, hybridoma, leukemia, myeloma, and fibroblast cells [50]. In contrast, increased levels of glutamine were detected in lung cancer tissue especially in NSCLC when compared to other types of cancer, such as colon or stomach cancer [47]. Glutamine dependency has been reported in H1299 and A549 cells [28]. Our findings strongly suggest the beneficial impact of δT on glutamine and related pathways in non-small cell lung cancer cells.

Considering metabolism of glutamine (Figure 5), one of its major roles in cancer cell proliferation is to replenish the TCA cycle intermediates removed by the process called glutaminolysis, and GSH synthesis [30,31]. In the process of glutaminolysis, the glutaminase enzyme (GLS1/2) catalyzes the conversion of glutamine to glutamic acid and the subsequent conversion of glutamate to α-ketoglutarate (2-oxoglutarate), catalyzed by glutamate dehydrogenase (GLUD) [32]. Aminotransferase also catalyzes the reaction from glutamate and oxaloacetate to aspartate or alanine and α-ketoglutarate. In this study, a significant reduction of glutamine, glutamate, and TCA cycle intermediates after treating with 30 μM of δT was observed, which is an indicator of reduced energy metabolism (Figure 5). In cancer cells, the enhanced production of 2-oxoglutarate and glutamate from glutamine metabolism can be observed, as it helps to maintain the citric acid cycle intermediate for energy production [32]. Glucose and glutamine provide substrates for macromolecular synthesis supplying both ATP and carbon skeletons in cancer cells [29]. This supports uncontrolled cell proliferation in cancer cells and requires a large number of macromolecules to create new biomass, including DNA, proteins, and lipids [28]. Our data suggests that by decreasing the availability of glutamine, δT retards this process, thereby leading to inhibition of uncontrolled cell proliferation in A549 and H1299 as reported in our previous studies [11,12,35].

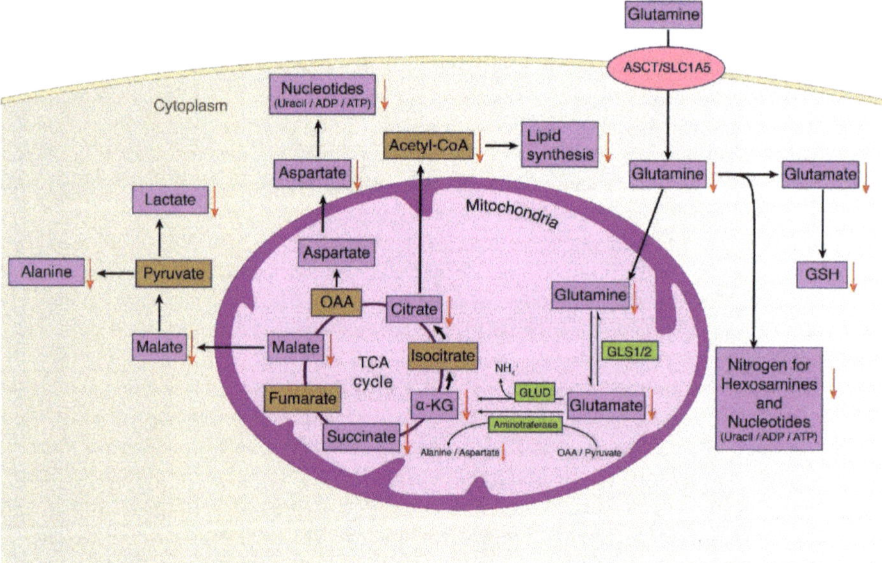


**Figure 5.** Glutamine metabolism and the effect of δT on glutamine metabolism in A549 and H1299 cells. Glutamine mainly replenishes the TCA cycle intermediates and GSH synthesis in cancer cell proliferation. In the process, glutaminase enzymes (GLS1/2) catalyzes the conversion of glutamine to glutamic acid and the subsequent conversion of glutamate to α-ketoglutarate (α-kG), catalyzed by glutamate dehydrogenase (GLUD) and amino transferase. This process supports for uncontrolled cell proliferation in cancer cells and requires a large number of macromolecules to create new biomass, including DNA, proteins, and lipids. The black arrows indicate the pathway's direction, while the red downward arrows indicate the inhibition of metabolites as an effect of δT treatment.

Considering possible causes for the significant decrease in glutamine and its downstream metabolites, we hypothesized that it may be due to inhibition of glutamine transporters. We thus measured the protein levels of glutamine transporters, namely LAT1 and ASCT2, known to play a fundamental role in glutamine uptake process in normal cell physiology. LAT-1 facilitates glutamine efflux in exchange for the influx of leucine and other essential amino acids (EAA) across the cell membrane; similarly, ASCT2 mediates uptake of neutral amino acids including glutamine [51]. Our observations from western blot analysis established that $\delta$T treatments inhibit the expression of LAT-1 and ASCT2 (Figure 4). We also quantified detectable EAA including leucine in cell lysates, the concentration of which were decreased significantly after treating NSCLC cells with $\delta$T by NMR analysis. Inhibition of EAA in A549 and H1299 cells upon $\delta$T treatment reflects function of LAT-1 which facilitate glutamine efflux in exchange for the influx of leucine and other essential amino acids (EAA). This supports the beneficial effects of $\delta$T on LAT1 transporters inside A549 and H1299 cells. In addition to facilitating the transport of EAAs for protein synthesis, LAT1 and ASCT2 stimulate the growth of cancer cells via mTOR [27,52,53]. In head and neck squamous cell carcinoma cell lines, inhibition of the LAT-1 transporter using an inhibitor lowered the levels of phosphorylation of mTOR and its downstream signaling molecules [54]. Thus, if the inhibition of glutamine transporters and EAA uptake with $\delta$T treatment is valid, it is logical to expect inhibition or lower activation of mTOR pathway after treating with $\delta$T in NSCLC. Indeed, we observed lower activation of mTOR along with LAT-1 and ASCT2 after treating with $\delta$T, using Western blot analysis, which illustrates that inhibition of glutamine transporters affect the mTOR signaling pathway (Figure 4).

mTOR functions are mediated by two downstream proteins, the eukaryotic initiation factor 4E (eIF4E)-binding protein 1 (4E-BP1) and p70 ribosomal S6 kinase 1 (p70S6K1, S6K1) (Figure 4) [55]. For further confirmation, we tested the expression levels of downstream genes of mTOR namely P-4E-BP1. We observed the similar inhibitory effect on mTOR downstream proteins 4E-BP1suggesting an inhibitory effect of glutamine transporters passing through mTOR to downstream pathway (Figure 4). mTOR downstream proteins 4E-BP1 and S6K1 regulate F-actin reorganization, focal adhesion formation, and tissue remodeling through the proteolytic digestion of extracellular matrix via upregulation of matrix metalloproteinase 9 (MMP-9) [56]. Interestingly, in our previous study, we observed that $\delta$T reduced cell migration, invasion and adhesion in a dose- and time-dependent manner, and inhibited MMP-9 expressions in NSCLC cells [13,34], which is an additional supporting inhibitory function of $\delta$T.

Further, in the previous study, we demonstrated that $\delta$T induces apoptosis in a dose-dependent manner in NSCLC from Annexin based flow cytometry analysis and histone ELISA [12]. The current literature also provides evidence to support the relationship between GSH and apoptosis. For instance, GSH depletion in cancer cells induces apoptosis in vitro and in vivo [57]. Dalton TP et al. showed GSH-depleted knockout mouse of $\gamma$-GCS died from massive apoptotic cell death [58]. Elevated levels of GSH are also associated with apoptotic resistant phenotypes in several models of apoptosis in previously reported studies [59,60], and GSH depletion by itself has been observed to either induce or stimulate apoptosis [59,61]. GSH quantification, after treating with $\delta$T in A549 and H1299 cells, shows a clear decline in intercellular GSH levels in both cell lines (Table 1). The results reveal there may also be a possible association between GSH levels and induction of apoptosis in NSCLC cells after treating with $\delta$T.

## 4. Materials and Methods

### 4.1. Cell Culture and Treatment with $\delta$-T

NSCLC cell lines A549 and H1299 were cultured in RPMI medium (Mediatech, Manassas, VA, USA) supplemented with 10% fetal bovine serum and 1% penicillin and streptomycin in 5% $CO_2$ at 37 °C. The culture medium was renewed every 2 to 3 days. Adherent cells were detached by incubation with trypsin-EDTA and centrifuged at $80\times g$. The treatment media was prepared by mixing $\delta$T (<0.01%

DMSO as a vector) in the RPMI medium, whereas the control was treated only with RPMI media. Three $\delta$T solutions at concentrations of 10 $\mu$M, 20 $\mu$M, and 30 $\mu$M containing <0.01% DMSO were chosen as the treatment concentration based on our previous studies. $\delta$T was a gift from the American River Nutrition for this study.

### 4.2. Intracellular Metabolite Extraction and Determination

We used a modified method which is explained in Saadat et al., 2018 [62]. In brief A549 and H1299 lines were seeded at a density of $2 \times 10^6$ per 100-mm dish for 24 h, followed by replacement of media absent or supplemented with different $\delta$T concentrations (10 $\mu$M, 20 $\mu$M, and 30 $\mu$M) at 37 °C. Cells were then incubated for another 72 h before extracting metabolites. Before extracting intracellular metabolites, existing culture media was removed on ice followed by washing twice with ice-cold PBS. Two milliliters of ice-cold methanol was added while scraping with cell scrapers on ice. The Petri dish was shaken for 5 min at 4 °C and ice-cold methanol was transferred into Eppendorf tubes. The cell debris was removed by centrifugation and all the extraction solvents were readily removed before NMR analysis by a Speed Vac at room temperature. Subsequently, the intracellular metabolites powder was prepared by evaporating with methanol, and redissolving in 450 $\mu$L $D_2O$ containing 0.5 $\mu$M 2,2-Dimethyl-2-silapentane-5-sulfonic acid (DSS) as aspectral calibration standard and 10 $\mu$M imidazole as a pH indicator. An additional Petri dish was prepared for each treatment/control with the same conditions and cells collected from the additional petri dish were used for analyzing total protein. The total protein quantifications include control-A549 (1.283 mg), 30 $\mu$M-A549 (1.099 mg), control-H1299 (1.325 mg), and 30 $\mu$M-H1299 (1.276 mg). The intracellular metabolite powder was redissolved in $D_2O$ and normalized based on the total protein contained in additional petri with corresponding treatment before performing NMR. We made sure to maintain the final concentration of internal standards at aforementioned levels.

### 4.3. 1H-NMR Spectroscopy

High-resolution 1H-NMR spectra of intracellular metabolites were obtained on a Varian 600 spectrometer operating at 600 MHz after normalizing the samples by total protein concentrations using BCA Protein Assays (Thermos Fisher Scientific, Rockford, IL, USA). 1H-NMR spectra of intracellular extracts were acquired using a 6-kHz spectral width and 64 K data points. The acquisition time was 5.44 s and the relaxation delay was 14.56 s with 64 scans.

### 4.4. 1H-NMR Spectroscopy Processing

After NMR analysis, Free Induction Decay (FID) files were obtained and processed using NMR processing software ACD (Advanced Chemistry Development, Inc. Toronto, ON, Canada). NMR spectra of all the samples were stacked and processed simultaneously. First, FID files were Fourier-transformed to visualize spectra followed by phasing, baseline correction and binning with the auto option of the software. After completing these steps, the full spectra, as a batch, were divided into 1000 bins using the intelligent bucketing algorithm in ACD software, giving a numerical value for corresponding peaks, and converted into a data table. Intelligent bucketing in ACD is an algorithm that was designed to make decisions as to where a bucket division should be. Intelligent bucketing chooses integral divisions based on local minima and therefore avoids the reduction of data resolution, while aligning the spectra as a batch.

### 4.5. Quality Control

Relative standard deviation (RSD) values were calculated for each treatment group separately and Technical variation within metabolomics datasets, recorded using one dimensional NMR maintained less than <8% (reported as the median spectral RSD)

## 4.6. Multivariate Data Analysis: OPLS-DA

The processed, digitized NMR spectral data table from ACD software (version 10) was imported into the SIMCA (version 15) software (Sartorius Stadium Biotech, Germany for Multivariate data analysis (MVDA). The data table was transposed and labeled accordingly. The integrals corresponding to the spectral region from 4.5 to 6 ppm were excluded as this region contains water peaks and exchangeable protons. Spectral regions displaying no peaks, DMSO, and spectral regions of methanol to all the samples were also excluded from the dataset. PCA, OPLS-DA models were created by generating optimum number of principal components needed to fit the data, using the autofit option in the software. Each model's characteristics are described by how well it fits the data and its ability to predict new data accurately. Thus the value for R2 describes how well the data fits the model while the value of Q2 relates to the models ability to predict unknown data correctly. These are calculated by the for the purpose of evaluating and validating the models generated. The following cutoff criteria are used for validating the models that were generated. For NMR metabolomic data, it is recommended that the model generated has a Q2 > 0.5, a value of R2 higher than Q2 with the difference between them being no greater than 0.3. These criteria were adhered to for all the models utilized for the investigation. Samples were identified and distinguished by their respective labels and colored for visual convenience. The data was subjected to Pareto-scaling prior to analysis. The Hotelling $T^2$ test (based on the 95% confidence interval) and DMOD-X test (based on the distance from the model plane) was used to remove any statistically extreme outliers while maintaining a minimum of 4 replicates in each group. Initially, unsupervised Principal Component Analysis (PCA) was performed to view the clustering effects in the samples (Supplemental Materials). Subsequently, OPLS-DA, a supervised pattern recognition method, was performed to maximize the identification of variation between groups tested.

## 4.7. Metabolite Identification and Quantification from Chenomx NMR Suite

The metabolites were identified using Chenomx NMR suite (Chenomx Inc., Edmonton, AB, Canada). The fid files from the 1D 1H-NMR spectra were imported into the Chenomx software. This software has its own processing interface where spectra were Fourier-transformed and baseline corrected. Phasing was done using DSS reference peak at 0.0 ppm, and the water peak was also deleted. The processed spectra were analyzed in the profiler module of the software. The 600 MHz library with the corresponding pH was selected. Identification and concentrations of different metabolites were calculated by fitting the set of peaks for those compounds in the sample spectrum. If the area was crowded with many peaks, then multiple metabolites were adjusted at one time to match the reference spectrum closest to the sample spectrum. The identified and quantified compounds were then exported into an excel sheet.

## 4.8. Additional Multivariate Data Analysis and Metabolic Pathway Identification Using MetaboAnalyst 3.0 Software

MetaboAnalyst 3.0 software, a web-based metabolomics data processing tool [63], was used to statistically analyze the metabolites identified using Chenomx NMR suite. Quantified data from Chenomx NMR suite were scaled using range scaling algorithm. Clustering differences, heat maps, and a Random Forest analysis plot were generated. Further, the top 25 metabolites correlating with glutamine were identified using Pearson correlation analysis and the significant features were identified by Random Forest analysis. Additionally, quantified data from Chenomx NMR suite was transferred into an excel table which allowed us to perform a Student's *t*-test and calculate fold changes. A *p*-value of less than 0.05 was considered to be statistically significant for univariate analysis.

Metabolic pathway identification was performed with the pathway analysis option of Metaboanalyst 3.0 software. Briefly, the Homo Sapiens Pathway Library was selected as a reference, and the pathway analysis was performed to generate pathway analysis output on all matched pathways, based on the *p*-values from pathway enrichment analysis and pathway impact values from pathway topology analysis.

Further, metabolites that were changing most significantly between the control and 30 µM treatment were traced back to their origin, and the pathways were interpreted for metabolism changes using current biochemistry.

### 4.9. Western Blot for Protein Expression Analysis

One million cells of each of A549 and H1299 were seeded in 100-mm dishes and incubated for 24 h; then, the original media was replaced by media with/without δT and incubated for another 72 h. After 72 h incubation, cells were washed with ice-cold PBS and lysed in the cold 1X cell lysis buffer (Cell Signaling Technology, Danvers, MA, USA) for 30 min on ice with 1X protease inhibitor (Cell Signaling Technology, Danvers, MA, USA). The cell lysate was kept at −80 °C overnight before quantifying.

Protein concentrations were estimated using Pierce BCA Protein Assay kit (Bio-Rad Laboratories, Hercules, CA, USA). Total cell lysates (40 µg) were mixed with equal amounts of 6x laemmli buffer (Bio-Rad Laboratories, Hercules, CA, USA), followed by boiling at 100 °C for 5 min. Samples were loaded on 10% SDS-polyacrylamide gel electrophoresis, and then the gel was electrophoretically transferred to a nitrocellulose membrane (Whatman, Clifton, NJ, USA) in transfer buffer (25 mM Tris, 190 mM glycine, 20% methanol) using a Bio-Rad Trans-Blot® Turbo™ Transfer System (Hercules, CA, USA). The membranes were incubated for 1 h at room temperature with 5% BSA in 1x TBS buffer containing 0.1% Tween. After incubation, the membranes were incubated overnight at 4 °C with primary antibodies (1:1000). The following antibodies ASCT2, LAT-1, p-mTOR, mTOR, p-4EBP-1,4-EBP1, and B-actin (Cell Signaling Technology, Danvers, MA, USA) were used in the analysis. The membranes were washed three times with TBS-T and subsequently incubated with the secondary antibodies (1:5000) containing 2% BSA for 2 h at room temperature. The signal intensity was then measured by chemiluminescent imaging with ChemiDoc XRS (Bio-Rad Laboratories, Hercules, CA, USA).

## 5. Conclusions

In this work, the anticancer effects of δT on NSCLC cell lines A549 and H1229 were investigated and confirmed by 1H-NMR metabolomics analysis. A closer look into the intracellular metabolome of NSCLC cells revealed significant and potentially beneficial alterations in glutamine concentrations and related metabolism upon treatment with δT. The data purports that δT exerts its action by inhibiting glutamine uptake into proliferating cells by inhibition of glutamine transporters, thereby resulting in inhibition of cell proliferation and induction of apoptosis via downregulation of the mTOR pathway (Figures 4B and 5). Through this work, NMR-based cellular metabolomics helps provide possible opportunities for evaluating the therapeutic effect of phytochemicals and systemic changes in cancer metabolism.

**Supplementary Materials:** The following are available online at http://www.mdpi.com/2218-1989/9/3/50/s1. Table S1: List of metabolite concentrations determined using Chenomx NMR Suite in A549 cells. Table S2: List of metabolite concentrations determined using Chenomx NMR Suite in H1299 cells. Figure S1: Effects of δT on A549 (A) and H1299 (B) on the metabolome of lung cancer cell lines.

**Author Contributions:** Conceptualization, L.D.R. and S.V.G.; Methodology, L.D.R.; Software, L.D.R.; Formal Analysis, L.D.R. and M.H.; Investigation, S.V.G.; Resources, S.V.G.; Data Curation, L.D.R.; Writing—Original Draft Preparation, L.D.R. Writing—Review and Editing, L.D.R., M.H., and S.V.G; Supervision, S.V.G.; Funding Acquisition, S.V.G.

**Funding:** This research was conducted using intra-mural funding.

**Acknowledgments:** We thank Alexander Buko, Vice President Business and Product Development at Human Metabolome Technologies America for comments on results and language and the assistance with additional bioinformatics methods that greatly improved the manuscript. We are also immensely grateful to Bashar Ksebati for NMR instrument support.

**Conflicts of Interest:** The authors declare no conflicts of interest.

## References


1. Society, A.C. Lung Cancer (Non-Small Cell). Available online: http://www.cancer.org/acs/groups/cid/documents/webcontent/003115-pdf.pdf (accessed on 09 March 2019).
2. American Cancer Society. *Cancer Facts & Figures 2016*; American Cancer Society: Atlanta, GA, USA, 2016.
3. Kelsey, C.R.; Clough, R.W.; Marks, L.B. Local Recurrence Following Initial Resection of NSCLC: Salvage Is Possible with Radiation Therapy. *Cancer J.* **2006**, *12*, 283–288. [CrossRef] [PubMed]
4. Newman, D.J.; Cragg, G.M. Natural products as sources of new drugs over the 30 years from 1981 to 2010. *J. Nat. Prod.* **2012**, *75*, 311–335. [CrossRef] [PubMed]
5. Wang, J.L.; Gold, K.A.; Lippman, S.M. Natural-agent mechanisms and early-phase clinical development. *Top. Curr. Chem.* **2013**, *329*, 241–252. [CrossRef]
6. Aggarwal, B.B.; Shishodia, S. Molecular targets of dietary agents for prevention and therapy of cancer. *Biochem. Pharmacol.* **2006**, *71*, 1397–1421. [CrossRef]
7. Surh, Y.J. Cancer chemoprevention with dietary phytochemicals. *Nat. Rev. Cancer* **2003**, *3*, 768–780. [CrossRef] [PubMed]
8. Theriault, A.; Chao, J.-T.; Wang, Q.; Gapor, A.; Adeli, K. Tocotrienol: A review of its therapeutic potential. *Clin. Biochem.* **1999**, *32*, 309–319. [CrossRef]
9. De Silva, L.; Chuah, L.H.; Meganathan, P.; Fu, J.-Y. Tocotrienol and cancer metastasis. *BioFactors* **2016**, *42*, 149–162. [CrossRef]
10. Constantinou, C.; Papas, A.; Constantinou, A.I. Vitamin E and cancer: An insight into the anticancer activities of vitamin E isomers and analogs. *Int. J. Cancer* **2008**, *123*, 739–752. [CrossRef] [PubMed]
11. Rajasinghe, L.D. Anti-Cancer Effects of Tocotrienols in NSCLC. Ph.D. Thesis, Wayne State University, Detroit, MI, USA, 2017.
12. Ji, X.; Wang, Z.; Geamanu, A.; Sarkar, F.H.; Gupta, S.V. Inhibition of cell growth and induction of apoptosis in non-small cell lung cancer cells by delta-tocotrienol is associated with notch-1 down-regulation. *J. Cell. Biochem.* **2011**, *112*, 2773–2783. [CrossRef]
13. Rajasinghe, L.D.; Pindiprolu, R.H.; Gupta, S.V. Delta-tocotrienol inhibits non-small-cell lung cancer cell invasion via the inhibition of NF-κB, uPA activator, and MMP-9. *OncoTargets Ther.* **2018**, *11*, 4301–4314. [CrossRef]
14. Kwon, H.; Oh, S.; Jin, X.; An, Y.J.; Park, S. Cancer metabolomics in basic science perspective. *Arch. Pharm. Res.* **2015**, *38*, 372–380. [CrossRef]
15. Puchades-Carrasco, L.; Pineda-Lucena, A. Metabolomics Applications in Precision Medicine: An Oncological Perspective. *Curr. Top. Med. Chem.* **2017**, *17*, 2740–2751. [CrossRef]
16. Tran, Q.; Lee, H.; Park, J.; Kim, S.H.; Park, J. Targeting Cancer Metabolism—Revisiting the Warburg Effects. *Toxicol. Res.* **2016**, *32*, 177–193. [CrossRef] [PubMed]
17. Mohamed, A.; Deng, X.; Khuri, F.R.; Owonikoko, T.K. Altered glutamine metabolism and therapeutic opportunities for lung cancer. *Clin. Lung Cancer* **2014**, *15*, 7–15. [CrossRef] [PubMed]
18. Abubakar, I.B.; Lim, S.-W.; Loh, H.-S. Synergistic Apoptotic Effects of Tocotrienol Isomers and Acalypha wilkesiana on A549 and U87MG Cancer Cells. *Trop. Life Sci. Res.* **2018**, *29*, 229–238. [CrossRef] [PubMed]
19. Zhdanov, A.V.; Waters, A.H.C.; Golubeva, A.V.; Dmitriev, R.I.; Papkovsky, D.B. Availability of the key metabolic substrates dictates the respiratory response of cancer cells to the mitochondrial uncoupling. *Biochim. Biophys. Acta BBA-Bioenerg.* **2014**, *1837*, 51–62. [CrossRef]
20. Gonzalez Herrera, K.N.; Lee, J.; Haigis, M.C. Intersections between mitochondrial sirtuin signaling and tumor cell metabolism. *Crit. Rev. Biochem. Mol. Biol.* **2015**, *50*, 242–255. [CrossRef]
21. Robert, S.M.; Sontheimer, H. Glutamate Transporters in the Biology of Malignant Gliomas. *Cell. Mol. Life Sci. CMLS* **2014**, *71*, 1839–1854. [CrossRef]
22. Fuchs, B.C.; Finger, R.E.; Onan, M.C.; Bode, B.P. ASCT2 silencing regulates mammalian target-of-rapamycin growth and survival signaling in human hepatoma cells. *Am. J. Physiol. Cell Physiol.* **2007**, *293*, C55–C63. [CrossRef]
23. Shimizu, K.; Kaira, K.; Tomizawa, Y.; Sunaga, N.; Kawashima, O.; Oriuchi, N.; Tominaga, H.; Nagamori, S.; Kanai, Y.; Yamada, M.; et al. ASC amino-acid transporter 2 (ASCT2) as a novel prognostic marker in non-small cell lung cancer. *Br. J. Cancer* **2014**, *110*, 2030–2039. [CrossRef]

24. Conciatori, F.; Ciuffreda, L.; Bazzichetto, C.; Falcone, I.; Pilotto, S.; Bria, E.; Cognetti, F.; Milella, M. mTOR Cross-Talk in Cancer and Potential for Combination Therapy. *Cancers* **2018**, *10*, 23. [CrossRef] [PubMed]

25. Jeon, Y.J.; Khelifa, S.; Feng, Y.; Lau, E.; Cardiff, R.; Kim, H.; Rimm, D.L.; Kluger, Y.; Ronai, Z.e. Abstract 2440: RNF5 mediates ER stress-induced degradation of SLC1A5 in breast cancer. *Cancer Res.* **2014**, *74*, 2440. [CrossRef]

26. Shimizu, K.; Kaira, K.; Tomizawa, Y.; Sunaga, N.; Kawashima, O.; Oriuchi, N.; Kana, Y.; Yamada, M.; Oyama, T.; Takeyoshi, I. P0143 ASC amino acid transporter 2 (ASCT2) as a novel prognostic marker in non-small-cell lung cancer. *Eur. J. Cancer* **2014**, *50*, e49. [CrossRef]

27. Fuchs, B.C.; Bode, B.P. Amino acid transporters ASCT2 and LAT1 in cancer: Partners in crime? *Semin. Cancer Biol.* **2005**, *15*, 254–266. [CrossRef]

28. van den Heuvel, A.P.J.; Jing, J.; Wooster, R.F.; Bachman, K.E. Analysis of glutamine dependency in non-small cell lung cancer: GLS1 splice variant GAC is essential for cancer cell growth. *Cancer Biol. Ther.* **2012**, *13*, 1185–1194. [CrossRef] [PubMed]

29. Wang, Q.; Hardie, R.A.; Hoy, A.J.; van Geldermalsen, M.; Gao, D.; Fazli, L.; Sadowski, M.C.; Balaban, S.; Schreuder, M.; Nagarajah, R.; et al. Targeting ASCT2-mediated glutamine uptake blocks prostate cancer growth and tumour development. *J. Pathol.* **2015**, *236*, 278–289. [CrossRef] [PubMed]

30. Imai, H.; Kaira, K.; Oriuchi, N.; Shimizu, K.; Tominaga, H.; Yanagitani, N.; Sunaga, N.; Ishizuka, T.; Nagamori, S.; Promchan, K. Inhibition of L-type amino acid transporter 1 has antitumor activity in non-small cell lung cancer. *Anticancer Res.* **2010**, *30*, 4819–4828.

31. Fuchs, B.C.; Bode, B.P. Stressing out over survival: Glutamine as an apoptotic modulator. *J. Surg. Res.* **2006**, *131*, 26–40. [CrossRef] [PubMed]

32. Matés, J.M.; Segura, J.A.; Alonso, F.J.; Márquez, J. Pathways from glutamine to apoptosis. *Front. Biosci.* **2006**, *11*, 3164–3180. [CrossRef]

33. Rajasinghe, L.; Gupta, S. Tocotrienols suppress non-small lung cancer cells via downregulation of the Notch-1 signaling pathway (644.1). *FASEB J.* **2014**, *28*. [CrossRef]

34. Rajasinghe, L.; Pindiprolu, R.; Razalli, N.; Wu, Y.; Gupta, S. Delta Tocotrienol Inhibits MMP-9 Dependent Invasion and Metastasis of Non-Small Cell Lung Cancer (NSCLC) Cell by Suppressing Notch-1 Mediated NF-κb and uPA Pathways. *FASEB J.* **2015**, *29*. [CrossRef]

35. Rajasinghe, L.D.; Gupta, S.V. Tocotrienol-rich mixture inhibits cell proliferation and induces apoptosis via down-regulation of the Notch-1/NF-κB pathways in NSCLC cells. *Nutr. Diet. Suppl.* **2017**, *9*, 103–114. [CrossRef]

36. Rajasinghe, L.D.; Gupta, S.V. Delta Tocotrienal Inhibit mTOR Pathway by Modulating Glutamine Uptake and Transporters in Non-Small Cell Lung Cancer. *FASEB J.* **2016**, *30*. [CrossRef]

37. Jewell, J.L.; Kim, Y.C.; Russell, R.C.; Yu, F.-X.; Park, H.W.; Plouffe, S.W.; Tagliabracci, V.S.; Guan, K.-L. Differential regulation of mTORC1 by leucine and glutamine. *Science* **2015**, *347*, 194–198. [CrossRef] [PubMed]

38. Petronini, P.G.; Urbani, S.; Alfieri, R.; Borghetti, A.F.; Guidotti, G.G. Cell susceptibility to apoptosis by glutamine deprivation and rescue: Survival and apoptotic death in cultured lymphoma-leukemia cell lines. *J. Cell. Physiol.* **1996**, *169*, 175–185. [CrossRef]

39. Folkers, K.; Satyamitra, M.; Srinivasan, V. Delta-tocotrienol Mediates the Cellular Response to Radiation-Induced DNA Damage through Upregulation of Anti-Apoptotic Effectors in Human Lung Fibroblast Cell Lines. *FASEB J.* **2015**, *29*. [CrossRef]

40. Husain, K.; Centeno, B.A.; Coppola, D.; Trevino, J.; Sebti, S.M.; Malafa, M.P. δ-Tocotrienol, a natural form of vitamin E, inhibits pancreatic cancer stem-like cells and prevents pancreatic cancer metastasis. *Oncotarget* **2017**, *8*, 31554–31567. [CrossRef]

41. Montagnani Marelli, M.; Marzagalli, M.; Moretti, R.M.; Beretta, G.; Casati, L.; Comitato, R.; Gravina, G.L.; Festuccia, C.; Limonta, P. Vitamin E δ-tocotrienol triggers endoplasmic reticulum stress-mediated apoptosis in human melanoma cells. *Sci. Rep.* **2016**, *6*, 30502. [CrossRef] [PubMed]

42. Swift, S.N.; Pessu, R.L.; Chakraborty, K.; Villa, V.; Lombardini, E.; Ghosh, S.P. Acute toxicity of subcutaneously administered vitamin E isomers delta-and gamma-tocotrienol in mice. *Int. J. Toxicol.* **2014**, *33*, 450–458. [CrossRef] [PubMed]

43. Mahipal, A.; Klapman, J.; Vignesh, S.; Yang, C.S.; Neuger, A.; Chen, D.-T.; Malafa, M.P. Pharmacokinetics and safety of vitamin E δ-tocotrienol after single and multiple doses in healthy subjects with measurement of vitamin E metabolites. *Cancer Chemother. Pharmacol.* **2016**, *78*, 157–165. [CrossRef]

44. Shen, C.-L.; Wang, S.; Yang, S.; Tomison, M.D.; Abbasi, M.; Hao, L.; Scott, S.; Khan, M.S.; Romero, A.W.; Felton, C.K. A 12-week evaluation of annatto tocotrienol supplementation for postmenopausal women: Safety, quality of life, body composition, physical activity, and nutrient intake. *BMC Complement. Altern. Med.* **2018**, *18*, 198. [CrossRef] [PubMed]

45. Scalise, M.; Galluccio, M.; Console, L.; Pochini, L.; Indiveri, C. The Human SLC7A5 (LAT1): The Intriguing Histidine/Large Neutral Amino Acid Transporter and Its Relevance to Human Health. *Front. Chem.* **2018**, *6*. [CrossRef] [PubMed]

46. Shin, G.; Kang, T.-W.; Yang, S.; Baek, S.-J.; Jeong, Y.-S.; Kim, S.-Y. GENT: Gene Expression Database of Normal and Tumor Tissues. *Cancer Inform.* **2011**, *10*, CIN–S7226. [CrossRef] [PubMed]

47. Nakanishi, K.; Matsuo, H.; Kanai, Y.; Endou, H.; Hiroi, S.; Tominaga, S.; Mukai, M.; Ikeda, E.; Ozeki, Y.; Aida, S.; et al. LAT1 expression in normal lung and in atypical adenomatous hyperplasia and adenocarcinoma of the lung. *Virchows Archiv* **2006**, *448*, 142–150. [CrossRef] [PubMed]

48. Hassanein, M.; Hoeksema, M.D.; Shiota, M.; Qian, J.; Harris, B.K.; Chen, H.; Clark, J.E.; Alborn, W.E.; Eisenberg, R.; Massion, P.P. SLC1A5 mediates glutamine transport required for lung cancer cell growth and survival. *Clin. Cancer Res.* **2013**, *19*, 560–570. [CrossRef] [PubMed]

49. Masle-Farquhar, E.; Bröer, A.; Yabas, M.; Enders, A.; Bröer, S. ASCT2 (SLC1A5)-Deficient Mice Have Normal B-Cell Development, Proliferation, and Antibody Production. *Front. Immunol.* **2017**, *8*. [CrossRef]

50. Chen, L.; Cui, H. Targeting Glutamine Induces Apoptosis: A Cancer Therapy Approach. *Int. J. Mol. Sci.* **2015**, *16*, 22830–22855. [CrossRef]

51. Kanai, Y.; Hediger, M.A. The glutamate/neutral amino acid transporter family SLC1: Molecular, physiological and pharmacological aspects. *Pflugers Archiv* **2004**, *447*, 469–479. [CrossRef]

52. Petroulakis, E.; Mamane, Y.; Le Bacquer, O.; Shahbazian, D.; Sonenberg, N. mTOR signaling: Implications for cancer and anticancer therapy. *Br. J. Cancer* **2006**, *94*, 195–199. [CrossRef]

53. Shaw, R.J.; Cantley, L.C. Ras, PI (3) K and mTOR signalling controls tumour cell growth. *Nature* **2006**, *441*, 424–430. [CrossRef]

54. Yamauchi, K.; Sakurai, H.; Kimura, T.; Wiriyasermkul, P.; Nagamori, S.; Kanai, Y.; Kohno, N. System L amino acid transporter inhibitor enhances anti-tumor activity of cisplatin in a head and neck squamous cell carcinoma cell line. *Cancer Lett.* **2009**, *276*, 95–101. [CrossRef]

55. Laplante, M.; Sabatini, D.M. mTOR signaling at a glance. *J. Cell Sci.* **2009**, *122*, 3589–3594. [CrossRef] [PubMed]

56. Zhou, H.; Huang, S. Role of mTOR Signaling in Tumor Cell Motility, Invasion and Metastasis. *Curr. Protein Pept. Sci.* **2011**, *12*, 30–42. [PubMed]

57. Circu, M.L.; Aw, T.Y. Glutathione and modulation of cell apoptosis. *Biochim. Biophys. Acta* **2012**, *1823*, 1767–1777. [CrossRef]

58. Dalton, T.P.; Chen, Y.; Schneider, S.N.; Nebert, D.W.; Shertzer, H.G. Genetically altered mice to evaluate glutathione homeostasis in health and disease. *Free Radic. Biol. Med.* **2004**, *37*, 1511–1526. [CrossRef] [PubMed]

59. Friesen, C.; Kiess, Y.; Debatin, K.M. A critical role of glutathione in determining apoptosis sensitivity and resistance in leukemia cells. *Cell Death Differ.* **2004**, *11*, S73–S85. [CrossRef] [PubMed]

60. Cazanave, S.; Berson, A.; Haouzi, D.; Vadrot, N.; Fau, D.; Grodet, A.; Letteron, P.; Feldmann, G.; El-Benna, J.; Fromenty, B.; et al. High hepatic glutathione stores alleviate Fas-induced apoptosis in mice. *J. Hepatol.* **2007**, *46*, 858–868. [CrossRef]

61. Armstrong, J.S.; Steinauer, K.K.; Hornung, B.; Irish, J.M.; Lecane, P.; Birrell, G.W.; Peehl, D.M.; Knox, S.J. Role of glutathione depletion and reactive oxygen species generation in apoptotic signaling in a human B lymphoma cell line. *Cell Death Differ.* **2002**, *9*, 252–263. [CrossRef]

62. Saadat, N.; Liu, F.; Haynes, B.; Nangia-Makker, P.; Bao, X.; Li, J.; Polin, L.; Gupta, S.; Mao, G.; Shekhar, M.P. Nano-targeted Delivery of Rad6/Translesion Synthesis Inhibitor for Triple Negative Breast Cancer Therapy. *Mol. Cancer Ther.* **2018**. [CrossRef]

63. Xia, J.; Wishart, D.S. Web-based inference of biological patterns, functions and pathways from metabolomic data using MetaboAnalyst. *Nat. Protoc.* **2011**, *6*, 743–760. [CrossRef]

 *metabolites*

Review

# Breast Cancer Metabolomics: From Analytical Platforms to Multivariate Data Analysis. A Review

Catarina Silva [1], Rosa Perestrelo [1], Pedro Silva [1], Helena Tomás [1,2] and José S. Câmara [1,2,*]

[1]  CQM - Centro de Química da Madeira, Universidade da Madeira, Campus Universitário da Penteada, 9020-105 Funchal, Portugal; cgsluis@staff.uma.pt (C.S.); rmp@staff.uma.pt (R.P.); pedro_dasilva@hotmail.com (P.S.); lenat@staff.uma.pt (H.T.)

[2]  Faculdade de Ciências Exactas e Engenharia, Universidade da Madeira, Campus Universitário da Penteada, 9020-105 Funchal, Portugal

*  Correspondence: jsc@staff.uma.pt; Tel.: +351-291-705-5112

Received: 7 February 2019; Accepted: 17 May 2019; Published: 22 May 2019

**Abstract:** Cancer is a major health issue worldwide for many years and has been increasing significantly. Among the different types of cancer, breast cancer (BC) remains the leading cause of cancer-related deaths in women being a disease caused by a combination of genetic and environmental factors. Nowadays, the available diagnostic tools have aided in the early detection of BC leading to the improvement of survival rates. However, better detection tools for diagnosis and disease monitoring are still required. In this sense, metabolomic NMR, LC-MS and GC-MS-based approaches have gained attention in this field constituting powerful tools for the identification of potential biomarkers in a variety of clinical fields. In this review we will present the current analytical platforms and their applications to identify metabolites with potential for BC biomarkers based on the main advantages and advances in metabolomics research. Additionally, chemometric methods used in metabolomics will be highlighted.

**Keywords:** breast cancer; omics; analytical platforms; chemometric methods

---

## 1. Introduction

Cancer is a public health problem and causes a tremendous burden on patients, families and society creating a significant problem on global economy. Although has been extensively investigated, cancer still remains one of the leading causes of death in the world after coronary diseases [1]. Globally, breast cancer (BC) remains at the top of women's cancers worldwide followed by colorectal, lung, cervix, and stomach cancers according to GLOBOCAN series of the International Agency for Research on Cancer (IARC), contributing with more than 11.6% of all cancer types (Figure 1).

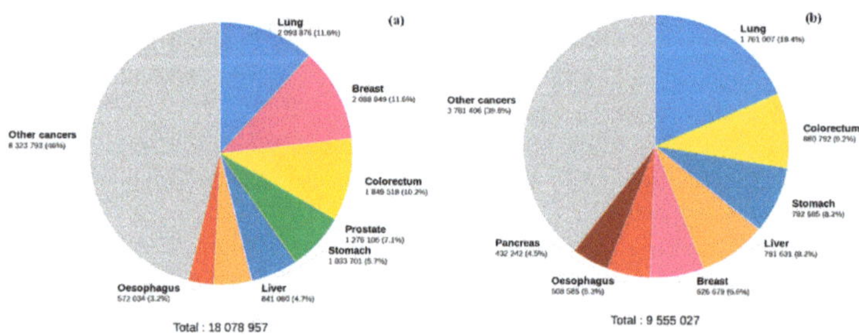

**Figure 1.** Estimated cancer incidence rates (**a**) and (**b**) estimated number of deaths worldwide for 2018. Adapted from GLOBOCAN [2].

In addition, around 2.1 million BC new cases were diagnosed in 2018 and occurred 630 thousand deaths (6.6% of all cancers) (Figure 1b).

The incidence rates are highest in North America, Australia and Europe and lowest in Asia. These differences might be related to societal changes, as result of industrialization, such as, unhealthy lifestyle, expressed by overweight and other symptoms, alcohol consumption, tobacco smoking, physical inactivity, early menarche, among others [2,3]. Although the incidence is high in some developed countries, mortality is higher in low and middle income countries [4]. The incidence of breast cancer increases with age and is usually diagnosed in the 50–60 age group. Moreover, the most aggressive type of the disease predominates in the younger age group (below 35 years) whereas in the older age group (above 75 years), the treatment cannot be so aggressive and has to be adjusted [5]. Concerning the incidence rates and mortality for breast cancer in Europe, it was observed that in 2012, the incidence of breast cancer was around 361,608 cases with 91,585 deaths. For 2020, around 400 thousand new cases will be diagnosed resulting in 100 thousand deaths according to International Agency for Research on Cancer (IARC). For Portugal and USA the expected number of breast cancer cases in 2020 will be nearly 6000 and 270 thousand resulting in around 1700 and 51,000 deaths, respectively as shown in Figure 2.

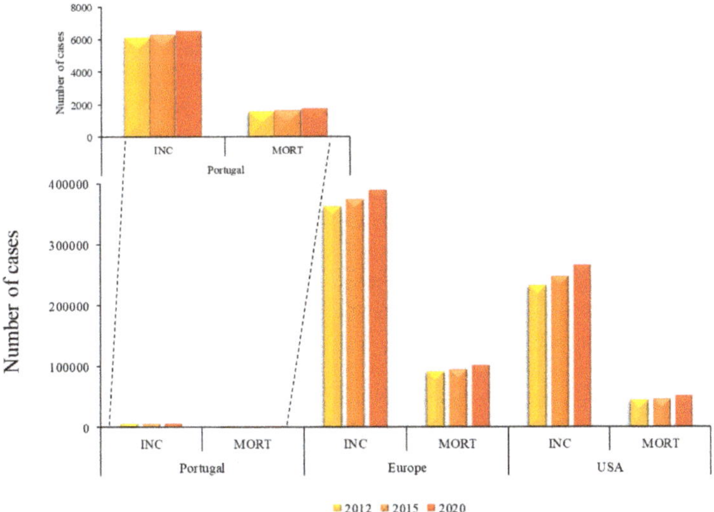


**Figure 2.** BC incidence and mortality rates in Portugal, Europe and USA from 2012, 2015 and expected rates for 2020. Data available at IARC. *Legend*: INC: incidence; MORT: mortality.

This trend might be as consequence by the availability of better screening procedures resulting in an early detection and also in the development of new treatments [3,6,7], which lead to an improved survival. Several risk factors associated with BC have been already recognized, namely epidemiological factors (e.g., age, reproductive factors, socioeconomic status, ethnicity), often using standard analysis approaches (e.g., logistic regression) with adjustment for multiple comparisons. Other factors as lifestyle (e.g., alcohol, tobacco, obesity, physical activity), and exposure to radiation [8] are also associated. The risk of developing BC increases with age being rare in women younger than 25 years, but tending to be more aggressive in younger people. The most common BC that occurs is the invasive type independently of age [9]. The highest risk of family history is associated with increasing number of first-degree relatives diagnosed with BC (age under 50 years). The risk is further increased when the affected relative is diagnosed in both breasts [10]. Particularly, the mutations in genes BRCA1, BRCA2 and TP53 are strongly associated with the development of BC [9], even if these mutations are low, accounting for a small portion of the total BC incidence [2]. Consistent physical activity has many benefits and greater activity has been related to lower BC risk by decreasing the circulating estrogen levels in postmenopausal women [11,12]. Extensive literature has linked alcohol consumption to BC risk and reveal the role of ethanol in carcinogenesis altering estrogen levels through acetaldehyde. Briefly ethanol is converted to acetaldehyde (AA) through alcohol dehydrogenase (ADH), that then binds to DNA interfering with the DNA synthesis and repair [13]. Obesity is another BC risk factor to take into account as it is involved in insulin resistance and hyperinsulinemia [14]. Insulin has anabolic effects on cellular metabolism and an overexpression of insulin receptor has been demonstrated in human cancer cells [9,15]. The involvement of insulin-like growth factor (IGFs) in carcinogenesis is attributed to their role in linking high energy intake, increased cell proliferation, and suppression of apoptosis to cancer risks [15,16]. With regard to obesity and BC risk, some studies indicate that is strongly associated with increased invasive BC risk in postmenopausal women particularly for estrogen receptor–positive cancers ($ER^+$) [17–19]. In clinical practice, there are nowadays several biomarkers routinely used for prognosis and identification of tumors, including the estrogen receptor (ER), progesterone receptor (PR) and the human epidermal growth factor receptor-2 (HER2) [20,21]. Another promising prognostic and predictive biomarker of BC is Ki-67 (present in dividing cells) as indicator of cell proliferation and also as an endpoint for neoadjuvant systemic therapy [20]. However there are other proposed markers of proliferation measured by immunohistochemistry (IHC), such as, cyclin D, cyclin E, p27, p21, among others that are used to determine the predictive and prognostic levels [22].

In the last years, metabolomics emerged as a powerful approach in the advanced disease biomarker discovery which includes the comprehensive study of metabolites that are present in biological samples [23]. The study of metabolome to search biomarkers for any disease involves the identification of endogenous metabolites that have the potential to discriminate between samples obtained from healthy subjects and diseased patients. Plasma, serum, urine, tissue and cerebrospinal fluid (CSF), are the most commonly used biological samples in metabolomic studies. These biological samples contain hundreds of metabolites that vary in chemical and physical properties and concentration levels. Metabolomic studies includes two main approaches – targeted and untargeted. The targeted analysis is focused in specific groups of chemical characterized and annotated metabolites and their related pathways, whereas in the untargeted analysis the study includes a comprehensive measurement of all metabolites present in samples [24,25].

The type of approach chosen will determine the experimental design, sample preparation, and which analytical techniques can be used to obtain the results. Both targeted and untargeted follow the similar pipeline. Briefly, the study design includes the population that will be part in the study and also the determination of the conditions that are relevant for the hypothesis in investigation, namely the sample size, randomization (as a study design consideration), storage (as a sample handling issue), freeze/thaw cycles and timing during sample preparation are the most common factors that should be taken into account to guarantee reproducible and successful experiments minimizing variability. There are three main analytical platforms frequently used in metabolomic studies, which include mass spectrometry (MS) and nuclear magnetic resonance (NMR) spectroscopy [26]. Moreover, after data acquisition, the obtained dataset, normally is subjected to statistical analysis (univariate and multivariate methods) to find significant variations that allow the discrimination of patients with a specific disease (in this case, BC) from a control group [27]. The most common approaches for the identification of important metabolites comprise the application of unsupervised methods, such as, principal component analysis (PCA), hierarchical cluster analysis (HCA), as well as supervised methods, like partial least squares discriminant analysis (PLS-DA), random forest (RF) and support vector machines (SVM) [26,28]. A training set is used to construct the multivariate analysis models (e.g., PCA or PLS-DA), followed by an external validation set to predict the new cohort of samples using the model constructed with the training model. Finally, the putative biomarkers can be placed in metabolic networks to allow the biological interpretation or which pathways are up- or down-regulated.

## 2. OMICS Science

The OMICs is a neologism broadly adopted in biomedical research, that comprises the dataset of genomics (DNA), transcriptomics (RNA), proteomics (proteins) and metabolomics (metabolites) based on the central dogma of molecular biology [29]. The purpose of OMICs science in cancer research is to discover cancer-specific biomarkers (diagnostic, prognostic and/or putative). The Food and Drug Administration (FDA) defined biomarkers as a "characteristic that is objectively measured and evaluated as an indicator of normal biological processes, pathogenic processes, or biological responses to a therapeutic intervention" [30]. Biomarkers are powerful tools, when used for the early cancer detection and selection of therapeutic strategy, thus improving the outcome of cancer treatment and reduce cancer-related mortalities.

One of the newest promising OMICs sciences is metabolomics being a suitable tool that provides state of the art of analytical instrumentation tandem with pattern recognition procedures and chemometric tools to discover new disease- biomarkers providing novel insights into disease etiology, and more robust assessment of etiological pathways [30,31]. Metabolomics studies the complex interaction in biological systems providing a comprehensive and detailed information of the phenotype and molecular physiology as result of environmental factors, genetic as well as exogenous and endogenous factors (e.g., age, gender, race, diet, drugs, exercise, gut microbiota) [31]. In addition, metabolomics can be used in early detection and diagnosis of cancer, in the assessment of therapies and medical interventions, since cancer is a disease that promotes changes in cellular metabolism [32–35]. This OMICs tool has been extensively applied in clinical health practice due to its ability to quickly analyze biological samples (e.g., blood, tissue, saliva and urine) with relatively simple sample preparation (10–30 min), cost-effective and high-throughput [30,36,37]. Nevertheless, metabolomics present several drawbacks resulting from biological and experimental features, such as sampling variability, inter- and intra-individual differences and a lack of validated protocols for biological samples handling which have a significant impact on the OMICs data approach [38,39].

The current review is focused in the metabolic profile of several biological samples, including lipidomics (lipids), labeled substrates (e.g., $^{13}$C labeled glucose), volatomic (volatile organic metabolites), and metabolites resulting from Krebs cycle [30,39] with the purpose of an early diagnosis, metabolic reprogramming, cancer typing, staging and therapeutic intervention response [29,37]. Regarding the Krebs/TCA cycle, there is evidence that the role of TCA for energy production and macromolecule synthesis by cancer cells, especially those with dysregulated oncogene and tumor suppressor expression [40–42]. Over the last years, there has been a rapidly growing number of metabolomic studies intended to discover new biomarkers or make disease diagnosis using different biological matrices, such as cell lines [43–46], blood [47], exhaled breath [48], plasma [33,49,50], saliva [51–54], tissues [55–57], serum [58] and urine [59]. In Table 1 are resumed the most common analytical approaches used in metabolomic studies grouped by type of biological sample and objective of the study. Interestingly, the main studies involve a diagnostic purpose using BC cell lines with the aim of search biomarkers, inspect the metabolome (endo- and exo-). Moreover, lipids as building blocks of cell membranes have their levels changed during the malignant transformation. Lipid metabolism plays a vital role in oxidative stress and is correlated with other parameters linked to BC risk (e.g., hormonal balance, body mass index, breast density, drug metabolism and growth of insulin levels) [43,60]. In addition, a summary of the total identified metabolites by analytical platform as well as the number of samples used for each biological specimen type is shown as Supplementary Materials (Figure S1).

**Table 1.** Summary of metabolomics studies performed in breast cancer biomarker discovery in different biological matrices.

| Biological Sample | Sample Groups | Aim | Analytical Approaches | Main Conclusions | References |
|---|---|---|---|---|---|
| Human cell lines | - | - | - | - | - |
| Diagnostic biomarkers | BC (ZR-75-1, T-74D, MCF7, MDA-MB-231, MDA-MB-453, MDA-MB-468, SK-BR-3, BT-474, BT-549), Control (MCF10A) | To compare the differences in the lipidomic compositions of human cell lines derived from normal and BC tissues, and tumor vs. normal tissues obtained after the surgery of BC patients. | LC-MS/MS, GC-MS | * 123 lipids were identified, and a differentiation was observed for MDA cells | [29,43] |
| Diagnostic biomarkers | BC (MDA-MB-231, -453, BT-474), Control (MCF-10A) | To determine endo- and exo-metabolite analysis of the BC cell lines | UPLC-MS/MS, LC-MS/MS | * Statistical analysis allowed a discrimination of the breast epithelial cells from the BC cell lines<br>* MDA-MB-231 showed an increase in nicotinamide levels, namely in 1-ribosyl-nicotinamide and NADþ | [46] |
| Diagnostic biomarkers | BC (T-47D, MDA-MB-231, MCF-7), Control (HMEC) | To establish the BC cell lines volatile metabolomic signature | GC–MS | * 60 VOMs were identified and six of them were detected only in the headspace of cancer cell lines | [44] |
| Diagnostic biomarkers | BC (MDA-MB-468, SKBR3, MCF-7) | To quantify specific metabolites in BC cell extracts | NMR | * Significantly differences were observed between cell lines, namely in the concentrations of 15 metabolites<br>* The current method represented a useful tool for the establishment of potential biomarkers | [61] |
| Diagnostic biomarkers | BC (Cal 51, SKBR3, MCF-7) | To measure the absolute metabolite concentrations in complex mixtures with a high precision in a reasonable time | NMR | * The proposed approach represented a powerful tool to quantify 14 metabolites (alanine, lactate, leucine, threonine, taurine, glutathione, glutamate, glutamine, choline, valine, isoleucine, myo-inositol, proline, and glucose) in cell extracts within 20 min | [45] |
| Diagnostic biomarkers | BC cell lines (MCF-7, HCC70, MDA-MB-231, MDA-MB-436, MDA-MB-468), BC patients ($n = 35$) | To investigate the metabolic profiles of human BC cell lines carrying BRCA1 pathogenic mutations | LC-MS/MS | * It was possible to collect differential metabolic signature for BC cells based on the BRCA1 functionality | [50] |
| Therapy response | BC cell line (MCF-7) | To develop a robust and highly sensitive platform to identify endogenous estrones in clinical specimens | MALDI-MS, LC-MS/MS | * The results suggested that MALDI-MS-based quantitative approach can be a broad method for the ketone-containing metabolites target analysis thus replicating the clinical stage. | [62] |
| Therapy response | BC tissue ($n = 40$), Blood ($n = 27$), BC cell lines ($n = 3$) | To detect alterations in metabolites and their linkage to metabolic processes in several pathological conditions including BC | NMR | * Functional of IP3Rs in causing metabolic disruption was observed in MCF-7 and MDA MB-231 cells<br>* The results offered new insights regarding the relationship of BC metabolites with IP3R. | [63] |

**Table 1.** *Cont.*

| Biological Sample | Sample Groups | Aim | Analytical Approaches | Main Conclusions | References |
|---|---|---|---|---|---|
| Metabolic reprogramming | MDA-MB-231, BC xenografts | To study toxic effects of bisphenol and the underlying mechanisms on tumor metastasis-related tissues | LC-MS/MS, MALDI-MS | * Metabolites-based studies might be suitable for BC diagnosis<br>* The data provided good indication for BPA screening secure option | [64] |
| Human Blood, plasma, serum | - | - | - | - | - |
| Diagnostic biomarkers | BC patients ($n = 258$), Benign mammary gland ($n = 159$), Control ($n = 78$) | To screen metabolite markers with BC diagnosis potentials | MS | * The method developed allowed the discrimination of BC from non-BC using six blood metabolites | [47] |
| Diagnostic biomarkers | Metastatic BC patients ($n = 95$), Early-stage BC patients ($n = 80$) | To explore whether serum metabolomic spectra could distinguish between early and metastatic BC patients and predict disease relapse | NMR | * Disease relapse was linked with lower and higher levels of histidine and glucose, respectively | [58] |
| Diagnostic biomarkers | BC patients (n= 132), Control ($n$= 76) | To develop a new computational method using personalized pathway dysregulation scores for disease diagnosis | LC-TOF-MS, GC-TOF-MS | * The method allowed to determine important metabolic pathways signature for BC diagnosis, representing a suitable tool for diagnostic and therapeutic interventions. | [65] |
| Diagnostic biomarkers | BC patients ($n = 45$), Control ($n = 45$) | To detect differences between BC and healthy individuals | UHPLC-MS, GC-MS | * 661 metabolites were detected, but only 338 metabolites were found in all samples, and 490 in more than 80% of samples. | [66] |
| Diagnostic biomarkers | BC patients ($n = 29$), Control ($n = 29$) | To establish a plasma metabolic fingerprint of Colombian Hispanic women with BC | LC-MS, GC-MS, NMR | * The current report showed the effectiveness of multiplatform strategies in metabolic/lipid fingerprinting works | [49] |
| Diagnostic biomarkers | BC patients ($n = 91$), Control ($n = 20$) | To explore whether serum metabolomic profile can discriminate the presence of human BC irrespective of the cancer subtype | LC-MS/MS | * From the 1269 metabolites identified in plasma from controls and patients; only 35 metabolites were related to BC. | [33] |
| Diagnostic biomarkers | BC patients ($n = 27$), control ($n = 30$) | To apply 1H NMR and DART-MS for the metabolomics analysis of serum samples from BC patients and healthy controls. | NMR, DART-MS | * The approach allowed the disease classification and the biochemical validation useful to identify the mechanisms associated to BC development. | [67] |
| Diagnostic biomarkers | Metastatic BC patients ($n = 39 + 51$ for validation), Early-stage BC patients ($n = 85 + 112$ for validation) | To distinguish between early and metastatic BC | NMR | * Metabolic phenotyping by NMR showed a robust potential for the diagnosis, prognosis, and management of BC cancer patients | [68] |

**Table 1.** *Cont.*

| Biological Sample | Sample Groups | Aim | Analytical Approaches | Main Conclusions | References |
|---|---|---|---|---|---|
| Diagnostic biomarkers | BC patients ($n = 40$) BE patients ($n = 40$) and healthy controls ($n = 34$). BE patients with fibroma ($n = 25$) and chronic fibroadenosis of breast ($n = 15$) | To investigate the free fatty acid (FFA) metabolic profiles to identify biomarkers that can be used to distinguish patients with BC (BC) from benign (BE) patients or healthy controls. | GC-MS | The FFA biomarkers proved to be helpful for the prevention and characterization of BC patients. | [69] |
| Therapy response | BC patients ($n = 19$) | To compare metabolite concentrations and Pearson's correlation coefficients to examine concomitant changes in metabolite concentrations and psychoneurologic symptoms before and after chemotherapy. | UPLC-MS/MS | * The post-chemotherapy global metabolites were characterized by higher and lower amounts of acetyl-L-alanine and indoxyl sulfate and 5-oxo-L-proline, respectively.<br>* Metabolomics was useful for further understanding of biological mechanisms associated with psychoneurologic symptoms. | [70] |
| Therapy response | BC patients ($n = 28$) | To identify potential biomarker candidates that can predict response to neoadjuvant chemotherapy for BC | LC-MS, NMR | * The concentrations of threonine, isoleucine, glutamine, linolenic acid had significantly different responses to chemotherapy<br>* The purposed approach clearly discriminates patients regarding the response to drugs providing a valuable tool for a non-invasive prognosis of the treatment strategy. | [71] |
| Endogenous factors | BC patients ($n = 206$), Control ($n = 396$) | To investigate whether plasma untargeted metabolomic profiles could contribute to predict the risk of developing BC | NMR | * The study contributed to the development of screening approaches for the identification of BC at-risk women. | [31] |
| Endogenous factors | BC patients ($n = 621$), Control ($n = 621$) | To evaluate associations of diet-related metabolites with the risk of BC in the prostate, lung, colorectal and ovarian cancer screening trial | GC-MS, LC-MS/MS | * The data obtained showed how nutritional metabolomics might identify diet-related exposures associated to cancer risk. | [72] |
| Human urine | - | - | - | - | - |
| Diagnostic biomarkers | BC patients ($n = 30$), CC ($n = 30$), Control ($n = 30$) | To discriminate different types of cancer based on urinary volatomic biosignature | GC-MS | * The butanoate metabolism was highly activated in studied cancers, as well as tyrosine metabolism, but in a reduced proportion<br>* Different clusters allowed to establish sets of VOMs fingerprints resulted in the discrimination of the studied cancers | [59] |
| Therapy response | BC patients ($n = 31$), Control ($n = 29$) | To identify metabolites which can be helpful in the understanding of metabolic alterations driven by BC as well as their potential usage as biomarkers | LC-MS, GC-MS | * The analytical multiplatform approach enabled a wide coverage of urine metabolites revealing significant alterations in BC samples | [73] |

**Table 1.** *Cont.*

| Biological Sample | Sample Groups | Aim | Analytical Approaches | Main Conclusions | References |
|---|---|---|---|---|---|
| Human Saliva | - | - | - | - | - |
| Diagnostic biomarkers | BC patients (primary, $n = 8$; relapse, $n = 22$), Control ($n = 14$) | To determine polyamines including N-acetylated forms in human saliva and the diagnostic approach to BC Patients | UPLC–MS/MS | * The increase on polyamines level in BC patients Ac-SPM, DAc-SPD, and DAc-SPM levels were significantly higher only in the relapsed patients | [52] |
| Diagnostic biomarkers | BC patients ($n = 30$), Control ($n = 25$) | To screen the potential salivary biomarkers for BC diagnosis, staging, and biomarker discovery. | UPLC-MS | * Saliva metabonomics approach may provide new insights into the discovery of BC diagnostic biomarkers. | [53] |
| Diagnostic biomarkers | BC patients ($n = 111$), Control ($n = 61$) | To determine of polyamines including their acetylated structures for the diagnosis of BC patients. | UPLC-MS/MS | * The ratio of N8-Ac-SPD/ (N1-Ac-SPD + N8-Ac-SPD) can be used as a health status index after the surgical treatment. | [54] |
| Diagnostic biomarkers | BC patients ($n = 66$), Control ($n = 40$) | To explore the potential of the volatile composition of saliva samples as biosignatures for BC non-invasive diagnosis | GC-MS | * This study defined an experimental layout appropriate for the characterization of volatile fingerprints from saliva as potential biosignatures for BC non-invasive diagnosis. | [51] |
| Human Exhaled breath | - | - | - | - | - |
| Diagnostic tool | BC patients ($n = 14$), Control ($n = 11$) | To detect and identify human exhaled BC–related volatile profile | MS | * Eight metabolites enabled a clear discrimination of exhaled breath of BC patients from controls. * The analytical technique provided a non-invasive strategy to detect VOMs for the BC diagnosis. | [48] |
| Human Tissues | - | - | - | - | - |
| Diagnostic biomarkers | BC patients ($n = 10$) | To establish a detailed lipidomic characterization with the goal to find the statistically differences between BC and normal tissues. | HPLC-MS | * Total concentrations for phosphatidylinositols, phosphatidylcholines, phosphatidylethanolamines and lysophosphatidylcholines were increased leading to a clear differentiation by PCA and OPLS-DA. | [55] |
| Diagnostic biomarkers | Paired tumor and non-tumor liver ($n = 60$), breast ($n = 130$) and pancreatic ($n = 76$) | To assess the metabolomic profiling as a novel tool for multiclass cancer characterization | GC-MS, LC-MS | * The findings provided a framework to validate cancer-type specific metabolite levels in tumor tissues. | [56] |
| Diagnostic biomarkers | BC patients ($n = 37$), Control ($n = 35$) | To identify potential biomarkers that differs TNBC from ER$^+$ BC | GC-MS, LC-MS/MS | * 133 metabolites presented significant differences between ER$^+$ and TNBC tumors * The metabolic pathway of tumors can provide new treatment targets. | [57] |

**Table 1.** *Cont.*

| Biological Sample | Sample Groups | Aim | Analytical Approaches | Main Conclusions | References |
|---|---|---|---|---|---|
| Diagnostic biomarkers | BC patients ($n = 47$), Control ($n = 35$) | To identify how TNBC differs from LABC subtypes within the African-American and Caucasian BC patients | HR-MAS-NMR | * Increased pyrimidine synthesis was related to TNBC in Caucasian women<br>* Novel treatment targets for TNBC could be explored through the metabolic changes | [74] |
| Diagnostic biomarkers | BC patients ($n = 228$) | To distinguish between tumor and non-involved adjacent tissue | HR-MAS-NMR | * Metabolic profiling of tumor tissues by NMR can be a suitable method for the analysis of the resection margins during BC surgery | [75] |
| Diagnostic biomarkers | BC patients ($n = 25$), Control ($n = 5$) | To establish metabolic profiles of ER$^+$ vs. ER$^-$ and of ER$^-$ subtypes linked to genetics | GC-MS, LC-MS | * Changes in the metabolic profile of ER$^-$ vs. ER$^+$ breast tumors were observed<br>* The data represents a potential tool for the hypothesis testing of tumor metabolism | [76] |
| Diagnostic biomarkers | BC patients ($n = 270$), Control ($n = 97$) | To quantify the dysregulation of the glutamate-glutamine equilibrium in BC | GC-TOFMS | * A positive correlation between glutamate and glutamine in normal breast tissues was observed, whereas a negative correlation was obtained for normal tissues | [77] |
| Diagnostic biomarkers | 95 OC (84 peritoneal, 11 pleural), 10 BC (7 pleural, 2 peritoneal, 1 pericardial), and 10 malignant mesotheliomas (6 peritoneal, 4 pleural) | To identify the metabolic differences between ovarian serous carcinoma effusions obtained pre- and post-chemotherapy and compare ovarian carcinoma (OC) effusions with breast carcinoma and malignant mesothelioma specimens. | 1H-NMR | * Differences in metabolic profiles of different malignant effusions were detected<br>* Metabolic characterization by NMR can be a technique to additional knowledge the mechanisms of effusion development | [78] |
| Therapy response | BC patients ($n = 122$) | To explore the effect of neoadjuvant therapy on metabolic profiles of BC tissues | HR-MAS-NMR | * Non-metastatic breast tumor tissue reflected different alterations in all patient groups after treatment.<br>* Metabolic profiles discriminated pNRs from pMRD patients thus complementing other molecular assays allowing the knowledge of the underlying mechanisms affecting the response. | [79] |
| Therapy response | BC patients ($n = 18$) | To study metabolite levels in human BC tissue, assessing, for instance, correlations with prognostic factors, survival outcome or therapeutic response | HR-MAS-NMR | * Significant changes between the tumors were identified, indicating that the intertumoral changes for numerous metabolites were greater than the intratumoral changes for these three tumors. | [80] |
| Therapy response | BC patients ($n = 37$) | To determine whether metabolic profiling of core needle biopsy (CNB) samples using HR-MAS-NMR could be used for predicting pathologic response to neoadjuvant chemotherapy (NAC) in patients with locally advanced BC | HR-MAS-NMR | * The purposed method can be applied to predict the pathologic response before neoadjuvant chemotherapy | [81] |

**Table 1.** *Cont.*

| Biological Sample | Sample Groups | Aim | Analytical Approaches | Main Conclusions | References |
|---|---|---|---|---|---|
| Therapy response | BC patients ($n = 271$) | To establish metabolic signatures for ER$^+$ vs. ER$^-$ BC | GC-TOFMS | Some metabolites levels were increased in ER$^-$ subtype, such as, beta-alanine, glutamate and xanthine. The down-regulation of the ABAT protein in ER$^-$ BC was confirmed by immunohistological analysis. | [82] |
| Mouse BC tissue | - | - | - | - | - |
| Metabolic reprogramming | MMTVPyMT, MMTV-PyMT-DB, MMTV-Wnt1, MMTV-Her2/neu, and C3(1)-SV40 T-antigen (C3-TAg) | To identify global metabolic profiles of breast tumors isolated from multiple transgenic mouse models and to identify unique metabolic signatures driven by these oncogenes | GC-MS, LC-MS/MS, CE-MS | * C3-TAg was the only cohort with a tumor metabolic signature composed of ten metabolites with significance prognostic value in BC patients | [83] |

ANOVA – Analysis of variance; AUC – Area under the curve; BC – BC; BFS– Bootstrap feature selection; CE-MS – Capillary electrophorese-mass spectrometer; DART-MS – Direct analysis in real time mass spectrometry; GC-MS – gas chromatography – mass spectrometry; GC-TOF-MS – Gas chromatography time-of-flight mass spectrometry; GGM – Gaussian graphical modelling; HCA – Hierarchical cluster analysis; HR-MAS-NMR - High resolution magic angle spinning nuclear magnetic resonance spectroscopy; LC-MS/MS – Liquid chromatography tandem with mass spectrometer; LC-TOF-MS – Liquid chromatography time-of-flight mass spectrometry; LDA – Linear discriminant analysis; MALDI-MS – Matrix-assisted laser desorption/ionization mass spectrometry; MCCV – Monte Carlo cross validation; MS – Mass spectrometry; MWT – Mann Whitney U test; NMR – Nuclear resonance magnetic; NRI – Net reclassification improvement; OPLS-DA – Orthogonal projections to latent structures discriminant analysis; OSC-PLS – Orthogonal signal correction partial least squares; PC – Pearson correlation; PCA – Principal component analysis; PEA – Pathway enrichment analysis; PLS-DA – Partial least squares discriminant analysis; RF – Random Forest classifier; ROC – Receiver operating characteristic; SCC – Spearman correlation coefficient; SVM – Support vector machine; TNBC – Triple negative BC; UPLC-MS/MS – Ultra performance liquid chromatography tandem mass spectrometer; VIP – variable importance in projection; VOMs – volatile organic metabolites.

In literature, the reports performed involving human cell lines focus mainly in diagnostic purpose. As for example in the volatile composition (VOMs) as described by Silva et al. [44] where the volatomic signature of BC cell lines was established, and based on the results, 2-pentanone, 2-heptanone, 3-methyl-3-buten-1-ol, ethyl acetate, ethyl propanoate and 2-methyl butanoate were detected only in cultured BC cell lines. These VOMs are formed endogenously or obtained from exogenous sources (e.g., environmental, lifestyle, biological agents) [51], and can be recognised as a useful tool to BC non-invasive diagnosis [44,51]. Other study by Willmann et al. [46] observed the changes of the exo- and endometabolite profiles in BC cell lines by LC-MS/MS and observed a clear discrimination of the breast epithelial from the BC cell lines through statistical tools. Moreover, a decrease on ratio of glutathione (GSH) and glutathione disulfide (GSSG) was observed in BC cell lines as a result of oxidative stress. The lipidomic profile of several BC cell lines was compared with normal cells obtained from non-cancerous tissues by LC-MS/MS and GC-MS that changes observed in breast tumor tissues were caused mainly by difference in lipidomic profiles of tumor cells and these alterations can be correlated with the lipidomic composition of the nine breast cancer cell lines. Furthermore, Martineau et al. [61], determined the absolute concentration of several metabolites (e.g., alanine, lactate, threonine, taurine, glutathione, glutamate, glutamine, choline, valine, isoleucine, myo inositol, serine, proline, aspartate and histidine), revealing the usefulness for the establishment of potential biomarkers. Also, BC cell lines with BRCA1 pathogenic mutations were investigated by LC-MS/MS in order to obtain their metabolic signature as possible diagnostic approach.

Regarding plasma, serum or blood, many studies have been conducted as observed in Table 1, with multiple aims as Cala et al. [49] that developed a pilot control case-study, where a metabolomic and lipidomic approach was performed in order to establish a plasma metabolic fingerprint of Colombian Hispanic women with BC. According to these authors, the plasma metabolites could contribute to an enhanced knowledge of the underlying metabolic shifts driven by BC in women of Colombian Hispanic origin. Moreover, despite racial differences, the mapped metabolic signatures in BC were comparable but not identical to those described for non-Hispanic women. Wang et al. [47] used a dried blood spot approach for rapid BC detection. In the first study, the target analytes were 23 amino acids and 26 acylcarnitines, and based on the results piperamide, asparagine, proline, tetradecenoylcarnitine/palmitoylcarnitine, phenylalanine/tyrosine, and glycine/alanine could be used as potential biomarkers to diagnose BC. Lyon et al. [70] established a serum metabolome analysis from the tryptophan pathway of 19 women with early-stage BC. The targeted analysis indicated higher kynurenine levels and kynurenine/tryptophan ratios post-chemotherapy. Also, the symptoms of pain and fatigue had association with several targeted metabolites. An improved metabolic profile of human serum samples was obtained using complementary thecniques, namely MS and NMR and this approach may be useful to achieve more accurate disease detection and gain more insights regarding disease mechanisms and biology [67].

Another study conducted by Lécuyer et al. [31] combined metabolomic and epidemiological approaches by NMR to investigate whether plasma untargeted metabolomic profiles could contribute to the identification of BC at-risk women, whereas Playdon et al. [72] focused on the evaluation of the associations of diet-related metabolites with the risk of breast cancer. It was possible to verify that the prediagnostic serum concentrations of metabolites related to alcohol, vitamin E, and animal fats were associated with ER$^+$ breast cancer risk.

Urine became a very interesting biological sample to investigate as diagnostic tool or as result of a treatment, as it is easy to collect, and also as ending point of all reactions that occur in the body. Furthermore, Porto-Figueira [59] established the urinary volatomic biosignature from breast (BC), and colon (CC) cancer patients as well as healthy individuals. This last work observed that several pathways are over activated in cancer patients, being phenylalanine pathway in BC and limonene and pinene degradation pathway in CC the most relevant. Yu et al. [84] explored the relationship between urinary metabolites and clinical chemotherapy response in BC. As results, chemotherapy-sensitive patients exhibited 30% of change in metabolite levels when compared to healthy individuals, while chemotherapy-insensitive patients showed only 9% of change in metabolite levels when compared to healthy people that presented recurrence.

Another explored biological fluid is saliva as described by Zhong et al. [53] that screened the putative salivary biomarkers for BC diagnosis, staging, and biomarker discovery. As a result, 18 biomarkers were identified, but only three up-regulated metabolites, displayed the area under the curve (AUC) values higher than 0.920, indicating the high accuracy to predict BC. Also, Cavaco et al. [51] screened salivary volatiles for a putative BC discrimination, and from metabolites identified, only 3-methyl-pentanoic acid, 4-methyl-pentanoic acid, phenol, p-tert-butyl-phenol, acetic, propanoic, benzoic acids, 1,2-decanediol, 2-decanone, and decanal were statistically relevant for the discrimination of BC patients in the populations analyzed. Another type of molecules, the polyamines were associated with tumor growth due to their biosynthesis and accumulation [54]. In this context, Tsutsui et al. [52] and Takayama et al. [54] determined polyamines including N-acetylated forms in saliva to diagnose BC. According to Tsutsui et al. [52], the level of polyamines increased in BC patients, and the levels of $N^1$-acetyl-spermine, $N^1N^8$-diacetyl-spermidine and $N,N$-diacetyl-spermine were significantly higher only in the relapsed patients. Takayama et al. [54] demonstrated that eight polyamines are strongly correlated with the BC patients. Furthermore, the ratio of $N^8$-acetyl-spermidine/ ($N^1$-acetylspermidine $+ N^8$-acetyl-spermidine) may be adopted as an index of the health status after the surgical treatment.

*In-vitro* analysis of BC tissues can be a valuable tool to inspect the metabolic differences between tissue classes, either using the hydrophilic or the lipophilic part. As a result, one might use the metabolomic profile as a novel tool for cancer characterization. Breast tissue is also an interesting biological sample used for diagnostic purposes and /or response to a treatment as demonstrated by Euceda et al. [79] that explored the effect of the antiangiogenic drug bevacizumab on metabolic profile from BC tissue. On the other hand, Budczies et al. [77] studied the glutamate enrichment as a new diagnostic opportunity in BC, and a positive correlation between glutamate and glutamine in normal breast tissues switched to negative correlation between glutamate and glutamine in BC tissues. Euceda et al. [79] observed a metabolic alteration indicating a decline in glucose consumption as an effect of chemotherapy. In addition, a lower glucose and higher lactate level was observed in patients (≥90% of tumor reduction) when compared to those with no response (≤10% of tumor reduction). In turn, Choi et al. [81] determined the metabolic profiling of core needle biopsy samples in order to predict pathologic response to neoadjuvant chemotherapy in patients with locally advanced BC. These authors observed that there was a trend of lower levels of phosphocholine/creatine ratio and choline-containing metabolite concentrations in the pathologic complete response group when compared to the non-pathologic complete response group. Most of the BC patients undergo a cycle or more of chemo being the general treatment that uses cancer-killing drugs before (neoadjuvant or preoperative therapy) and after (adjuvant therapy) surgery [31,36], Then, the therapeutic chemo effect may shift significantly between patients, as a result of BC phenotypes [37] of and intra- and inter- individual differences. For this reason, it is necessary to punctually and accurately evaluate the therapeutic effects of chemotherapy, which could help to adjust the chemotherapy regimen [71,84]. whereas the advances in treatment increased significantly the survival rates for women with BC, as women often report psychoneurologic symptoms (e.g., pain, fatigue, depression) during and after chemotherapy cycles.

Regarding exhaled breath a less explored biological sample in terms of BC diagnostic purpose. In a study performed by Martinez-Lozano Sinues et al. [48] who developed a pilot study to identify cancer–related volatile profile in exhaled breath of BC patients. Concerning exhaled breath and the possible mechanisms involved in the production of endogenous VOMs, in Figure 3 is represented a schematic illustration about the possible pathways.

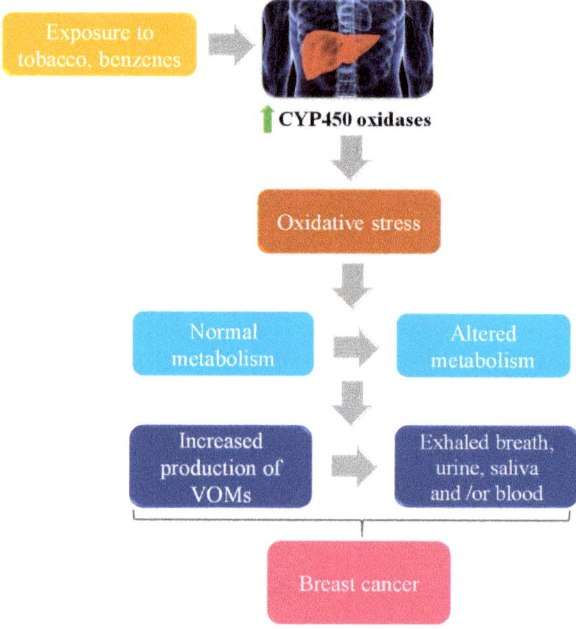


**Figure 3.** Schematic illustration of possible origin of some VOMs.

The principle behind this is based on the fact that the cancer growth is promoted by the progressive accumulation of genetic and epigenetic changes leading to cellular oxidative stress, which in turn increases the liver's production of cytochrome P-450 (CYP450) oxidase enzymes to take into account with stress. Both processes affect the abundance of VOMs in breath once oxidative stress causes lipid peroxidation of polyunsaturated fatty acids (PUFA) in membranes, producing alkanes and methylalkanes which are catabolized by CYP450 [85].

## 3. Analytical Approaches

Metabolomics encompasses targeted and non-targeted analysis of endogenous and exogenous metabolites (<1500 Da), such as lipids, amino acids, hormonal steroids, peptides, nucleic acids, organic acids, vitamins, thiols and carbohydrates, which represent a promising tool for biomarker discovery [86,87]. The complexity of the metabolome, the metabolites properties and their concentration levels in biological samples complicates the separation and detection on a single analytical platform. For this fact, the integration of high resolution analytical frameworks, mass spectrometry (MS) and nuclear magnetic resonance (NMR), appear as an outcome in metabolomics studies, providing sensitive, reliable detection and quantification of thousands of metabolites in a biological sample and related metabolic pathways within a few minutes [27,86,87] as shown in Figure 4.

This review will provide an update of the most commonly used analytical methods in metabolomics, namely MS- and NMR- based metabolomics [27].

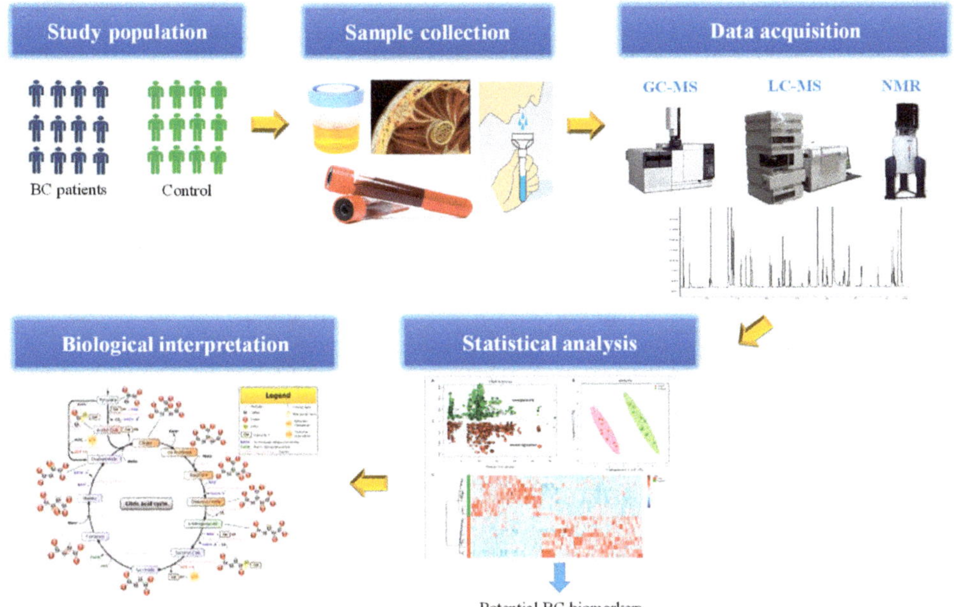

**Figure 4.** General flowchart in targeted/untargeted metabolomic approaches.

## 3.1. MS–Based Metabolomics

MS is an analytical tool extensively used in metabolomics applications, ranging from understanding the structural characterization of important metabolites to biomarker discovery [86]. Metabolic fingerprinting is general obtained by MS direct-injection, but this approach presents several drawbacks namely co-suppression and low ionization efficiency. Thus, generally MS based metabolomics includes a separation step, based on gas chromatography (GC–MS) [43,44,51,59,65,66,77,82], liquid chromatography (LC–MS) [33,43,46,50,52–55,70] or capillary electrophoresis (CE-MS) [83,84], to solve the co-suppression and to decrease the complexity of the biological sample. The integration of MS with a chromatographic technique (GC, LC) and capillary electrophoresis showed high sensitivity, speed, selectivity and improves the accuracy of compound identification, detection and quantification. In addition, GC-, LC- and CE-MS are destructive methods, requires sample preparation and are expensive, being these facts the main drawbacks of these hyphenated frameworks [86,88,89].

### 3.1.1. Gas Chromatography-mass Spectrometry (GC-MS) - Based Metabolomics

In the last decades, MS and chromatography have been broadly developed, and GC-MS becomes a core and reliable separation, detection and identification analytical framework on metabolomic analysis [43,44,51,59,65,66,77,82]. After sample collection and metabolite extraction, a small volume of sample is commonly injected in splitless mode, once the metabolites are in trace levels, to improve the sensitivity and the carrier gas propels the sample through the high-resolution capillary column (30 or 60 m columns with 5–50% phenyl stationary phases). The separation in GC occurs in an oven at high temperatures, and the metabolites need to be thermally stable and volatile (e.g., aldehydes, ketones, alkanes, organic acids) or non-volatile metabolites requiring derivatization (e.g., amino acids, sugars, phosphorylated metabolites, amines, lipids) [86,88,89]. The samples are ionized by electron-impact (EI) or chemical ionization (CI) for MS detection, being EI the most used since it provides molecular ion fragmentation to obtain a mass spectrum revealing of the metabolite's structure [88]. The MS employed influences the sensitivity of detection, being the quadrupole (q), time-of-flight (TOF) and ion trap the most usually applied in metabolomics. GC-qMS was used to screen salivary volatiles for putative BC as an exploratory study involving geographically distant populations [51], also to establish the metabolomic signature of human BC cell lines [44] and to discriminate different types of cancer based on urinary volatomic biosignature [59], among other examples reported in Table 1. In the first study, up to 120 volatiles from distinct chemical classes, with significant variations among the groups, were identified [51], whereas Silva et al. [44] and Porto-Figueira et al. [59] identified 60 and 130 volatiles in BC cell lines and urine, respectively. On the other hand, Budczies et al. [77,82] used GC-TOFMS framework to evaluate the glutamate enrichment as new diagnostic opportunity in BC and to accomplish a comparative metabolomics of estrogen receptor positive (ER$^+$) and estrogen receptor negative (ER$^-$) in BC. Budczies et al. [82] identified 19 metabolites BC tissues revealed significantly differences in central metabolism in ER$^-$ when compared to the ER$^+$ type. The affected metabolic pathways included the metabolism of glutamine with a decrease in concentration of glutamine and an increase glutamate and 2-hydroxyglutaric acid [82]. In turn, Dougan et al. [66] used GC-MS to evaluate the detectability, reliability, and distribution of metabolites measured in pre-diagnostic plasma samples in a pilot study of women listed in the Northern California site of the BC Family Registry. In this study. 661 metabolites were detected, 338 (51%) of them were found in all samples, and 490 (74%) in more than 80% of samples.

The main advantages of GC-MS-based metabolomics are sensitivity, specificity, high-throughput technology to handle a large volume of samples and reproducible. Nevertheless, this hyphenated technique has limited in mass range ($m/z$ 30–550), the molecular ion is often not detected owing to fragmentation, which makes more difficult the identification of unknown metabolites and the metabolites need be volatile and thermally stable [89,90].

### 3.1.2. Liquid Chromatography-Mass Spectrometry (LC-MS) - Based Metabolomics

Currently, liquid chromatography (LC) in particular high-performance liquid chromatography-mass spectrometry (HPLC-MS, LC-MS) represents an easy-going tool on separation and characterization of a metabolites pool, namely salts, acids, bases, hydrophilic and hydrophobic metabolites. The versatility of LC-MS is due to the several separation procedures and wide-ranging mass analyzers [90]. Contrarily to GC-MS, HPLC-MS is not limited to volatile and thermo stable metabolites and it is a promising tool for global metabolomics and the establishment of disease biomarkers.

Basically, the metabolites are eluted through a column based on their selective partition between a stationary phase (column material) and a mobile liquid phase. The metabolites according to the type of stationary phase can be eluted based on their charge, size, hydrophobicity and molecular weight [91]. Nowadays, the evolution of the HPLC is focused in miniaturization, smaller columns and low solvent volumes to attain a faster separation of metabolites. Ultra-high performance chromatography (UHPLC) appears as solution, since compared to HPLC promotes the resolution within a low analysis time and requires low volumes of solvent [92,93]. UHPLC columns are packed with 2 µm particles and the system operates at higher pressures (1000 bar) and tandem with MS, results in higher peak capacity, resolution, specificity and high-throughput abilities (reduced run time per sample) compared with HPLC [86,90,92–94].

Furthermore, Willmann et al. [46] analyzed the endo- and exometabolite of the BC cell lines MDA-MB-231, -453 and BT-474 as well as the breast epithelial cell line MCF-10A through two different analytical platforms: UHPLC-ESI-QTOF and HPLC-ESI-QqQ, which resulted in the identification of 92 annotated exometabolites and 58 endometabolites. In turn, Jové [33] used LC-ESI-QTOFMS/MS to establish the metabolomic profile of BC, whereas HPCL-ESI-MS was used to determine the determine the lipidomic differences between human BC and the surrounding normal tissues [55]. UHPLC tandem with MS was applied to explore novel blood plasma biomarkers associated to the BRCA1-mutated phenotype of BC [50], to determine polyamines including N-acetylated forms in saliva [52,54], and to screen the potential salivary biomarkers for BC diagnosis, staging, and biomarker discovery [53].

### 3.2. NMR–Based Metabolomics

NMR spectroscopy has been announced as a promising tool of metabolomics, providing a comprehensive view of metabolite fingerprinting, profiling and metabolic flux analysis under specific conditions, despite its inherent lower sensitivity compared to MS, limiting its skill with trace level metabolites. The main advantages of NMR are automation, requires low or no sample preparation, high reproducibility, non-destructive, non-selectivity in metabolite detection and the ability to simultaneously quantify multiple classes of metabolites [29,87].

The principle of NMR spectroscopy is based on the fact that the nucleic of many isotopes (e.g., $^1H$, $^{13}C$, $^{14}N$, $^{15}N$, $^{17}O$), when placed in a magnetic field, absorb radiation at a specific frequency [90]. The result is a NMR spectrum which corresponds to a unique metabolite pattern and provides structural information that can simplify the identification of unknown metabolites [86,89]. A fast identification of metabolite results from a combination of chemical shifts, spin–spin coupling, and relaxation or diffusion information [86,89]. Jobard et al. [68] reported a $^1H$ NMR-based metabolic phenotyping study aiming the identification of metabolic serum changes associated with advanced metastatic BC (MBC) in comparison to the localized early disease (EBC). Histidine, acetoacetate, glycerol, pyruvate, glycoproteins (N-acetyl), mannose, glutamate and phenylalanine were the metabolites that allowed the discrimination between MBC and EBC groups. NMR was also used by Tenori et al. [58] to explore whether serum metabolomic spectra could distinguish between early and metastatic BC patients and predict disease relapse, whereas Singh et al. [63] used NMR to detect alterations in metabolites and their linkage to metabolic processes in a number of pathological conditions including BC. In the last study, the authors observed an increase in lipoprotein, lactate, lysine and alanine level and a decrease in the levels of pyruvate and glucose in serum of inositol 1, 4, 5-trisphosphate (IP3R) receptor group patients when compared to control. In addition, NMR offers the possibility to study tissue through high-resolution magic angle spinning (HR-MAS) to reduce line widths in NMR spectra of tissue samples [74,75,79–81]. Tayyari et al. [74] performed the metabolomic analysis of triple-negative and luminal A BC subtypes in African-American using HR-MAS-NMR. A total of 27 metabolites were assigned and the metabolic profiles of these subtypes were also distinct from those revealed in Caucasian women. In turn, the feasibility of HR-MAS-NMR of small tissue biopsies to distinguish between tumor and non-involved adjacent tissue was investigated by Bathen et al. [75]. The results showed that the levels of glucose were higher in samples with low tumor content, whereas samples with high tumor content presented higher levels of ascorbate, lactate, creatine, glycine, taurine and the choline-containing metabolites. Euceda et al. [79] evaluate the metabolomic changes during neoadjuvant chemotherapy combined with bevacizumab in BC using HR-MAS-NMR. According to these authors, despite metabolic profiles not being able to predict the pathological complete response (pCR) prior to treatment, a significant metabolic difference in $pCR^+$ patients compared to $pCR^-$ was detected after neoadjuvant chemotherapy.

### 3.3. Comprehensive Analytical Frameworks on Metabolomics Approach

Comprehensive analytical frameworks have gained popularity on metabolomics field [86], being hundreds of metabolites detected simultaneously through analytical frameworks such as GCxGC-MS, HPLC-CE-MS, LCxLC-MS, LC-MS-NMR, MALDI-FT-ICR-MS, LC-FT-ICR-MS, among others. In the last decade, two dimension (2D) liquid-liquid chromatography (LCxLC) as well as gas-gas chromatography (GCxGC) have been gained increasing attention since overcome overlapping of metabolites by diverting each peak from a GC or LC column to a second GC or LC column, improve sensitivity and complementary selectivity being a promising tool in metabolomics field [95]. Nevertheless, other comprehensive analytical framework has been purposed in metabolomic field, in this context LC-MS-NMR platform is used in the identification of unknown metabolites in biological samples at trace levels, providing sample efficiency higher than the conventional flow injection methods [86]. In this sense, Reichenbach and co-workers [96] developed a suitable approach based on GCxGC-HRMS to analyse a cohort of 18 samples from BC tumors. This approach avoided the intractable problem of comprehensive peak matching, through a few reliable peaks for alignment and peak-based retention-plane windows to define comprehensive features that can be consistently matched for cross-sample analysis. In addition, a clear discrimination was achieved between sample of different grades and establish potential BC biomarkers. On the other hand, Yu et al. [97] optimized GC×GC-MS for robust BC cells, tissue, serum and urine metabolite profiling. GC×GC-MS analysis revealed detection around 600 molecular features from which 165 were characterized representing different chemical groups, such as amino acids, fatty acids, lipids, carbohydrates, nucleosides and small polar components of glycolysis and the Krebs cycle using EI spectrum matching. NanoLC-FT-ICR MS was used to analyse protein digests of ~3000 laser capture microdissection (LCM)-derived tumor cells from breast carcinoma tissue, corresponding to ~300 ng of total protein [98].

## 4. Data Analysis

Data analysis is crucial in metabolomics, being indispensable in every step of research, namely in sampling and experiment designs, data pre-processing and metabolite identification, as well in variables selection, classification modeling and validation procedures. The great challenge of data analysis in metabolomics is the high dimensionality and complexity of datasets under analysis. Several chemometric tools and statistical softwares are used in order to attribute value for high-dimensional metabolomic information obtained previously by the analytical tools [99,100]. Normally, a complete data analysis procedure in metabolomics is based on the following steps: dataset pre-treatment (centering, scaling, normalization), pre-processing (exploratory projection, variables selection), processing (predictive models), validation (model verification) and post-processing (pathway analysis) [101]. However, data analysis is dependent on the objective of the study and may be a simple exploratory research or complex discovery of biomarkers and metabolic pathways, for this reason not all steps are always present or are not followed in the same order. The data analysis procedures of recent metabolomics studies in BC are described in Table 2.

**Table 2.** Summary of the main chemometric methods applied to metabolomic studies.

| Biological Sample | Data Pre-Treatment | Pre-Processing | Processing | Validation | Post-Processing | Reference |
| --- | --- | --- | --- | --- | --- | --- |
| | | **Data Analysis** | | | | |
| Diagnostic tool | - | - | - | - | - | - |
| Human BC cell lines | Scaling (Pareto scaled), Transformation (log transformed) | PCA, HCA | OPLS-DA | LOOCV, ROC | none | [43] |
| | Centering (mean centered), Scaling (autoscaled) | ANOVA, PCA, HCA, Pearson correlation | PLS-DA | LOOCV | none | [46] |
| | Experimental correction (sample weight corrected) | PCA | none | none | none | [61] |
| | none | none | none | none | none | [45] |
| | none | ANOVA, PCA | PLS, LDA | K-CV | none | [44] |
| Human blood | none | T-test | PLS-DA, LRA | ROC, Permutation test | none | [47] |
| | Scaling (total intensity value scaled) | Wilcoxon test | RF | ROC, Bootstrapping | none | [58] |
| Human Exhaled breath | Transformation (quantile transformed) | T-test | RF, SVM | LOOCV, ROC, Bootstrapping | none | [48] |
| Human plasma | none | Correlation feature selection (CFS) | LRA, SVM, RF | K-CV, ROC | Pathway-based metabolite sets analysis (pathifier) | [65] |
| | Scaling (median value scaled), Transformation (log transformed) | ANOVA, PCA | none | none | none | [66] |
| | none | T-test, PCA, HCA | PLS-DA, RF | K-CV, ROC | Pathway enrichment analysis (metaboanalyst) | [33] |

**Table 2.** *Cont.*

| Biological Sample | Data Analysis | | | | | |
|---|---|---|---|---|---|---|
| | Data Pre-Treatment | Pre-Processing | Processing | Validation | Post-Processing | Reference |
| Human BC cell lines, plasma | none | KS-test, T-test, PCA | none | none | none | [50] |
| Human saliva | none | none | none | none | none | [52] |
| | Scaling (Pareto and total intensity value scaled) | T-test, PCA | PLS-DA | ROC, Permutation test | none | [53] |
| | none | none | LDA | K-CV, ROC | none | [54] |
| | Scaling (autoscaled and median value scaled), Transformation (cubic root transformed) | MW-test, HCA | PLS-DA, OPLS-DA | MCCV, Permutation test | none | [51] |
| | Experimental correction (internal standard corrected) | MW-test, PCA | PLS-DA, SVM, LRA | K-CV, ROC | none | [102] |
| Human tissues | Scaling (Pareto scaled) | PCA | OPLS | K-CV | none | [55] |
| | none | PCA, HCA | none | none | none | [56] |
| | Scaling (median scaled), Transformation (log transformed) | T-test | none | none | none | [57] |
| | Scaling (total intensity value scaled) | T-test | PLS-DA | LOOCV, ROC | Pathway enrichment analysis (metaboanalyst) | [74] |
| | Scaling (median scaled) | PCA | PLS-DA | LOOCV | none | [75] |
| | Scaling (median scaled) | T-test, PCA | PLS-DA | LOOCV | none | [78] |
| | Transformation (log transformed) | T-test, Pearson correlation, HCA | none | none | none | [76] |
| | none | T-test, Pearson correlation | PLS-DA | K-CV, ROC | none | [77] |

**Table 2.** *Cont.*

| | Data Analysis | | | | | |
| --- | --- | --- | --- | --- | --- | --- |
| **Biological Sample** | **Data Pre-Treatment** | **Pre-Processing** | **Processing** | **Validation** | **Post-Processing** | **Reference** |
| Human serum | Centering (mean centered), Scaling (total intensity value scaled) | PCA | PLS-DA, OPLS-DA | K-CV, ROC | none | [67] |
| | Centering (mean centered), Scaling (total intensity value scaled) | T-test, PCA, ANOVA | OPLS | K-CV, ROC, Bootstrapping | none | [68] |
| | Transformation (log transformed), Experimental correction (internal standard corrected) | ANOVA, PCA | PLS-DA, LRA | K-CV, ROC | none | [69] |
| Human urine | Scaling (autoscaled and median value scaled), Transformation (cubic root transformed) | T-test, HCA | PLS-DA, SVM, RF | MCCV, ROC | Pathway enrichment analysis (metaboanalyst) | [59] |
| Drug therapy | - | - | - | - | - | - |
| BC cell line | none | T-test | none | none | none | [62] |
| Human blood | Transformation (log transformed) | T-test, Pearson correlation | none | none | none | [70] |
| BC tissues | Centering (mean centered), Transformation (log transformed - only in univariate analysis) | T-test, Pearson correlation, PCA | PLS-DA | K-CV, Permutation test | none | [79] |
| | Scaling (mean scaled - only in PCA) | ANOVA, Spearman correlation, PCA | RF | K-CV, Bootstrapping, Permutation test | none | [80] |
| | Scaling (total intensity value scaled) | MW-test | OPLS-DA | LOOCV | none | [81] |
| | none | Spearman correlation | none | none | none | [82] |
| Serum | Scaling (total intensity value scaled) | T-test | PLS, PLS-DA | LOOCV, ROC | none | [71] |

**Table 2.** *Cont.*

| | Data Analysis | | | | | |
|---|---|---|---|---|---|---|
| **Biological Sample** | **Data Pre-Treatment** | **Pre-Processing** | **Processing** | **Validation** | **Post-Processing** | **Reference** |
| Serum, tissues, cell lines | none | T-test, ANOVA, PCA | PLS-DA | K-CV, ROC | Pathway enrichment analysis (metaboanalyst) | [63] |
| Urine | Scaling (total intensity value scaled) | KS-test, L-test, SW-test, T-test, PCA | OPLS-DA | K-CV, ROC | Pathway enrichment analysis (metaboanalyst) | [73] |
| | none | T-test, PCA | PLS-DA | K-CV | none | [84] |
| Metabolic reprogramming | - | - | - | - | - | - |
| Human BC cell lines, BC xenografts | none | ANOVA, PCA | PLS-DA | K-CV | none | [64] |
| Mouse BC tissue | Scaling (median scaled) | ANOVA, PCA | none | none | none | [83] |
| Endogenous factors | - | - | - | - | - | - |
| Human plasma | none | T-test, Spearman correlation, PCA | LRA | ROC | none | [31] |
| Human serum | Transformation (log transformed) | Pearson correlation, PCA | LRA | none | none | [72] |

ANOVA – Analysis of variance; ROC – Receiver operating characteristic; LOOCV – leave-one-out-cross validation; AUC – Area under the curve; BFS– Bootstrap feature selection; GGM–Gaussian graphical modelling; HCA – Hierarchical cluster analysis; LDA – Linear discriminant analysis; MCCV – Monte Carlo cross validation; MWT – Mann Whitney U test; NRI – Net reclassification improvement; OPLS-DA – Orthogonal projections to latent structures discriminant analysis; LRA - logistic regression analysis; OSC-PLS – Orthogonal signal correction partial least squares; PC – Pearson correlation; PCA – Principal component analysis; PEA – Pathway enrichment analysis; PLS-DA – Partial least squares discriminant analysis; RF – Random Florest classifier; SCC – Spearman correlation coefficient; SVM – Support vector machine; VIP – variable importance in projection.

### 4.1. Dataset Pre-Treatment

Dataset pre-treatment is the initial step in data analysis, being extensively used in metabolomics to resolve the heteroscedasticity of high-dimensional datasets. Commonly, pre-treatment in BC metabolomics is done through normalization of dataset based on the centering, scaling, transformation and/or experimental corrections of variables values [103–105]. Centering is performed when the data analysis is focused on the differences between variables, where all measurements (e.g., concentrations, areas) are converted to values around zero based on variation measures. Mean [46,67,68,79] is the measure normally used in centering. Scaling is used to adjust the variables measurements based on a scaling factor, converting the measurements of all variables into values relative to the scaling factor. The scaling factor selected can be a dispersive measure (e.g., standard deviation) or size measure (e.g., mean). The main scaling approaches based on dispersive measures are autoscaling (standard deviation) [46,51,59] and pareto scaling (square root of the standard deviation) [43,53,55]. On the other hand, the most of size measure approaches uses scaling factors based on the mean [80], median [51,57,59,66,75,78,83] or total intensity value [53,58,67,68,71,73,74,81]. Transformations are mathematical approaches used to decrease the heteroscedasticity of dataset, which the variability between variables is dramatically reduce. Log [43,57,66,69,70,72,76,79] is the main transformation in BC metabolomics. However, cubic root [51,59] and quantile [48] transformations are also used. Other normalization approaches based on experimental corrections are also used in metabolomics, such as internal standards [102,106] and sample weight [61]. Internal standards normalization assumes that the heteroscedasticity of all variables is systematic and can be corrected by variance of internal standards. Sample weight normalization is the direct correction of variables values by experimental sample measures (e.g., volume and weight).

### 4.2. Pre-Processing

Pre-processing methods are performed to obtain an exploratory projection of dataset or an overview of variables importance prior to prediction models processing. Primarily, normality tests are used to determine if the data distribution is normal (parametric) or not normal (non-parametric). The most commonly used are Kolmogorov-Smirnov test (KS-test) [50,73], Shapiro-Wilk test (SW-test) [73] and Lilliefors test (L-test) [73]. Two types of approaches are normally used in exploratory projections/variables importance ranking of BC metabolomics datasets: univariate and multivariate analysis. Univariate statistical methods are used to analyzed only one variable at a time, being useful to easily discover significant differences or measure correlations between samples groups. The differentiation is based on variance between groups by rejection of the null hypothesis or acceptation the alternate hypothesis [101,107,108]. The most common methods used when the data is parametric are T-tests [31,47,48,50,53,57,59,62,68,70,71,73,74,76–79,84] and ANOVA [33,44,46,63,64,66,68,69,72,83] T-tests, such as Student and Welch's tests, are recommended to analyze differences between two groups, and ANOVA-based methods, such as one-way ANOVA, two-way ANOVA, factorial ANOVA and MANOVA are used to evaluate more than two groups. Alternative univariate methods are implemented when the assumption of the normal distribution is non-parametric, such as Mann-Whitney test (MW-test) [51,81,102] and Wilcoxon test (W-test) [58]. In addition, univariate methods are also widely used to measure the correlations between continuous variables and response. The Pearson correlation [46,70,72,76,77,79] is the preferred option for linear relationships in populations with normal distribution. On the other hand, the Spearman correlation [31,80,82] is usually used in non-parametric datasets. More complex correlation methods are also used in data analysis, such as Correlation Feature Selection (CFS) [65], where the appropriate correlation measure and a heuristic search strategy are performed by experiments on artificial and natural datasets based on algorithms.

Similarly, the multivariate methods are also widely used for exploratory studies to obtain dataset patterns based on relationships between groups, being divided into two sub-groups, unsupervised and supervised methods. Unsupervised methods are the preferential option for exploratory studies, where the modeling process is based only on the explanatory variables, without external intervention of user (Yi et al., 2016). The most commons are principal component analysis (PCA) [56,57,60,62,70,75,76,80,92,93] and hierarchical cluster analysis (HCA) [33,43,51,56,59,76] PCA provides the projection of dataset into low dimensional based on orthogonal transformation, converting the variables variability from a set of observations into score vectors and loadings, called principal components [100,109]. HCA methods are used to form subsets of samples at ordered levels based on variables similarities/dissimilarities (such as distances or correlations) and can be performed in agglomerative mode (samples are aggregate into clusters) or divisive mode (complete dataset is divided into clusters). In both modes, the linkage criterion need to be selected, being that the most commonly used are single-linkage clustering (the minimum of distances) and complete linkage clustering (the maximum of distances) [110,111].

### 4.3. Processing Methods

After the explorative studies and variable selection, the next step is the processing of dataset in order to create a predictive response model to classification of new samples (ex. diagnostic tools), identification of valuable variables (ex. biomarkers) or exploring the mechanisms of metabolomic studies (ex. metabolic pathways). In this stage, the supervised methods are the preferential choice, where the response models are mainly based on two types, continuous (regression) and discrete (classification) [100,103]. The main methods for continuous response are based on multiple linear regression (MLR), sometimes called ordinary least squares (OLS). MLR is performed to predict the values of a dependent variable (response) based on a set of continuous explanatory variables, assuming a linear combination of the explanatory variables [109]. The most applied MLR-based method in metabolomics is partial least squares (PLS) [44,55,68,71]. Unlike PCA, which uses only the variables variation, PLS is a predictive and supervised method that use an informative response to maximize the covariance between the explanatory variables and the response, producing score vectors and loading vectors. The prediction model is based on interaction between the variables and response, ignoring the variables with irrelevant importance. The importance of each variable is defined according the PLS-based criteria, such as loading weights, variable importance on projection scores, regression coefficient, target projection and selectivity ratio [100,101,109]. However, when categorical variables are introduced, the discrete models should be used. Discrete models provide a predictive classification of response based on continuous and categorical variables, being classified into linear or non-linear. In linear methods, the classification is performed by highest probability based on linear relationships between explanatory variables, where exist a grouping variable (categorical). Linear discriminant analysis (LDA) [44,54] is the preferential method to classification models of discrete responses. LDA perform linear transformations of explanatory variables to create discriminant functions that will maximize the separation between multiple classes of samples (groups) based on the information of the categorical variables [109]. Among the various LDA-based methods, PLS-DA [33,46,47,51,53,59,63,64,67,69,71,74,75,77,79,84,102] is most widely used in metabolomics studies. PLS-DA is a successful combination of PLS and LDA that provides a visual low-dimensional pattern of samples discrimination based on the analysis of relationships between continuous and categorical variables [101,109]. Recently, some extensions of PLS-DA were used in BC metabolomics, namely the OPLS-DA [43,51,67,73,81]. OPLS-DA separates out response orthogonal variations in rotations of the original component [109].

On the other hand, non-linear methods are used when metabolomics dataset follow a non-linear response. The most applied non-linear methods are support vector machines (SVM) [48,59,65,102], random forests (RF) [33,48,58,59,65,80] and logistic regression analysis (LRA) [31,47,65,69,72,102]. SVM is a kernel-based model used for regression and classification of non-linear datasets, transforming the non-linear data into more general spaces (linear) by algorithm based on kernels functions. SVM perform the mapping of dataset into a high-dimensional space through kernels functions for the separation of two groups of samples into distinctive regions. The separation is based on support vectors, which are points (samples) on the boundary or on the incorrect side of the margin supporting the separation. SVM is a versatile method that transforms non-linear complex datasets into a high-dimensional space where classes are linearly separable [100,101,109]. RF is a non-linear method for regression and classification of high-dimensional datasets, where a large number of classification and regression trees are created by bootstrapping (replacement) based on random selection of a training samples from the original dataset. Afterwards, bootstrapping is performed systematically to build a large group of simple trees that are used to estimate classification accuracy of the model [100,101]. Another non-linear predictive method widely used is LRA, which is similar to linear regression, but with a binomial response variable. LRA is used to explain the relationship between one dependent binary variable and one or more nominal, ordinal, interval or ratio-level independent variables [112].

### 4.4. Model Validation

The validation of predictive models is a key step in data analysis of metabolomics studies. Validation process analyzes the performance/ability of model to predict correctly the hypothesized relationships between variables and responses [101]. Several validation methods have been used in BC metabolomics. The coefficient of determination ($R^2$) is the simplest method to evaluate the ability of predictive model, being used for continuous responses. The $R^2$ is expressed as the ratio between 0 and 1, where a value of 1 indicates the perfect prediction. However, this validation is recommended for small datasets, due to fact that the $R^2$ value tends to be increased when a predictor variable is added to the model [113]. However, in validation of predictive models used to high-dimensional and complex datasets, as the case of metabolomics studies, the cross validation (CV) methods are the preferential option. CV provides qualitative and quantitative analysis of the model ability to model's ability to predict new independent samples without collecting additional data. During the CV, the available data are split into two sets, where one set is used to create a predictive model using the values of continuous and predictor variables (training set). The second set is used to test the performance of predictive model (validation set) [100]. The most applied CV procedure is k-fold (K-CV) [33,44,54,55,63–65,67–69,73,77,79,80,84,102]. K-CV processing is based on random partition of original dataset into equal sized subsamples (k). A single k subsample is used as the validation set for testing the model, and the remaining k -1 subsamples are used as a training set. This process is then repeated k times (folds), with each of the k subsamples being used exactly one time as the validation set [106]. One special type of K-CV is the leave-one-out cross validation (LOOCV) (Bathen et al., 2013; Choi et al., 2013; Cífková et al., 2017; Martinez-Lozano Sinues et al., 2015; Tayyari et al., 2018; R. Vettukattil et al., 2013; Willmann et al., 2016), where the number of folds equals the number of k subsamples. LOOCV is considered an exhaustive CV, being recommend for small datasets [106,113]. Another type of CV is the Monte Carlo cross validation (MCCV) [51,59]. Although less used in metabolomics than LOOCV, MCCV is asymptotically consistent and showed better prediction ability. In MCCV proceeding, significant part of dataset is leaved out at a time during model validation, repeating systematically this procedure several times [114,115]. The $Q^2$ value, which is the equivalent $R^2$ value, is the preferential coefficient of determination for CV procedures.

A visual and easy model validation method is the receiver operating characteristic (ROC) curve [31,33,43,47,48,53,54,58,59,63,65,67–69,71,73,74,77,102] which the prediction ability of a model is validated considering the specificity (ratio of the correctly predicted negatives) and sensitivity (ratio of correctly predicted positives). The ROC curve is given by plotting the sensitivity versus (1 - specificity) across a series of cutoff points. The area under curve (AUC) is a quantitative measure (between 0 and 1) of the ability of predictive model, where a AUC value close to 1 indicates a nearly perfect prediction response [100,113].

Random resampling-based methods are a robust alternative for model validation. The most used in BC metabolomics are bootstrapping [48,58,68,80] and permutation tests [47,51,53,79,80]. Bootstrapping is a model validation based on replacement of samples, which can be considered non-parametric when the replacement is from the original dataset, or parametric when random noise is added from a recognized distribution to the dataset to estimate underlying sampling distribution or establish robust confidence intervals. Normally, in metabolomics studies the common approach is non-parametric bootstrapping [113,116]. Permutation tests provide the exact control of false positives from a predictive model (linear or non-linear), under minimal assumptions, based on differences between the randomly permuted response variables model and the original model. Permutation tests are based on a repeatedly permuting (repetitive reordering) of the N entries in the response variable. Permuted vectors containing integers between 1 and N are produced in a random number generator, creating new scrambled response variables only by switching their internal positions. The scrambled vectors are modelled one by one, where for every test, the $R^2$ and $Q^2$ values are calculated and saved. After, these values are compared with the values calculated from the original data. The results of permutation tests are displayed as a percentage overlap between the real and permuted $R^2$ and $Q^2$ values, where a 0% of overlap is the optimal result [109,117].

*4.5. Post-Processing*

The post-processing step consists in the interpretation of metabolomic responses from original dataset. Normally, pathway analysis is the most used strategy to provide an overview of association/relationship between identified metabolites and metabolic pathways and other general biological networks. Pathifier [65] and metaboanalyst [33,59,63,73,74] are the most used software for this propose in metabolomics.

## 5. Future Directions

The advances in analytical techniques and chemometric methods in metabolomics have been growing rapidly becoming possible the identification of potential biomarkers. Furthermore, the integration of analytical platforms increases the comprehensive analysis of metabolites in biological samples. In this context, metabolites became valuable identifications, regardless their hierarchical source, enabling the phenotypic properties in a biological system. Additionally, the identification of key metabolic pathways from which significant metabolites are linked, it is possible to reveal potential targets for cancer therapy.

Also, standard procedures for sample collection, data analysis and shared in repositories have potential to be adopted by both researchers and medical communities.

Since the metabolome instantly responds to environmental stimuli including therapeutic or surgical intervention, could be also used to monitor the metabolic status of the individual and indicate any possible toxic effects. Moreover, metabolomics may help in the detection of potential cancer biomarkers, being useful for example in the development of different devices, including biosensors, that can significantly improve the cancer diagnosis. These devices include a biorecognition element within a biosensor system. The biorecognition molecules interact with the target, which is then converted into a measurable signal by a transducer. Basically, these molecules, usually enzymes or antibodies, can be immobilized on the transducer surface and interact with the target (biomarker) to produce a signal is interpreted, providing information about the disease and their possible recurrence after therapy.

**Supplementary Materials:** The following is available online at http://www.mdpi.com/2218-1989/9/5/102/s1, Figure S1: (a) Total identified metabolites by analytical technique and (b) number of samples used by each type of biological specimen.

**Author Contributions:** C.S. conception of the review, drafting of manuscript, performed the introduction, pathophysiology, risk factors, diagnostic and screening tools, future directions. R.P. performed the omics science and analytical platforms. P.S. performed the data analysis on chemometric tools. H.T. performed the supervision and revision of the manuscript. J.S.C. performed the supervision and revision of the manuscript. All authors read and approved the final version of manuscript.

**Funding:** This work was supported by FCT-Fundação para a Ciência e a Tecnologia (project PEst-OE/QUI/UI0674/2019, CQM, Portuguese Government funds, and INNOINDIGO/0001/2015), Madeira 14-20 Program (project PROEQUIPRAM - Reforço do Investimento em Equipamentos e Infraestruturas Científicas na RAM - M1420-01-0145-FEDER-000008) and by ARDITI-Agência Regional para o Desenvolvimento da Investigação Tecnologia e Inovação through the project M1420-01-0145-FEDER-000005 - Centro de Química da Madeira - CQM+ (Madeira 14-20).

**Acknowledgments:** The authors also acknowledge the FCT for the Ph.D. grant SFRH/BD/97039/2013 and Post-Doctoral fellowship SFRH/BPD/97387/2013 given to Catarina L. Silva and Rosa Perestrelo. Pedro Silva acknowledge ARDITI through the Ph.D. grant under the M1420 Project - 09-5369-FSE-000001.

**Conflicts of Interest:** The authors declare no conflict of interest.

## References

1. Donepudi, M.S.; Kondapalli, K.; Amos, S.J.; Venkanteshan, P. Breast cancer statistics and markers. *J. Cancer Res. Ther.* **2014**, *10*, 506–511. [PubMed]

2. Bray, F.; Ferlay, J.; Soerjomataram, I.; Siegel, R.L.; Torre, L.A.; Jemal, A. Global cancer statistics 2018: GLOBOCAN estimates of incidence and mortality worldwide for 36 cancers in 185 countries. *CA Cancer J. Clin.* **2018**, *68*, 394–424. [CrossRef] [PubMed]

3. Siegel, R.L.; Miller, K.D.; Jemal, A. Cancer statistics, 2018. *CA Cancer J. Clin.* **2018**, *68*, 7–30. [CrossRef]

4. Ferlay, J.; Soerjomataram, I.; Dikshit, R.; Eser, S.; Mathers, C.; Rebelo, M.; Parkin, D.M.; Forman, D.; Bray, F. Cancer incidence and mortality worldwide: Sources, methods and major patterns in GLOBOCAN 2012. *Int. J. Cancer* **2015**, *136*, E359–E386. [CrossRef]

5. Allison, K.H. Molecular Pathology of Breast Cancer. *Am. J. Clin. Pathol.* **2012**, *138*, 770–780. [CrossRef]

6. DeSantis, C.E.; Bray, F.; Ferlay, J.; Lortet-Tieulent, J.; Anderson, B.O.; Jemal, A. International Variation in Female Breast Cancer Incidence and Mortality Rates. *Cancer Epidemiol. Biomarkers Prev.* **2015**, *24*, 1495–1506. [CrossRef]

7. Ghoncheh, M.; Pournamdar, Z.; Salehiniya, H. Incidence and Mortality and Epidemiology of Breast Cancer in the World. *Asian Pac. J. Cancer Prev.* **2016**, *17*, 43–46. [CrossRef] [PubMed]

8. Verma, R.; Bowen, R.L.; Slater, S.E.; Mihaimeed, F.; Jones, J.L. Pathological and epidemiological factors associated with advanced stage at diagnosis of breast cancer. *Br. Med. Bull.* **2012**, *103*, 129–145. [CrossRef]

9. Shah, R.; Rosso, K.; Nathanson, S.D. Pathogenesis, prevention, diagnosis and treatment of breast cancer. *World J. Clin. Oncol.* **2014**, *5*, 283–298. [CrossRef]

10. Libson, S.; Lippman, M. A review of clinical aspects of breast cancer. *Int. Rev. Psychiatry* **2014**, *26*, 4–15. [CrossRef]

11. Maruti, S.S.; Willett, W.C.; Feskanich, D.; Rosner, B.; Colditz, G.A. A prospective study of age-specific physical activity and premenopausal breast cancer. *J. Natl. Cancer Inst.* **2008**, *100*, 728–737. [CrossRef] [PubMed]

12. Kyu, H.H.; Bachman, V.F.; Alexander, L.T.; Mumford, J.E.; Afshin, A.; Estep, K.; Veerman, J.L.; Delwiche, K.; Iannarone, M.L.; Moyer, M.L.; et al. Physical activity and risk of breast cancer, colon cancer, diabetes, ischemic heart disease, and ischemic stroke events: systematic review and dose-response meta-analysis for the Global Burden of Disease Study 2013. *BMJ* **2016**, *354*, i3857. [CrossRef] [PubMed]

13. McDonald, J.A.; Goyal, A.; Terry, M.B. Alcohol Intake and Breast Cancer Risk: Weighing the Overall Evidence. *Curr. Breast Cancer Rep.* **2013**, *5*. [CrossRef] [PubMed]

14. Maskarinec, G.; Jacobs, S.; Park, S.-Y.; Haiman, C.A.; Setiawan, V.W.; Wilkens, L.R.; Le Marchand, L. Type II Diabetes, Obesity, and Breast Cancer Risk: The Multiethnic Cohort. *Cancer Epidemiol. Biomarkers Prev.* **2017**, *26*, 854–861. [CrossRef]

15. Heidegger, I.; Ofer, P.; Doppler, W.; Rotter, V.; Klocker, H.; Massoner, P. Diverse Functions of IGF/Insulin Signaling in Malignant and Noncancerous Prostate Cells: Proliferation in Cancer Cells and Differentiation in Noncancerous Cells. *Endocrinology* **2012**, *153*, 4633–4643. [CrossRef] [PubMed]

16. Djiogue, S.; Nwabo Kamdje, A.H.; Vecchio, L.; Kipanyula, M.J.; Farahna, M.; Aldebasi, Y.; Seke Etet, P.F. Insulin resistance and cancer: the role of insulin and IGFs. *Endocr. Relat. Cancer* **2013**, *20*, R1–R17. [CrossRef]

17. Neuhouser, M.L.; Aragaki, A.K.; Prentice, R.L.; Manson, J.E.; Chlebowski, R.; Carty, C.L.; Ochs-Balcom, H.M.; Thomson, C.A.; Caan, B.J.; Tinker, L.F.; et al. Overweight, Obesity, and Postmenopausal Invasive Breast Cancer Risk. *JAMA Oncol.* **2015**, *1*, 611. [CrossRef] [PubMed]

18. Picon-Ruiz, M.; Morata-Tarifa, C.; Valle-Goffin, J.J.; Friedman, E.R.; Slingerland, J.M. Obesity and adverse breast cancer risk and outcome: Mechanistic insights and strategies for intervention. *CA Cancer J. Clin.* **2017**, *67*, 378–397. [CrossRef]

19. Moore, S.C.; Playdon, M.C.; Sampson, J.N.; Hoover, R.N.; Trabert, B.; Matthews, C.E.; Ziegler, R.G. A Metabolomics Analysis of Body Mass Index and Postmenopausal Breast Cancer Risk. *JNCI J. Natl. Cancer Inst.* **2018**, *6*, 588–597. [CrossRef]

20. Kos, Z.; Dabbs, D.J. Biomarker assessment and molecular testing for prognostication in breast cancer. *Histopathology* **2016**, *68*, 70–85. [CrossRef]

21. Duffy, M.J.; Harbeck, N.; Nap, M.; Molina, R.; Nicolini, A.; Senkus, E.; Cardoso, F. Clinical use of biomarkers in breast cancer: Updated guidelines from the European Group on Tumor Markers (EGTM). *Eur. J. Cancer* **2017**, *75*, 284–298. [CrossRef] [PubMed]

22. Harris, L.N.; Ismaila, N.; McShane, L.M.; Andre, F.; Collyar, D.E.; Gonzalez-Angulo, A.M.; Hammond, E.H.; Kuderer, N.M.; Liu, M.C.; Mennel, R.G.; et al. Use of Biomarkers to Guide Decisions on Adjuvant Systemic Therapy for Women With Early-Stage Invasive Breast Cancer: American Society of Clinical Oncology Clinical Practice Guideline. *J. Clin. Oncol.* **2016**, *34*, 1134–1150. [CrossRef] [PubMed]

23. Jasbi, P.; Wang, D.; Cheng, S.L.; Fei, Q.; Cui, J.Y.; Liu, L.; Wei, Y.; Raftery, D.; Gu, H. Breast cancer detection using targeted plasma metabolomics. *J. Chromatogr. B* **2019**, *1105*, 26–37. [CrossRef] [PubMed]

24. Clish, C.B. Metabolomics: An emerging but powerful tool for precision medicine. *Cold Spring Harb. Mol. Case Stud.* **2015**, *1*, a000588. [CrossRef]

25. Roberts, L.D.; Souza, A.L.; Gerszten, R.E.; Clish, C.B. Targeted metabolomics. *Curr. Protoc. Mol. Biol.* **2012**, *98*, 30.2.1–30.2.24. [CrossRef]

26. Klupczyńska, A.; Dereziński, P.; Kokot, Z.J. Metabolomics in medical sciences–trends, challenges and perspectives. *Acta Pol. Pharm.* **2015**, *72*, 629–641. [PubMed]

27. Alonso, A.; Marsal, S.; Julià, A. Analytical methods in untargeted metabolomics: State of the art in 2015. *Front. Bioeng. Biotechnol.* **2015**, *3*, 23. [CrossRef]

28. Cho, K.; Mahieu, N.G.; Johnson, S.L.; Patti, G.J. After the feature presentation: Technologies bridging untargeted metabolomics and biology. *Curr. Opin. Biotechnol.* **2014**, *28*, 143–148. [CrossRef]

29. Claudino, W.M.; Goncalves, P.H.; di Leo, A.; Philip, P.A.; Sarkar, F.H. Metabolomics in cancer: A bench-to-bedside intersection. *Crit. Rev. Oncol. Hematol.* **2012**, *84*, 1–7. [CrossRef]

30. Beger, R.D. A review of applications of metabolomics in cancer. *Metabolites* **2013**, *3*, 552–574. [CrossRef]

31. Lécuyer, L.; Victor Bala, A.; Deschasaux, M.; Bouchemal, N.; Nawfal Triba, M.; Vasson, M.-P.; Rossary, A.; Demidem, A.; Galan, P.; Hercberg, S.; et al. NMR metabolomic signatures reveal predictive plasma metabolites associated with long-term risk of developing breast cancer. *Int. J. Epidemiol.* **2018**, *47*, 484–494. [CrossRef] [PubMed]

32. Gu, H.; Gowda, G.A.N.; Raftery, D. Metabolic profiling: Are we en route to better diagnostic tests for cancer? *Future Oncol.* **2012**, *8*, 1207–1210. [CrossRef]

33. Jové, M.; Collado, R.; Quiles, J.L.; Ramírez-Tortosa, M.-C.; Sol, J.; Ruiz-Sanjuan, M.; Fernandez, M.; de la Torre Cabrera, C.; Ramírez-Tortosa, C.; Granados-Principal, S.; et al. A plasma metabolomic signature discloses human breast cancer. *Oncotarget* **2017**, *8*, 19522–19533. [CrossRef]

34. Fan, Y.; Zhou, X.; Xia, T.-S.; Chen, Z.; Li, J.; Liu, Q.; Alolga, R.N.; Chen, Y.; Lai, M.-D.; Li, P.; et al. Human plasma metabolomics for identifying differential metabolites and predicting molecular subtypes of breast cancer. *Oncotarget* **2016**, *7*, 9925–9938. [CrossRef] [PubMed]

35. Bain, J.R.; Stevens, R.D.; Wenner, B.R.; Ilkayeva, O.; Muoio, D.M.; Newgard, C.B. Metabolomics Applied to Diabetes Research: Moving From Information to Knowledge. *Diabetes* **2009**, *58*, 2429–2443. [CrossRef]

36. Günther, U.L. Metabolomics Biomarkers for Breast Cancer. *Pathobiology* **2015**, *82*, 153–165. [CrossRef] [PubMed]

37. Hadi, N.I.; Jamal, Q. "OMIC" tumor markers for breast cancer: A review. *Pakistan J. Med. Sci.* **2015**, *31*, 1256–1262.

38. Cappelletti, V.; Iorio, E.; Miodini, P.; Silvestri, M.; Dugo, M.; Daidone, M.G. Metabolic Footprints and Molecular Subtypes in Breast Cancer. *Dis. Markers* **2017**, *2017*, 1–19. [CrossRef]

39. McCartney, A.; Vignoli, A.; Biganzoli, L.; Love, R.; Tenori, L.; Luchinat, C.; Di Leo, A. Metabolomics in breast cancer: A decade in review. *Cancer Treat. Rev.* **2018**, *67*, 88–96. [CrossRef]

40. Anderson, N.M.; Mucka, P.; Kern, J.G.; Feng, H. The emerging role and targetability of the TCA cycle in cancer metabolism. *Protein Cell* **2018**, *9*, 216–237. [CrossRef]

41. Ryan, D.G.; Murphy, M.P.; Frezza, C.; Prag, H.A.; Chouchani, E.T.; O'Neill, L.A.; Mills, E.L. Coupling Krebs cycle metabolites to signalling in immunity and cancer. *Nat. Metab.* **2019**, *1*, 16–33. [CrossRef] [PubMed]

42. Ciccarone, F.; Vegliante, R.; Di Leo, L.; Ciriolo, M.R. The TCA cycle as a bridge between oncometabolism and DNA transactions in cancer. *Semin. Cancer Biol.* **2017**, *47*, 50–56. [CrossRef] [PubMed]

43. Cífková, E.; Lísa, M.; Hrstka, R.; Vrána, D.; Gaték, J.; Melichar, B.; Holčapek, M. Correlation of lipidomic composition of cell lines and tissues of breast cancer patients using hydrophilic interaction liquid chromatography/electrospray ionization mass spectrometry and multivariate data analysis. *Rapid Commun. Mass Spectrom.* **2017**, *31*, 253–263. [CrossRef] [PubMed]

44. Silva, C.L.; Perestrelo, R.; Silva, P.; Tomás, H.; Câmara, J.S. Volatile metabolomic signature of human breast cancer cell lines. *Sci. Rep.* **2017**, *7*, 43969. [CrossRef]

45. Le Guennec, A.; Tea, I.; Antheaume, I.; Martineau, E.; Charrier, B.; Pathan, M.; Akoka, S.; Giraudeau, P. Fast Determination of Absolute Metabolite Concentrations by Spatially Encoded 2D NMR: Application to Breast Cancer Cell Extracts. *Anal. Chem.* **2012**, *84*, 10831–10837. [CrossRef]

46. Willmann, L.; Schlimpert, M.; Hirschfeld, M.; Erbes, T.; Neubauer, H.; Stickeler, E.; Kammerer, B. Alterations of the exo- and endometabolite profiles in breast cancer cell lines: A mass spectrometry-based metabolomics approach. *Anal. Chim. Acta* **2016**, *925*, 34–42. [CrossRef]

47. Cao, Y.; Wang, Q.; Gao, P.; Dong, J.; Zhu, Z.; Fang, Y.; Fang, Z.; Sun, X.; Sun, T. A dried blood spot mass spectrometry metabolomic approach for rapid breast cancer detection. *Onco. Targets. Ther.* **2016**, *9*, 1389. [CrossRef] [PubMed]

48. Martinez-Lozano Sinues, P.; Landoni, E.; Miceli, R.; Dibari, V.F.; Dugo, M.; Agresti, R.; Tagliabue, E.; Cristoni, S.; Orlandi, R. Secondary electrospray ionization-mass spectrometry and a novel statistical bioinformatic approach identifies a cancer-related profile in exhaled breath of breast cancer patients: A pilot study. *J. Breath Res.* **2015**, *9*, 31001. [CrossRef]

49. Cala, M.P.; Aldana, J.; Medina, J.; Sánchez, J.; Guio, J.; Wist, J.; Meesters, R.J.W. Multiplatform plasma metabolic and lipid fingerprinting of breast cancer: A pilot control-case study in Colombian Hispanic women. *PLoS ONE* **2018**, *13*, e0190958. [CrossRef]

50. Roig, B.; Rodríguez-Balada, M.; Samino, S.; Lam, E.W.-F.; Guaita-Esteruelas, S.; Gomes, A.R.; Correig, X.; Borràs, J.; Yanes, O.; Gumà, J. Metabolomics reveals novel blood plasma biomarkers associated to the BRCA1-mutated phenotype of human breast cancer. *Sci. Rep.* **2017**, *7*, 17831. [CrossRef]

51. Cavaco, C.; Pereira, J.A.M.; Taunk, K.; Taware, R.; Rapole, S.; Nagarajaram, H.; Câmara, J.S. Screening of salivary volatiles for putative breast cancer discrimination: An exploratory study involving geographically distant populations. *Anal. Bioanal. Chem.* **2018**, *410*, 1–10. [CrossRef] [PubMed]

52. Tsutsui, H.; Mochizuki, T.; Inoue, K.; Toyama, T.; Yoshimoto, N.; Endo, Y.; Todoroki, K.; Min, J.Z.; Toyo'oka, T. High-Throughput LC–MS/MS Based Simultaneous Determination of Polyamines Including N-Acetylated Forms in Human Saliva and the Diagnostic Approach to Breast Cancer Patients. *Anal. Chem.* **2013**, *85*, 11835–11842. [CrossRef] [PubMed]

53. Zhong, L.; Cheng, F.; Lu, X.; Duan, Y.; Wang, X. Untargeted saliva metabonomics study of breast cancer based on ultra performance liquid chromatography coupled to mass spectrometry with HILIC and RPLC separations. *Talanta* **2016**, *158*, 351–360. [CrossRef]

54. Takayama, T.; Tsutsui, H.; Shimizu, I.; Toyama, T.; Yoshimoto, N.; Endo, Y.; Inoue, K.; Todoroki, K.; Min, J.Z.; Mizuno, H.; et al. Diagnostic approach to breast cancer patients based on target metabolomics in saliva by liquid chromatography with tandem mass spectrometry. *Clin. Chim. Acta* **2016**, *452*, 18–26. [CrossRef] [PubMed]

55. Cífková, E.; Holčapek, M.; Lísa, M.; Vrána, D.; Gaték, J.; Melichar, B. Determination of lipidomic differences between human breast cancer and surrounding normal tissues using HILIC-HPLC/ESI-MS and multivariate data analysis. *Anal. Bioanal. Chem.* **2015**, *407*, 991–1002. [CrossRef] [PubMed]

56. Budhu, A.; Terunuma, A.; Zhang, G.; Hussain, S.P.; Ambs, S.; Wang, X.W. Metabolic profiles are principally different between cancers of the liver, pancreas and breast. *Int. J. Biol. Sci.* **2014**, *10*, 966–972. [CrossRef] [PubMed]

57. Kanaan, Y.M.; Sampey, B.P.; Beyene, D.; Esnakula, A.K.; Naab, T.J.; Ricks-Santi, L.J.; Dasi, S.; Day, A.; Blackman, K.W.; Frederick, W.; et al. Metabolic profile of triple-negative breast cancer in African-American women reveals potential biomarkers of aggressive disease. *Cancer Genom. Proteom.* **2014**, *11*, 279–294.

58. Tenori, L.; Oakman, C.; Morris, P.G.; Gralka, E.; Turner, N.; Cappadona, S.; Fornier, M.; Hudis, C.; Norton, L.; Luchinat, C.; et al. Serum metabolomic profiles evaluated after surgery may identify patients with oestrogen receptor negative early breast cancer at increased risk of disease recurrence. Results from a retrospective study. *Mol. Oncol.* **2015**, *9*, 128–139. [CrossRef]

59. Porto-Figueira, P.; Pereira, J.A.M.; Câmara, J.S. Exploring the potential of needle trap microextraction combined with chromatographic and statistical data to discriminate different types of cancer based on urinary volatomic biosignature. *Anal. Chim. Acta* **2018**, *1023*, 53–63. [CrossRef]

60. Thomson, C.A.; Thompson, P.A. Dietary patterns, risk and prognosis of breast cancer. *Futur. Oncol.* **2009**, *5*, 1257–1269. [CrossRef]

61. Martineau, E.; Tea, I.; Akoka, S.; Giraudeau, P. Absolute quantification of metabolites in breast cancer cell extracts by quantitative 2D 1H INADEQUATE NMR. *NMR Biomed.* **2012**, *25*, 985–992. [CrossRef]

62. Kim, K.-J.; Kim, H.-J.; Park, H.-G.; Hwang, C.-H.; Sung, C.; Jang, K.-S.; Park, S.-H.; Kim, B.-G.; Lee, Y.-K.; Yang, Y.-H.; et al. A MALDI-MS-based quantitative analytical method for endogenous estrone in human breast cancer cells. *Sci. Rep.* **2016**, *6*, 24489. [CrossRef] [PubMed]

63. Singh, A.; Sharma, R.K.; Chagtoo, M.; Agarwal, G.; George, N.; Sinha, N.; Godbole, M.M. 1H NMR Metabolomics Reveals Association of High Expression of Inositol 1, 4, 5 Trisphosphate Receptor and Metabolites in Breast Cancer Patients. *PLoS ONE* **2017**, *12*, e0169330. [CrossRef] [PubMed]

64. Zhao, C.; Xie, P.; Wang, H.; Cai, Z. Liquid chromatography-mass spectrometry-based metabolomics and lipidomics reveal toxicological mechanisms of bisphenol F in breast cancer xenografts. *J. Hazard. Mater.* **2018**. [CrossRef]

65. Huang, S.; Chong, N.; Lewis, N.E.; Jia, W.; Xie, G.; Garmire, L.X. Novel personalized pathway-based metabolomics models reveal key metabolic pathways for breast cancer diagnosis. *Genome Med.* **2016**, *8*, 34. [CrossRef] [PubMed]

66. Dougan, M.M.; Li, Y.; Chu, L.W.; Haile, R.W.; Whittemore, A.S.; Han, S.S.; Moore, S.C.; Sampson, J.N.; Andrulis, I.L.; John, E.M.; et al. Metabolomic profiles in breast cancer: A pilot case-control study in the breast cancer family registry. *BMC Cancer* **2018**, *18*, 532. [CrossRef]

67. Gu, H.; Pan, Z.; Xi, B.; Asiago, V.; Musselman, B.; Raftery, D. Principal component directed partial least squares analysis for combining nuclear magnetic resonance and mass spectrometry data in metabolomics: Application to the detection of breast cancer. *Anal. Chim. Acta* **2011**, *686*, 57–63. [CrossRef] [PubMed]

68. Jobard, E.; Pontoizeau, C.; Blaise, B.J.; Bachelot, T.; Elena-Herrmann, B.; Trédan, O. A serum nuclear magnetic resonance-based metabolomic signature of advanced metastatic human breast cancer. *Cancer Lett.* **2014**, *343*, 33–41. [CrossRef] [PubMed]

69. Lv, W.; Yang, T. Identification of possible biomarkers for breast cancer from free fatty acid profiles determined by GC–MS and multivariate statistical analysis. *Clin. Biochem.* **2012**, *45*, 127–133. [CrossRef]

70. Lyon, D.E.; Starkweather, A.; Yao, Y.; Garrett, T.; Kelly, D.L.; Menzies, V.; Dereziński Pawełand Datta, S.; Kumar, S.; Jackson-Cook, C. Pilot Study of Metabolomics and Psychoneurological Symptoms in Women With Early Stage Breast Cancer. *Biol. Res. Nurs.* **2018**, *20*, 227–236. [CrossRef]

71. Wei, S.; Liu, L.; Zhang, J.; Bowers, J.; Gowda, G.A.N.; Seeger, H.; Fehm, T.; Neubauer, H.J.; Vogel, U.; Clare, S.E.; et al. Metabolomics approach for predicting response to neoadjuvant chemotherapy for breast cancer. *Mol. Oncol.* **2013**, *7*, 297–307. [CrossRef]

72. Playdon, M.C.; Ziegler, R.G.; Sampson, J.N.; Stolzenberg-Solomon, R.; Thompson, H.J.; Irwin, M.L.; Mayne, S.T.; Hoover, R.N.; Moore, S.C. Nutritional metabolomics and breast cancer risk in a prospective study. *Am. J. Clin. Nutr.* **2017**, *106*, 637–649. [CrossRef] [PubMed]

73. Cala, M.; Aldana, J.; Sánchez, J.; Guio, J.; Meesters, R.J.W. Urinary metabolite and lipid alterations in Colombian Hispanic women with breast cancer: A pilot study. *J. Pharm. Biomed. Anal.* **2018**, *152*, 234–241. [CrossRef] [PubMed]

74. Tayyari, F.; Gowda, G.A.N.; Olopade, O.F.; Berg, R.; Yang, H.H.; Lee, M.P.; Ngwa, W.F.; Mittal, S.K.; Raftery, D.; Mohammed, S.I. Metabolic profiles of triple-negative and luminal A breast cancer subtypes in African-American identify key metabolic differences. *Oncotarget* **2018**, *9*, 11677–11690. [CrossRef]

75. Bathen, T.F.; Geurts, B.; Sitter, B.; Fjøsne, H.E.; Lundgren, S.; Buydens, L.M.; Gribbestad, I.S.; Postma, G.; Giskeødegård, G.F. Feasibility of MR Metabolomics for Immediate Analysis of Resection Margins during Breast Cancer Surgery. *PLoS ONE* **2013**, *8*, e61578. [CrossRef] [PubMed]

76. Tang, X.; Lin, C.-C.; Spasojevic, I.; Iversen, E.S.; Chi, J.-T.; Marks, J.R. A joint analysis of metabolomics and genetics of breast cancer. *Breast Cancer Res.* **2014**, *16*, 415. [CrossRef]

77. Budczies, J.; Pfitzner, B.M.; Györffy, B.; Winzer, K.-J.; Radke, C.; Dietel, M.; Fiehn, O.; Denkert, C. Glutamate enrichment as new diagnostic opportunity in breast cancer. *Int. J. Cancer* **2015**, *136*, 1619–1628. [CrossRef]

78. Vettukattil, R.; Hetland, T.E.; Flørenes, V.A.; Kærn, J.; Davidson, B.; Bathen, T.F. Proton magnetic resonance metabolomic characterization of ovarian serous carcinoma effusions: chemotherapy-related effects and comparison with malignant mesothelioma and breast carcinoma. *Hum. Pathol.* **2013**, *44*, 1859–1866. [CrossRef]

79. Euceda, L.R.; Haukaas, T.H.; Giskeødegård, G.F.; Vettukattil, R.; Engel, J.; Silwal-Pandit, L.; Lundgren, S.; Borgen, E.; Garred, Ø.; Postma, G.; et al. Evaluation of metabolomic changes during neoadjuvant chemotherapy combined with bevacizumab in breast cancer using MR spectroscopy. *Metabolomics* **2017**, *13*, 37. [CrossRef]

80. Gogiashvili, M.; Horsch, S.; Marchan, R.; Gianmoena, K.; Cadenas, C.; Tanner, B.; Naumann, S.; Ersova, D.; Lippek, F.; Rahnenführer, J.; et al. Impact of intratumoral heterogeneity of breast cancer tissue on quantitative metabolomics using high-resolution magic angle spinning 1H NMR spectroscopy. *NMR Biomed.* **2018**, *31*, e3862. [CrossRef]

81. Choi, J.S.; Baek, H.-M.; Kim, S.; Kim, M.J.; Youk, J.H.; Moon, H.J.; Kim, E.-K.; Nam, Y.K. Magnetic resonance metabolic profiling of breast cancer tissue obtained with core needle biopsy for predicting pathologic response to neoadjuvant chemotherapy. *PLoS ONE* **2013**, *8*, e83866. [CrossRef]

82. Budczies, J.; Brockmöller, S.F.; Müller, B.M.; Barupal, D.K.; Richter-Ehrenstein, C.; Kleine-Tebbe, A.; Griffin, J.L.; Orešič, M.; Dietel, M.; Denkert, C.; et al. Comparative metabolomics of estrogen receptor positive and estrogen receptor negative breast cancer: alterations in glutamine and beta-alanine metabolism. *J. Proteomics* **2013**, *94*, 279–288. [CrossRef]

83. Dai, C.; Arceo, J.; Arnold, J.; Sreekumar, A.; Dovichi, N.J.; Li, J.; Littlepage, L.E. Metabolomics of oncogene-specific metabolic reprogramming during breast cancer. *Cancer Metab.* **2018**, *6*, 5. [CrossRef]

84. Yu, L.; Jiang, C.; Huang, S.; Gong, X.; Wang, S.; Shen, P. Analysis of urinary metabolites for breast cancer patients receiving chemotherapy by CE-MS coupled with on-line concentration. *Clin. Biochem.* **2013**, *46*, 1065–1073. [CrossRef] [PubMed]

85. Krilaviciute, A.; Heiss, J.A.; Leja, M.; Kupcinskas, J.; Haick, H.; Brenner, H. Detection of cancer through exhaled breath: A systematic review. *Oncotarget* **2015**, *6*, 38643–38657. [CrossRef] [PubMed]

86. Zhang, A.; Sun, H.; Wang, P.; Han, Y.; Wang, X. Modern analytical techniques in metabolomics analysis. *Analyst* **2012**, *137*, 293–300. [CrossRef]

87. Issaq, H.J.; Van, Q.N.; Waybright, T.J.; Muschik, G.M.; Veenstra, T.D. Analytical and statistical approaches to metabolomics research. *J. Sep. Sci.* **2009**, *32*, 2183–2199. [CrossRef] [PubMed]

88. Dunn, W.B.; Bailey, N.J.C.; Johnson, H.E. Measuring the metabolome: Current analytical technologies. *Analyst* **2005**, *130*, 606. [CrossRef] [PubMed]

89. Wang, J.H.; Byun, J.; Pennathur, S. Analytical approaches to metabolomics and applications to systems biology. *Semin. Nephrol.* **2010**, *30*, 500–511. [CrossRef]

90. Sas, K.M.; Karnovsky, A.; Michailidis, G.; Pennathur, S. Metabolomics and diabetes: Analytical and computational approaches. *Diabetes* **2015**, *64*, 718–732. [CrossRef] [PubMed]

91. Ahad, T.; Jasia Nissar, I.; Tehmeena Ahad, C.; Nissar, J. Division of food science and technology, Skuast-k Fingerprinting in determining the adultration of food. *J. Pharmacogn. Phytochem. JPP* **2017**, *6*, 1543–1553.

92. Narwate, B.M.; Ghule, P.J.; Ghule, A.V.; Darandale, A.S.; Wagh, J.G. Ultra performance liquid chromatography: A new revolution in liquid chromatography. *Int. J. Pharm. Drug Anal.* **2014**, *2*.

93. Yandamuri, N.; Srinivas Nagabattula, K.R.; Swamy Kurra, S.; Batthula, S.; S Nainesha Allada, L.P.; Bandam, P. Comparative Study of New Trends in HPLC: A Review. *Int. J. Pharm. Sci. Rev. Res.* **2013**, *23*, 52–57.

94. De Vos, J.; Broeckhoven, K.; Eeltink, S. Advances in Ultrahigh-Pressure Liquid Chromatography Technology and System Design. *Anal. Chem.* **2016**, *88*, 262–278. [CrossRef]

95. Cacciola, F.; Farnetti, S.; Dugo, P.; Marriott, P.J.; Mondello, L. Comprehensive two-dimensional liquid chromatography for polyphenol analysis in foodstuffs. *J. Sep. Sci.* **2017**, *40*, 7–24. [CrossRef]

96. Reichenbach, S.E.; Tian, X.; Tao, Q.; Ledford, E.B.; Wu, Z.; Fiehn, O. Informatics for cross-sample analysis with comprehensive two-dimensional gas chromatography and high-resolution mass spectrometry (GCxGC–HRMS). *Talanta* **2011**, *83*, 1279–1288. [CrossRef] [PubMed]

97. Yu, Z.; Huang, H.; Reim, A.; Charles, P.D.; Northage, A.; Jackson, D.; Parry, I.; Kessler, B.M. Optimizing 2D gas chromatography mass spectrometry for robust tissue, serum and urine metabolite profiling. *Talanta* **2017**, *165*, 685–691. [CrossRef] [PubMed]

98. Umar, A.; Luider, T.M.; Foekens, J.A.; Paša-Tolić, L. NanoLC-FT-ICR MS improves proteome coverage attainable for ~3000 laser-microdissected breast carcinoma cells. *Proteomics* **2007**, *7*, 323–329. [CrossRef]

99. Hendriks, M.M.W.B.; van Eeuwijk, F.A.; Jellema, R.H.; Westerhuis, J.A.; Reijmers, T.H.; Hoefsloot, H.C.J.; Smilde, A.K. Data-processing strategies for metabolomics studies. *TrAC Trends Anal. Chem.* **2011**, *30*, 1685–1698. [CrossRef]

100. Yi, L.; Dong, N.; Yun, Y.; Deng, B.; Ren, D.; Liu, S.; Liang, Y. Chemometric methods in data processing of mass spectrometry-based metabolomics: A review. *Anal. Chim. Acta* **2016**, *914*, 17–34. [CrossRef]

101. Gromski, P.S.; Muhamadali, H.; Ellis, D.I.; Xu, Y.; Correa, E.; Turner, M.L.; Goodacre, R. A tutorial review: Metabolomics and partial least squares-discriminant analysis—A marriage of convenience or a shotgun wedding. *Anal. Chim. Acta* **2015**, *879*, 10–23. [CrossRef]

102. Sugimoto, M.; Wong, D.T.; Hirayama, A.; Soga, T.; Tomita, M. Capillary electrophoresis mass spectrometry-based saliva metabolomics identified oral, breast and pancreatic cancer-specific profiles. *Metabolomics* **2010**, *6*, 78–95. [CrossRef] [PubMed]

103. van den Berg, R.A.; Hoefsloot, H.C.; Westerhuis, J.A.; Smilde, A.K.; van der Werf, M.J. Centering, scaling, and transformations: Improving the biological information content of metabolomics data. *BMC Genom.* **2006**, *7*, 142. [CrossRef]

104. Sysi-Aho, M.; Katajamaa, M.; Yetukuri, L.; Orešič, M. Normalization method for metabolomics data using optimal selection of multiple internal standards. *BMC Bioinform.* **2007**, *8*, 93. [CrossRef] [PubMed]

105. Kohl, S.M.; Klein, M.S.; Hochrein, J.; Oefner, P.J.; Spang, R.; Gronwald, W. State-of-the art data normalization methods improve NMR-based metabolomic analysis. *Metabolomics* **2012**, *8*, 146–160. [CrossRef] [PubMed]

106. Xi, B.; Gu, H.; Baniasadi, H.; Raftery, D. Statistical analysis and modeling of mass spectrometry-based metabolomics data. *Methods Mol. Biol.* **2014**, *1198*, 333–353.

107. Zhang, A.; Sun, H.; Qiu, S.; Wang, X. Metabolomics in noninvasive breast cancer. *Clin. Chim. Acta* **2013**, *424*, 3–7. [CrossRef]

108. Xia, J.; Broadhurst, D.I.; Wilson, M.; Wishart, D.S. Translational biomarker discovery in clinical metabolomics: An introductory tutorial. *Metabolomics* **2013**, *9*, 280–299. [CrossRef]

109. Liland, K.H. Multivariate methods in metabolomics—From pre-processing to dimension reduction and statistical analysis. *TrAC Trends Anal. Chem.* **2011**, *30*, 827–841. [CrossRef]

110. Köhn, H.-F.; Hubert, L.J. Hierarchical Cluster Analysis. In *Wiley StatsRef: Statistics Reference Online*; John Wiley & Sons, Ltd.: Chichester, UK, 2015; pp. 1–13.

111. Jain, A.K. Data clustering: 50 years beyond K-means. *Pattern Recognit. Lett.* **2010**, *31*, 651–666. [CrossRef]

112. Sperandei, S. Understanding logistic regression analysis. *Biochem. Med.* **2014**, *24*, 12–18. [CrossRef] [PubMed]

113. Ivanescu, A.E.; Li, P.; George, B.; Brown, A.W.; Keith, S.W.; Raju, D.; Allison, D.B. The importance of prediction model validation and assessment in obesity and nutrition research. *Int. J. Obes.* **2016**, *40*, 887–894. [CrossRef] [PubMed]

114. Xu, Q.-S.; Liang, Y.-Z.; Du, Y.-P. Monte Carlo cross-validation for selecting a model and estimating the prediction error in multivariate calibration. *J. Chemom.* **2004**, *18*, 112–120. [CrossRef]

115. Haddad, K.; Rahman, A.; A Zaman, M.; Shrestha, S. Applicability of Monte Carlo cross validation technique for model development and validation using generalised least squares regression. *J. Hydrol.* **2013**, *482*, 119–128. [CrossRef]

116. Jaki, T.; Su, T.-L.; Kim, M.; Van Horn, M.L. An evaluation of the bootstrap for model validation in mixture models. *Commun. Stat. -Simul. Comput.* **2018**, *47*, 1028–1038. [CrossRef] [PubMed]

117. Lindgren, F.; Hansen, B.; Karcher, W.; Sjöström, M.; Eriksson, L. Model Validation by Permutation tests: Applications to VariableSselection. *J. Chemom.* **1996**, *10*, 521–532. [CrossRef]

 *metabolites* 

*Review*

# HR-MAS NMR Based Quantitative Metabolomics in Breast Cancer

**Mikheil Gogiashvili [1,*], Jessica Nowacki [1], Roland Hergenröder [1], Jan G. Hengstler [2], Jörg Lambert [1,†] and Karolina Edlund [2,†]**

[1] Leibniz Institut für Analytische Wissenschaften, ISAS, e.V., 44139 Dortmund, Germany; jessica.nowacki@tu-dortmund.de (J.N.); roland.hergenroeder@isas.de (R.H.); joerg.lambert@isas.de (J.L.)
[2] Leibniz Research Centre for Working Environment and Human Factors (IfADo), 44139 Dortmund, Germany; Hengstler@ifado.de (J.G.H.); Edlund@ifado.de (K.E.)
* Correspondence: mikheil.gogiashvili@isas.de; Tel.: +492311392231; Fax.: +4923192120
† These authors contributed equally to this work.

Received: 16 December 2018; Accepted: 18 January 2019; Published: 22 January 2019

**Abstract:** High resolution magic-angle spinning (HR-MAS) nuclear magnetic resonance (NMR) spectroscopy is increasingly used for profiling of breast cancer tissue, delivering quantitative information for approximately 40 metabolites. One unique advantage of the method is that it can be used to analyse intact tissue, thereby requiring only minimal sample preparation. Importantly, since the method is non-destructive, it allows further investigations of the same specimen using for instance transcriptomics. Here, we discuss technical aspects critical for a successful analysis—including sample handling, measurement conditions, pulse sequences for one- and two dimensional analysis, and quantification methods—and summarize available studies, with a focus on significant associations of metabolite levels with clinically relevant parameters.

**Keywords:** NMR; HR MAS; breast cancer; metabolomics

---

## 1. Introduction

The improved understanding of breast cancer has been supported by the development of omics-based technologies. Transcriptomics has made key contributions, for instance by delineating clinically relevant subtypes based on gene expression patterns [1–3]. Moreover, gene expression-based assays are now used for the assessment of recurrence risk in the clinical setting [4–6]. As a relative newcomer to the omics field, metabolomics offers the potential to further reveal alterations that underlie breast cancer development and progression, as well as the discovery of novel therapeutic targets and biomarkers for improved diagnostics and prediction of prognosis as well as response to therapy.

Nuclear magnetic resonance (NMR) spectroscopy and mass spectrometry (MS) are the two most widely used methods for quantitative metabolomic analysis of tumour tissue. Their respective advantages and disadvantages have been extensively reviewed elsewhere, for example, by Wishart et al. [7] and Nagana Gowda & Raftery [8]. Importantly, the higher sensitivity of MS allows the quantification of a much larger number of metabolites compared to NMR. However, despite this disadvantage of NMR spectroscopy, certain characteristics make this method indispensable in metabolomics. Properties of NMR spectroscopy, such as the capacity to provide information about the number of chemically identical atoms, the chemical shift of individual atomic groups and the spin-spin coupling with the resulting signal splitting pattern contribute to the high selectivity of this method [9]. Consequently, one-dimensional NMR spectroscopy can often be sufficient, without a need of using two dimensional (2D) NMR, to perform the reliable identification and also reproducible quantification of small molecules. Moreover, NMR does not require chemical separation of analytes prior to analysis, which is one of the reasons for its excellent technical reproducibility [8]. A particular advantage of NMR spectroscopy is the possibility to directly analyse intact tissue in a non-destructive manner, while MS requires an extraction step, thus destroying tissue integrity. Most studies use the $^1$H nucleus for sensitivity reasons. For the same reason, techniques employing the $^{13}$C nucleus have not found wide applications in tissue analysis and metabolomics studies that use $^{13}$C usually require labelled samples. $^{31}$P HR-MAS NMR, however, is a valuable technique for the analysis of tissue specimens, as the sensitivity is just one order of magnitude lower than that of $^1$H NMR and potential applications to both phospholipid and energy metabolism studies are evident.

Magnetic resonance spectroscopy (MRS) provides spatially resolved information about the chemical composition of tissue in vivo [10]. If combined with ex vivo NMR studies, in vivo MRS can be used to non-invasively detect biomarkers that were identified in previous ex vivo NMR studies. A disadvantage of in vivo MRS, however, is its poor spectral resolution. The ex vivo analysis of intact tissue specimens also suffers from poor resolution compared to conventional liquid NMR techniques. In liquid samples, the dipole-dipole couplings, that is, the through-space interactions between protons, as well as the anisotropy, that is, the orientational dependence of the chemical shift, are completely averaged out due to the high molecular mobility. Under these conditions, only the isotropic chemical shift and the coupling through bonds remain, giving rise to well-resolved signals with small (1–2 Hz) linewidths. For $^1$H-NMR investigations of tissue, the anisotropy of the chemical shift can be neglected but the dipole-dipole coupling, albeit small (20–50 Hz) is non-negligible [11]. Since the limited mobility of molecules in tissue impedes a full averaging of the dipolar couplings, the resulting spectra are poorly resolved. This makes quantification difficult, as the superposition of numerous broad signals from a mixture of metabolites can no longer be easily disentangled. Moreover, the semisolid character of tissue gives rise to local magnetic field gradients, caused by spatial variations of the magnetic bulk susceptibility, which also leads to spectral broadening [12,13]. Both the anisotropic part of the latter interaction [11,13] and the dipole-dipole coupling follow a [3cos$^2$($\Theta$)-1] dependence on the angle $\Theta$ between the B$_0$ field and the distance vector between interacting nuclei [11]. If the sample is rotated at the so-called "magic angle" of $\Theta$ = 54.7°, the [3cos$^2$($\Theta$) - 1] term becomes zero and a high-resolution NMR spectrum is obtained [14,15] (Figure 1).

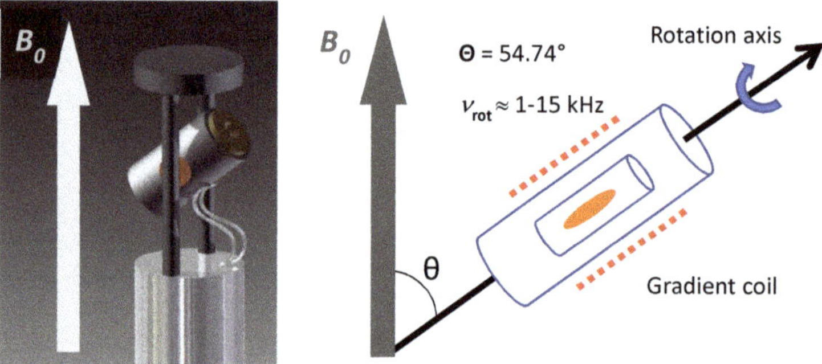


**Figure 1.** Representation of an HR-MAS NMR probe head with an orientation of the stator at an angle ($\theta$) of 54.7° between axis of rotation and B$_0$. The rotation speed ($\nu_{rot}$) of the sample reaches up to 15 kHz. A gradient coil is arranged around the rotor.

One of the earliest applications of high-resolution magic angle spinning (HR-MAS) NMR spectroscopy was the study of polymer beads, which are synthetic organic materials used in the preparation of combinatorial chemistry libraries [16]. Towards the end of the 1990s, the examination of intact tissue became possible [17–20], enabled by the development of high-field magnets, improved probe heads with ultra-low-inductance decoupling coils and symmetric spinner drive designs that led to a reduction of spinning-, decoupling- and variable temperature-induced thermal gradients [21,22].

The alignment of the z-axis gradient along the magic angle gave access to a wide variety of experimental NMR techniques, such as gradient enhanced solvent suppression and to the full range of 1D and 2D homo- and hetero-nuclear experiments. In this way, HR-MAS NMR spectra, with a typical linewidth of 1–2 Hz, became comparable in quality to spectra from liquid samples [23,24]. As already mentioned, HR-MAS NMR allows metabolite quantification directly from intact tissue specimens, which abolishes the need for an extraction step and thereby avoids one potential source of poor reproducibility. In addition, the risk of partial extraction of certain metabolites, as for instance reported for choline [19], is minimized. Also, the non-destructive nature of the method allows the analysis of metabolite levels in tissue to be combined with other analytical techniques based on the same tissue specimen, which may be important when tissue availability is limited.

HR-MAS NMR has been used to study metabolites in breast cancer tissue and to correlate metabolite concentrations with clinically relevant parameters. Here, we summarize the HR-MAS NMR-based studies of breast cancer tissue available until December 2018, focusing on analytical aspects, including measurement conditions, pulse sequences used for 1D and 2D NMR, quantification methods and the numbers and identities of the reported metabolites. Moreover, we summarize findings based on quantitative HR-MAS NMR with regard to significant associations with clinically relevant factors in breast cancer.

## 2. Preanalytical Factors and Measurement Conditions

Reliable and reproducible analysis of tissue samples using HR-MAS NMR requires a robust and standardized protocol that considers factors both before and during the measurement. In order to standardize the analytical conditions, various time windows should be considered, including sampling, storage, sample preparation and measurement (Figure 2). A brief overview of potential issues related with the analysis of tissue specimens using HR-MAS NMR was recently published by Esteve and colleagues [25]. However, as of yet, only few studies directly examined the impact of time and temperature during the different steps of the analytical pipeline in the analysis of breast tumour tissue [26,27].

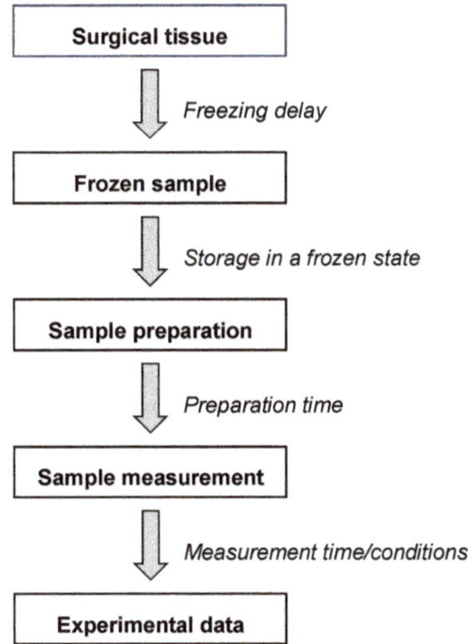


**Figure 2.** Different steps and time windows between the tissue sampling and the final HR-MAS NMR data.

Any omics analysis represents a snapshot of the tissue at one point in time. The time delay after the tissue has been removed from the circulation during surgery until freezing should therefore be kept as short as possible or ongoing biochemical processes in the tissue may alter the metabolite content. Haukaas et al. studied the influence of the length of the time interval before the tissue was snap-frozen in liquid nitrogen (0, 15, 30, 60, 90 and 120 min) on the metabolic profile of breast cancer xenografts [27]. A significant difference between metabolite levels in samples frozen directly after surgical removal and frozen after 60 min was shown for ascorbate ($-25\%$), choline ($+56\%$) and creatine ($-28\%$) and after 90 min for glutathione ($-35\%$). The freezing delay did not have a statistically significant influence on the eleven additionally studied metabolites but ranged from $-15\%$ to $+11\%$ and $-20\%$ to $+31\%$ after 30 and 60 min, respectively. Notably, for all time points, increased metabolite levels were observed for certain metabolites, while the levels of other metabolites decreased. Moreover, the pattern across the increasing time delays varied for the individual metabolites, including consistently increasing or decreasing metabolite levels, as well as more difficult to interpret mixed patterns. Nevertheless, no significant alterations in metabolite levels are expected at freezing delay times less than 30 min. The impact of the duration of the ischemic period before freezing was also studied in other tissue types. For instance, in rat brain, statistically significant changes were observed for glucose (down) as well as alanine, $\gamma$-aminobutyric acid (GABA) and lactate (up) after 30 min and additionally for glutamine, myo-inositol, GPC and total choline (down) and acetate, creatine and glycine (up) after three hours [28]. We are unaware of any study that directly compares the effect of ischemia/freezing delay time between different tissue types, but the occurrence of tissue-specific differences can probably not be excluded.

In the above-mentioned study by Haukaas et al., the authors also compared metabolite levels between samples that were immediately snap-frozen and thereafter analysed ($n = 6$) and samples that were analysed directly, without being frozen ($n = 6$) [27]. Importantly, significantly higher mean metabolite levels were observed for 12 of 16 metabolites in the snap-frozen tissues (20–60% increase after freezing compared to fresh tissue). Noticeable differences could also be observed between the metabolites with regard to the level of variability between the replicates (20–100%). This study is in agreement with the results of an early study of rat kidney from 1998, where considerably increased signal intensities were observed after freezing for several metabolites, including alanine (> 100%), glutamine (> 40%) and glycine (> 100%) [29]. Middleton et al. assigned these changes to the release of metabolites that were bound to macromolecules and therefore are invisible for HR-MAS NMR. This can happen due to freezing-induced cellular disruption and/or the precipitation of non-freezing resistant proteins, in both cases leading to fewer available non-specific binding sites for small molecules. Increased levels of several metabolites in rat kidney after snap-freezing were also observed by Waters et al., including leucine, isoleucine, valine, alanine and glycine, when compared to fresh tissue that was kept on ice for up to five hours before analysis [30]. In addition, decreased signals of choline, glycerophosphocholine, glucose, myo-inositol, trimethylamine N-oxide (TMAO) and taurine were found after snap-freezing. However, the statistical significance and magnitude of these changes were not reported. Interestingly, much fewer changes were observed in liver compared to kidney, indicating tissue-specific differences [30]. All in all, despite the observed freezing-induced changes, freezing the tissue is likely to remain a standard approach for practical reasons, such as the distance between the surgical unit and the laboratory, as well as programs for prospective tissue collection and biobanking for later analysis.

After snap-freezing, the storage of biological material at $-80\,^{\circ}\text{C}$ until analysis is standard [31]. Only one published study thus far investigated the impact of storage time at $-80\,^{\circ}\text{C}$ on the metabolic profile of human breast cancer tissue [26]. In this study, samples were snap-frozen after being kept for approximately 30 min on ice and analysed using HR-MAS NMR after 1, 6 and 12 months. It was reported that the levels of choline in healthy breast tissue increased ($p < 0.000001$) with longer storage time, while phosphocholine decreased ($p < 0.000001$), which could be due to the breakdown of phosphocholine to choline. Lower phosphocholine levels were also observed in breast tumour tissue

($p < 0.0002$), together with increased levels of lactate ($p < 0.05$). The concentrations of nine other metabolites showed no significant changes during the one year storage period. Further studies would be required to assess the impact of storage at $-80\,°C$ for even longer time periods, which usually is the case when studying, for instance, the association of metabolite concentrations with cancer survival in retrospective frozen tissue collections. Findings by Jordan et al. of no significant storage time-associated effect on metabolite levels, evaluating human prostate cancer tissue after three years of storage at $-80\,°C$ [32], support the conclusion that the influence of the storage time in a low-temperature freezer is most likely minor.

Before HR-MAS NMR, preparation of the sample for analysis includes punching or cutting the tissue to fit into an insert, placing it in the rotor, weighing, and adding the internal standard. These preparatory steps are commonly performed at room temperature, with the specimen kept on ice to avoid extensive thawing. The usage of a cooled workstation has also been reported [33,34]. Another option is to prepare the sample at $-10\,°C$ in a closed glovebox under nitrogen atmosphere [35,36]. This would, in addition, mitigate the possible influence of condensation of ambient water from the air, which may distort the sample weight and in turn affect the quantification. However, the impact of factors during the sample preparation step on final metabolite concentrations has not been systematically studied until now.

Finally, different conditions during the measurement, with regard to temperature, analysis time and rotation frequency, were used in the thus far published HR-MAS NMR studies of breast cancer tissue. To avoid line broadening and to achieve a high resolution MAS NMR spectrum, the sample must be thawed and the measurement performed at a temperature above $0\,°C$. To minimize the risk of temperature-induced changes during the measurement, the temperature after methanol calibration is usually adjusted to approximately $5\,°C$. In brain tissue, it was shown that the rate of degradation of N-acetyl aspartate (NAA) to acetate was four times higher at $20\,°C$ than at $2\,°C$ [19]. Most measurements of breast cancer tissue reported in the literature were performed at $4\,°C$ [37–41], while others were performed at $5\,°C$ [27,42–45], $6\,°C$ [41,46], $19\,°C$ [47,48], $20\,°C$ [49] or $26\,°C$ [50], with an approximate mean measurement time of 19 min (range: 3 min 7 sec [27] to 1 h 5 min [51]). Haukaas et al. reported that a prolonged HR-MAS NMR measurement time of 1.5 h at $5\,°C$ influenced the concentration of certain metabolites, including a significant increase of glucose, glycine and choline and a decrease of glycerophosphocholine (GPC) [27]. Similar observations were made in brain tissue, with decreasing levels of N-acetyl aspartate (NAA) and increasing levels of acetate in spectra collected at $20\,°C$ during the course of 24 h [19]. Similarly as for the freezing delay time, different tolerance to thawing, as well as multiple freeze-thaw cycles, can probably be expected in various tissue types, so it is difficult to assess to what extent findings in one tissue type can be extrapolated to other. In studies such as those above, it is also difficult to separate the effect of the extended time at a temperature above $0\,°C$ from that of prolonged rotation at a high frequency. Indeed, the rotation of tissue at a high frequency during HR-MAS NMR has been reported to impact the morphology of the specimen [38,41,52] as well as the metabolite content [28]. A high speed of rotation can destroy cell and tissue structures. For instance, analysing cells, two hours of rotation at 2.5 kHz destroyed approximately 20% of adipocytes [53] and cell lysis was observed in erythrocytes at 4 kHz MAS rotation [54]. In human prostate tissue, distortion of ductal structures occurred after one hour spinning at 3 kHz; whereas, no obvious morphological alterations were observed after 45 min spinning at 600 Hz followed by 15 min at 700 Hz [52], indicating that preservation of tissue integrity could be achieved by slower rotation. However, the centrifugal forces are only reduced by two orders of magnitude when the spin rate is reduced by a factor of ten [55]. The results of one study of rat brain tissue suggest that mechanical stress due to prolonged spinning at $4\,°C$ may have a larger impact on the metabolic profile than the delay in the freezing of the sampled tissue sample in liquid nitrogen, for instance leading to increased creatine levels, possibly because of the tissue-damage associated release from initially undetectable creatine stores [28]. In the published studies of breast cancer tissue using HR-MAS NMR, a rotation frequency of 5 kHz was applied, except in some publications spinning the sample at 2 kHz [47,48,50,51,56], 2.5 kHz [49] and 6 kHz [24]. As of yet, no study directly compared the impact of rotation frequency on tissue morphology and metabolite concentrations in breast cancer tissue. Renault et al. compared HR-MAS measurements of liver tissue at different rotations frequencies (150 Hz, 500 Hz, 4000 Hz) and found that the presence and intensity of sidebands is strongly dependent on the sample preparation (position and shape of the sample, presence of air bubbles) [57]. The authors claim that sideband free spectra can be obtained at rotation frequencies as low as 500 Hz by minimizing the volume of the sample chamber and by positioning the sample chamber at the coil centre with an insert located at the top of the rotor.

In summary, further studies are warranted to better understand the impact of conditions during sample procurement, storage and analysis of breast cancer tissue on metabolite levels analysed by HR-MAS NMR. Only one study of breast cancer tissue comprehensively examined the impact of the freezing delay time after surgery on metabolite concentrations. Given the practical difficulties of tissue collection immediately after surgery, further validation that no significant changes in metabolite concentrations occur during the first 30–60 min would be important to provide additional confidence in current protocols for tissue procurement. The impact of the rotation frequency and measurement time on metabolite concentrations have not yet been systematically examined in breast cancer tissue and should be addressed in future studies.

## 3. NMR Techniques Employed in Tissue Analysis

Several common NMR techniques can be employed in the analysis of intact tissue specimens to provide specific information about the metabolite and lipid content. A successful analysis also depends on techniques to suppress unwanted signals that originate from, for instance, tissue water and the macromolecular background of lipids and proteins that would otherwise impede the analysis. Below, the presaturation approach to suppress solvent signals, as well as common pulse sequences used for one and two dimensional NMR experiments, are described.

### 3.1. Water Suppression

Water is a dominating component of biological samples, as a solute or as a solvent. Because of the high concentration of water protons, the $^1$H-NMR signal of water exceeds the metabolite signals by several orders of magnitude. If not suppressed, the water signal will saturate the analogue to digital converter (ADC), that is, it uses the full dynamic range of the ADC, leaving just a few bits for the metabolite signals, which impedes a correct quantification of the latter. Moreover, the strong water signal distorts the baseline of the spectrum. The $^1$H-NMR spectrum of a sample with high water content is displayed in Figure 3A.

Presaturation is the most simple and most widely used technique to suppress solvent signals [58] and can be easily combined with most pulse sequences used in NMR. It uses an extended period of a weak continuous wave irradiation at the frequency of the water signal at 4.7 ppm (Figure 3B). This irradiation results in an equal population of the two energy levels of the water hydrogen spins. Hence, if spin-lattice relaxation is neglected, there is no longitudinal magnetization of the water protons and the subsequent excitation pulse exclusively excites the signals of the metabolites. The power level employed for the presaturation must be chosen to only saturate the water signal and no metabolite signal in its vicinity. The determination of the presaturation power level is performed once for every sample in a series of experiments, where the power level is incremented in small steps and the outcome of the presaturation is evaluated by the operator. This evaluation has to take into account the intensities of the water signal and of the metabolite signals in its vicinity, with and without presaturation, as well as recommendations of the manufacturer concerning the maximum power level. Another important factor for the successful presaturation is the homogeneity of the $B_0$ field, because signal broadening caused by inhomogeneity cannot always be suppressed by presaturation [58]. Figure 3C shows the spectrum of the same sample as in Figure 3A but here the water signal is suppressed using the described presaturation technique. The intensity of the residual water signal now is at the same level as the metabolite signal intensities and does not impede the quantification of the metabolite signals.

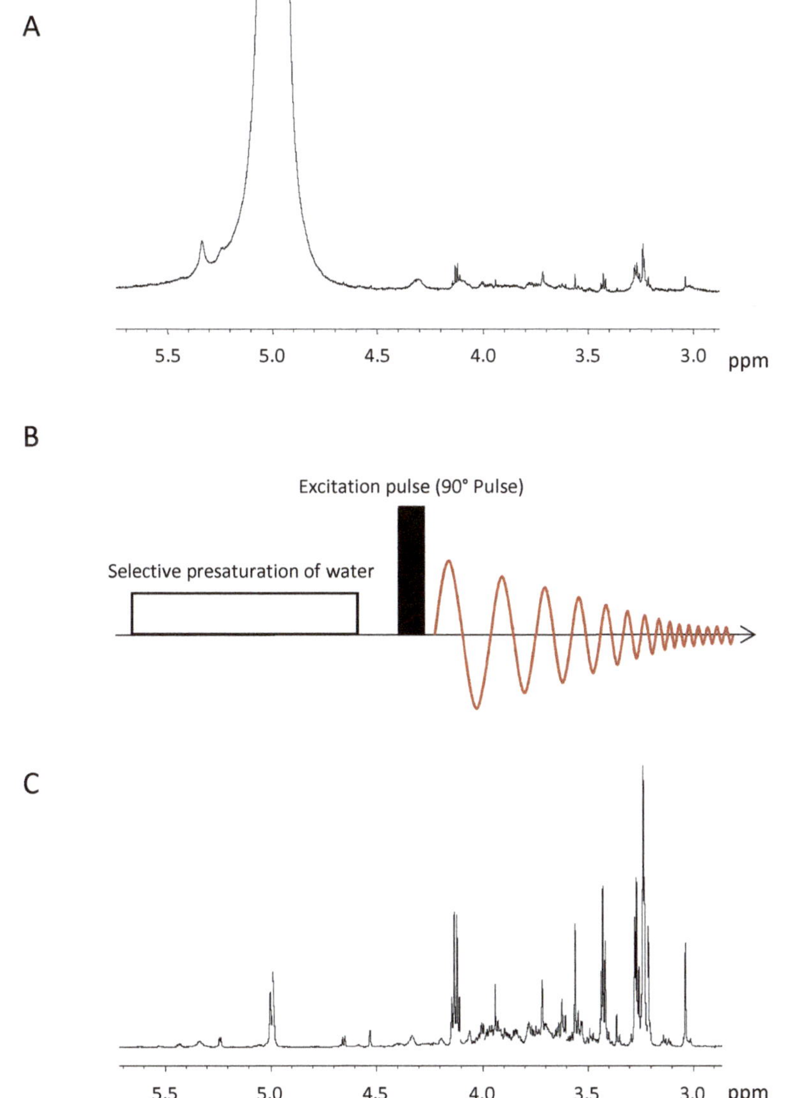



**Figure 3. (A)** $^1$H NMR spectrum of a tumour sample without water suppression. The strong water signal (0 ppm) saturates the analogue to digital converter and leaves just a few bits for the metabolite signals, which impedes the quantification of the latter. **(B)** Schematic representation of the presaturation pulse sequence [58], which consists of a selective presaturation at the water frequency, the excitation of the sample and data acquisition. **(C)** $^1$H NMR spectrum of a breast cancer specimen with an almost completely suppressed water signal at 5 ppm. As a result, the metabolite signals of the sample can be quantified.

### 3.2. Pulse Sequences for 1D-NMR

The 1D Nuclear Overhauser-Effect Spectrometry (1D NOESY) technique is employed when the goal is to observe both the signals of low-molecular weight compounds (metabolites) and macromolecules (lipids, proteins) in a $^1$H-NMR spectrum [23]. This experiment is helpful in breast cancer studies to estimate the amount of lipids in relation to the metabolite content. The 1D NOESY pulse sequence [RD-90°-t1-90°-tm-90°-ACQ] (Figure 4A) [23] starts with a presaturation irradiation of the water signal during the relaxation delay time (English "relaxation delay"—RD). The first 90° pulse hence produces transverse magnetization only for the metabolites, since the signal of water has previously been saturated. This is followed by a short interval $t_1$ of approximately 3 µs, which serves as a switching time for the phases of the pulses. The subsequent second 90° pulse rotates the magnetization of the metabolites to the z-direction. During the time $t_m$ (duration approximately 10 ms) after the second 90° pulse, presaturation is switched on again to once more saturate the water magnetization that has relaxed during the course of the experiment. The metabolite magnetization is not affected by this saturation because it is longitudinal at this time. Finally, the third 90° pulse rotates the magnetization of the metabolites back to the transverse plane. During the acquisition time (here denoted ACQ) the transverse magnetization is sampled. The magnetization of the water is saturated at this time and therefore provides no signal after the third 90° pulse.

If the aim is to observe the metabolite signals only, while the macromolecular background of lipids and proteins is suppressed, the Carr-Purcell Meiboom-Gill (CPMG) sequence is employed. This sequence, like the 1D NOESY sequence, has an integrated one-dimensional water presaturation interval. It uses the pulse train [RD-90 °-(τ1-180°-τ1)n-ACQ] (Figure 4B) [23], where (τ1-180°-τ1)n acts as a $T_2$ filter to suppress signals from macromolecules and other substances with short $T_2$ times. As a result, the $^1$H-NMR spectrum of tissue samples obtained with this pulse sequence only consists of small molecule signals from metabolites, which have relatively long $T_2$ times. After presaturation, the excitation starts with a 90° pulse, which generates transverse magnetization. The subsequent precession of the magnetization during the delay time D2 (typically 1 ms for CPMG) is refocused by a 180° pulse followed by another delay D2. This spin-echo sandwich is repeated n times, followed by the acquisition after the n-th echo [23]. If the lipid content of the tissue is high, the echo time n*τ can be prolonged to suppress the lipid signals by spin-spin relaxation. Depending on the type of tissue, the total echo time can be between 30 ms [59] and 720 ms [60]. For some tissues, such as brain tumours, which have lower lipid content, the echo times are short and only vary between 30 and 150 ms [59,61]. Tissue with higher lipid content requires longer echo times to suppress the lipids. In breast cancer samples echo times as long as 580 ms have been used [24,60]. The CPMG pulse sequence is an indispensable tool in NMR studies of breast cancer tissue, as lipid signal suppression is essential.

Finally, it should be noted that the above-mentioned pulse techniques were originally designed for liquid state NMR measurements. When applied to HR-MAS NMR, the timing of the pulse sequence must be synchronized with the rotation period, so that the interpulse spacings are equal to multiples of the rotor cycle time [52].

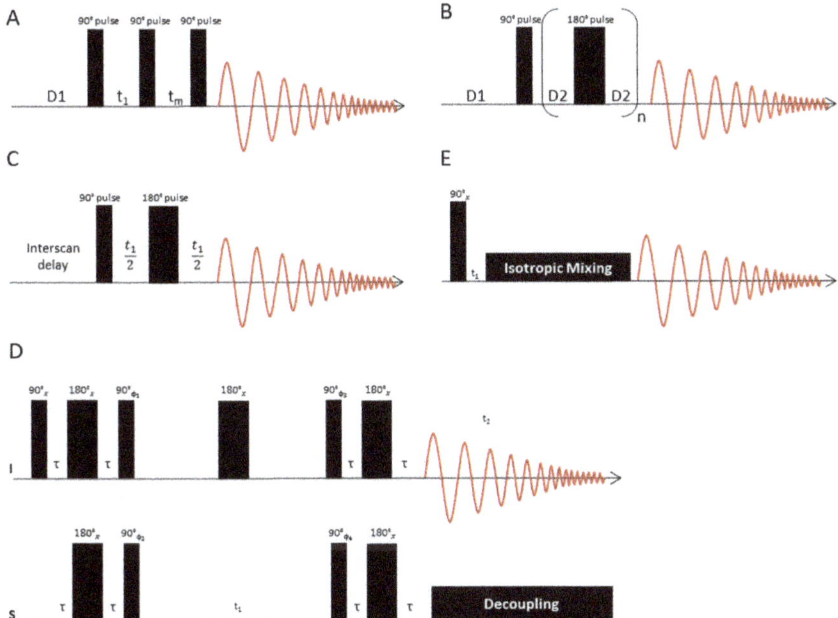

**Figure 4. (A)** The 1D NOESY pulse sequence consists of a presaturation interval (D1), an excitation pulse, which after a short time $t_1$ (phase switching time) is followed by another 90° pulse. After another time $t_m$, the spin system is excited with another 90° pulse. During the time $t_m$, the water saturation is switched on again to ensure complete water suppression. D1 denotes the relaxation delay time. **(B)** Representation of the CPMG pulse sequence: After a presaturation interval (D1), the spin system is excited with a 90° pulse. This is followed by a train of n 180° pulses, each pulse bracketed by two delay times D2, in which the spins refocus. Subsequently, the signal is recorded as an FID. The total delay time n*τ after n spin echo periods τ = (D2-180°-D2) is chosen so as to suppress the signals of fast relaxing molecules like lipids. **(C)** JRES pulse sequence to separate $^1$H chemical shifts and $^1$H,$^1$H couplings into separate dimensions of a 2D display. The chemical shifts are displayed in the horizontal dimension, while the multiplet patterns show up in the vertical dimension. If strong coupling artefacts can be neglected, the horizontal dimension corresponds to a "broadband" decoupled proton spectrum. **(D)** HSQC pulse sequence to correlate proton and carbon chemical shift information. The sequence starts with an INEPT block, which transfers proton magnetization to $^{13}$C. The carbon magnetization is then labelled with $^{13}$C chemical shift information via a spin echo sequence, the duration of which is incremented in subsequent experiments. A reverse INEPT transfer brings back $^{13}$C magnetization to the proton channel, where it is recorded under $^{13}$C broadband decoupling. **(E)** TOCSY pulse sequence to correlate $^1$H chemical shifts that are part of a spin-spin coupling network. The sequence starts with proton magnetization, which is labelled with its precession frequency during $t_1$. In a subsequent isotropic mixing step, a net magnetization transfer to coupled protons is performed. The extent of this magnetization transfer can be steered by adjustment of the length of the mixing delay. Values of 80 ms typically give spectra where magnetization has been transferred to all coupling partners of the excited spin, so that the whole spin system can be traced out. The signals of the coupling partners appear on horizontal lines in the spectrum, which eases analysis of metabolite spin systems.

## 3.3. Pulse Sequences for 2D-NMR

In the presence of strong peak overlaps, which are typical for complex mixtures such as those encountered in metabolomics, special measurement techniques are required to untangle the overlapping peaks and to assist in peak assignment. Peaks that overlap in the 1D NMR spectra can often be resolved in two dimensional (2D) NMR spectra.

Two dimensional J-resolved (2D JRES) NMR spectroscopy (Figure 4C) [62] is helpful for the analysis of metabolite mixtures, as it allows the recording of a second spectral dimension with relatively little overlap of signals. In the 2D JRES NMR experiment, the $^1$H spectrum is presented in the horizontal dimension and the coupling pattern of each signal is displayed in the vertical dimension. The number of identifiable metabolites is, however, strongly limited by the 2D JRES spectral resolution as well as by strong coupling effects, which hamper the applicability of this method especially at low field strengths.

The use of 2D $^1$H-$^{13}$C heteronuclear NMR (Figure 4D) in metabolomics may be advisable to identify metabolites if the $^1$H-NMR spectrum is heavily congested. Among the 2D NMR methods, the 2D Heteronuclear Single Quantum Coherence (HSQC) technique [63] offers a very high resolution by incorporation of $^{13}$C chemical shift information. The HSQC pulse sequence correlates proton and carbon chemical shifts. The sequence starts with an Insensitive Nuclei Enhanced by Polarization Transfer (INEPT) block [64], which performs a polarization transfer of proton magnetization to the $^{13}$C channel. In a subsequent spin echo sequence, the carbon magnetization is labelled with $^{13}$C chemical shift information. The duration of the spin echo sequence is incremented in subsequent experiments. Finally, a reverse INEPT transfer brings back $^{13}$C magnetization to the proton channel. During the acquisition, $^{13}$C broadband decoupling is switched on to collapse the $^{13}$C, $^1$H-couplings. In many cases, however, HSQC is generally not sensitive enough for metabolomics studies but the acquisition of an HSQC spectrum is of particular importance if new metabolites have to be identified or if additional evidence is sought for signal assignments obtained from $^1$H-NMR measurements, as exemplified for breast cancer tissue in Reference [35].

A disadvantage of both HSQC- and of 2D-JRES spectra is the missing spin system information, as the cross-peaks are all independent of each other. This shortcoming is avoided by the 2D $^1$H-$^1$H Total Coherence Spectroscopy (TOCSY) experiment [65], which permits the identification of individual $^1$H spin systems that can be assigned to the various mixture components. The TOCSY pulse sequence (Figure 4E) correlates distinct $^1$H-NMR signals, which are part of a network of spin-spin couplings (usually denoted as a "spin system"). An excitation pulse creates proton magnetization that is labelled with its precession frequency during the delay time $t_1$, which is incremented in subsequent experiments. During an isotropic mixing step, a net magnetization transfer to coupled protons occurs, the extent of which is controlled by adjustment of the length of the mixing delay. Values of 80 ms typically give spectra where magnetization has been transferred to all coupling partners of the excited spin, so that the whole spin system can be traced out. The signals of the coupling partners appear on horizontal lines in the spectrum. TOCSY is particularly well suited for computational analysis, since each cross-section of signals represents the 1D spectrum of the whole spin system, that is, the signals of a metabolite. Notably, for HR-MAS NMR, adiabatic mixing sequences [66] are recommended to perform the isotropic mixing. The composite pulse mixing sequences commonly employed in liquid state NMR introduce modulations of the effective field in the presence of both radiofrequency field and $B_0$ inhomogeneities, when applied to rotating samples [66]. These modulations compromise the performance of composite pulse mixing sequences and introduce a sensitivity of the signal intensities to the sample spinning speed. Adiabatic mixing sequences are less susceptible to such modulations and perform better [66].

As a final point, 2D NMR techniques that use pulse sequences adopted from liquid state NMR require a rotor synchronization [67], that is, all delays used in the pulse sequence must be integer multiples of the rotor period. Rotor synchronization helps to eliminate both residual anisotropic interactions and the effect of radial inhomogeneities of the radiofrequency field. Rotor synchronization is of particular importance for TOCSY experiments, where the lengths of every basic cycle of the isotropic mixing sequence as well as the trim pulses have to be integer multiples of the rotor period [67].

## 4. Metabolites Identified with HR-MAS NMR in Breast Tumour Tissue

Two complementary strategies are used in metabolomics: the targeted and the non-targeted approach. With a targeted approach, a preselected subset of metabolites is measured, usually based on an a priori hypothesis. Contrarily, with a non-targeted approach, the number of metabolites to be measured is not predetermined; rather the aim is to capture as much information as possible. To date, 27 publications using HR-MAS NMR spectroscopy report a total of 46 metabolites to be detectable in breast cancer tissue (tumour tissue from patients and/or xenografts) (Table 1). Cheng et al. published the first report using HR-MAS NMR to analyse breast cancer tissue already in 1998 [49], followed by the milestone work of Sitter et al., who identified more than 30 metabolites, in 2002 [24]; however, no quantification was performed in these early studies. Table 1 gives an overview of metabolites detected in breast cancer tissue analysed by HR-MAS [1]H-NMR spectroscopy and in addition indicates if metabolite concentrations were determined, for example, in $\mu$mol/g tissue, using an internal- (TSP) or external standard (ERETIC, PULCON) (twelve publications) or if relative quantification based on integrated peak areas was performed (eight publications).

The most comprehensive quantitative investigations of breast cancer tissue using HR-MAS NMR were those by Park et al. and Yoon et al., where 34 metabolites were reported in each study [50,68]. The exact inclusion criterion for HR-MAS NMR-based identification of metabolites in breast cancer tissue is, however, rarely stated in the literature. In the work of Sitter et al., reporting concentrations for nine metabolites, a signal-to-noise ratio (SNR) >10 for the creatine singlet was applied [41]. In a recent publication, we quantified all metabolites that had a baseline-separated signal with a SNR >3 for at least one peak used for quantification, thus reporting concentrations for 32 metabolites [35].

As indicated in Table 1, the quantification of approximately 40 metabolites is what can be maximally achieved by this method in breast cancer tissue with a non-targeted approach. A substantial further increase of the number of metabolites to be quantified would require higher $B_0$ field strengths at the expense, however, of proportionally increasing rotation speeds to shift the spinning sidebands out of the spectral window. The concomitant increase in centrifugal forces acting on the fragile samples would compromise the non-destructive nature of NMR. Currently, HR-MAS spectra are acquired at a maximum [1]H frequency of 600 MHz (14.1 T), which gives sideband-free spectra at a rotation speed of 5 kHz. There are, however, approaches to use slow-spinning HR-MAS techniques [55] that employ a sideband suppression like PASS [69], PHORMAT [70] or PROJECT [71], which suggest that high quality HR-MAS spectra could be obtained at higher field strengths than 14.1 T. However, the application of PASS and PHORMAT to tissue analysis is hampered, not only by sensitivity and resolution problems but especially by the fact that the extent of the side band pattern affects the intensity of the isotropic peak. This feature impedes quantification, as the extent of the sideband pattern may vary from sample to sample [55]. A suitable approach to slow MAS in metabolomics seems to be the PROJECT pulse sequence but so far spectra devoid of spinning sidebands were only observed at 400 Hz rotation speed in case of favourable conditions, such as a spherical sample (for instance fish eggs), minimal $B_1$ inhomogeneities, small-volume rotors and a sample composition close to that of an isotropic liquid [55].

Finally, the quantification of metabolites from 1D [1]H-NMR can be difficult due to the complexity of the spectrum. For instance, in 1D HR-MAS [1]H-NMR spectra, choline- and ethanolamine-containing metabolite signals are superimposed and the weak signals from phosphoethanolamine and glycerophosphoethanolamine are difficult to detect. HR-MAS [31]P-NMR spectra have higher chemical shift dispersion than their [1]H-NMR counterparts, which helps to separate the signals of choline- and ethanolamine-containing metabolites. The quantification of the signals in the HR-MAS [31]P-NMR spectra is fairly straightforward, as the signal of phosphocholine shows up in both types of spectra ([1]H and [31]P) and can be used for cross-calibration. Glycerophosphoethanolamine and glycerol-3-phosphate were only reported in breast cancer tissue in the one study that used HR-MAS [31]P-NMR [72].

**Table 1.** Overview of metabolites reported in HR-MAS NMR breast cancer tissue studies. Q: absolute quantification with concentrations given in e.g. µmol/g tissue; R: relative quantification; I: identified but not quantified; [#]patient tumour tissue; [$]xenograft tissue; (A) ERETIC; (B) TSP; (C) PULCON.

| | Gogiashvili et al. 2018[#][A] | Tayyari et al. 2018[#] | Euceda et al. 2017a[#] | Euceda et al. 2017b[$] | Park et al. 2016[#][B] | Yoon 2016[#][B] | Haukaas 2016a[$] | Haukaas 2016b[#] | Chae 2016[#][B] | Cao et al. 2014[#] | Grinde et al. 2014[$][C] | Choi et al. 2013[#][B] | Borgan et al. 2013[$][A] | Bathen et al. 2013[#] | Giskeødegård et al. 2012[#] | Cao et al. 2012b[#] | Cao et al. 2012a[#][A] | Choi et al. 2012[#][B] | Li et al. 2011[#] | Sitter et al. 2010[#][A] | Moestue et al. 2010[#][$][A] | Borgan et al. 2010[#] | Giskeødegård et al. 2010[#] | Bathen et al. 2007[#] | Sitter et al. 2006[#][B] | Sitter et al. 2002[#] | Cheng et al. 1998[#] |
|---|---|---|---|---|---|---|---|---|---|---|---|---|---|---|---|---|---|---|---|---|---|---|---|---|---|---|---|
| 3-Hydroxybutyrate | Q | | | | | | | | | | | | | | | | | | | | | | | | | | |
| Acetate | Q | | | | Q | Q | | R | | | | I | | R | | | | | | | | | | | | I | |
| Adipate | | | | | Q | | | | | | | | | | | | | | | | | | | | | | |
| Alanine | Q | R | R | R | Q | Q | R | R | Q | R | | Q | | R | I | I | | Q | I | | | I | I | | | I | I |
| Arginine | | R | | | Q | Q | | | I | | | | | | | | | | | | | | | | | | |
| Ascorbate | Q | | R | R | | | R | R | | | | | | R | I | | | | | | | | | | I | | |
| Asparagine | | | | | Q | Q | | | I | | | | | | | | | | | | | | | | I | | |
| Aspartate | Q | | | | Q | Q | | | | | | | | I | | | | | I | | | | | | | I | I |
| ATP | | R | | | | | | | | | | | | | | | | | | | | | | | | | |
| Betaine | Q | | | | Q | Q | | | | | | | | | | | | | | | | | | | I | | |
| Choline | Q | R | R | R | Q | Q | R | R | Q | R | Q | Q | Q | R | I | R | Q | Q | I | Q | Q | I | I | I | Q | I | I |
| Creatine | Q | R | R | R | Q | Q | R | R | Q | R | I | Q | Q | R | I | I | I | Q | I | Q | Q | I | I | I | Q | I | I |
| Ethanol | | | | | Q | | | | | | | | | | | | | | | | | | | | | | |
| Ethanolamine | | | | | Q | Q | | | | | | | | | | | | | | | | | | | | | |
| Formate | | | | | | | | | | | | | I | | | | | | | | | | | | | | |
| Fumarate | | | | | Q | Q | | | | | | | | | | | | | | | | | | | | | |
| Glucose | Q | R | R | R | Q | Q | R | R | | I | | I | | R | I | R | I | | | Q | | I | | | I | Q | I |
| Glutamate | Q | R | R | R | Q | Q | R | R | I | R | I | I | | R | I | | I | I | | | | | | | | I | I |
| Glutamine | Q | R | R | R | Q | Q | R | R | I | R | I | | | R | I | | | | | | | | | | | I | I |
| Glutathione | Q | R | R | R | | | R | R | | | | | | | | | | | | | | | | | | | |
| Glycerol | | | | | Q | Q | | | I | | | | | | | | | | | | | | | | I | | |
| Glycine | Q | R | R | R | Q | Q | R | R | Q | R | I | Q | Q | R | R | R | Q | Q | I | Q | Q | I | I | I | Q | I | I |
| GPC | Q | | R | R | Q | Q | R | R | Q | I | Q | Q | Q | R | I | R | Q | Q | | Q | Q | I | I | I | Q | I | |
| Histidine | | | | | Q | Q | | | | | | | | | | | | | | | | | | | | I | |
| Inosine | Q | | | | | | | | | | | | | | | | | | | | | | | | | I | |
| Isoleucine | Q | | | | Q | Q | | | I | | | I | | | | | | | I | | | | | | I | | |
| Lactate | Q | R | R | R | Q | Q | R | R | I | R | I | I | Q | R | R | R | I | I | I | Q | | I | I | I | I | I | I |

**Table 1.** *Cont.*

| | Gogiashvili et al. 2018 #(A) | Tayyari et al. 2018 # | Euceda et al. 2017a # | Euceda et al. 2017b $ | Park et al. 2016 #(B) | Yoon 2016 #(B) | Haukaas 2016a $ | Haukaas 2016b # | Chae 2016 #(B) | Cao et al. 2014 # | Grinde et al. 2014 $(C) | Choi et al. 2013 #(B) | Borgan et al. 2013 $(A) | Bathen et al. 2013 # | Giskeødegård et al. 2012 # | Cao et al. 2012b # | Cao et al. 2012a #(A) | Choi et al. 2012 #(B) | Li et al. 2011 # | Sitter et al. 2010 #(A) | Moestue et al. 2010 #$(A) | Borgan et al. 2010 # | Giskeødegård et al. 2010 # | Bathen et al. 2007 # | Sitter et al. 2006 #(B) | Sitter et al. 2002 # | Cheng et al. 1998 # |
|---|---|---|---|---|---|---|---|---|---|---|---|---|---|---|---|---|---|---|---|---|---|---|---|---|---|---|---|
| Leucine | Q | R | | | Q | Q | | | I | | | I | | | | | | | | | | I | | | | I | |
| Lysine | Q | R | | | Q | Q | | | I | | | I | | | | | | | | | | | | | | I | I |
| Methionine | Q | R | | | Q | Q | | | | | | | | | | | | | | | | | | | | | |
| *myo*-Inositol | Q | R | R | R | Q | Q | R | R | Q | I | I | Q | | R | | | | Q | I | Q | | I | | I | Q | I | |
| O-Phosphocholine | Q | R | R | R | Q | Q | R | R | Q | R | Q | Q | Q | R | I | R | Q | Q | I | Q | Q | I | I | I | Q | I | I |
| O-Phosphoethanolamine | Q | | | | Q | Q | | | I | | I | | | | | | | | | | | | | | | I | I |
| Phenylalanine | Q | R | | | Q | Q | | | I | | | | | | | | | | | | | | | I | | | |
| Proline | Q | | | | Q | Q | | | | | | | | | | | | | | | | | | | | | |
| *scyllo*-Inositol | Q | | | | | | | R | | I | | | | R | | | Q | | | | | I | | | I | I | |
| Serine | Q | | | | Q | Q | | | I | | | | | | | | | | | | | | | | | | |
| Succinate | Q | | R | R | | | R | R | Q | R | | I | | | | | Q | | | | | | | | | I | |
| Taurine | Q | R | R | R | Q | Q | R | R | Q | I | I | Q | Q | R | I | R | Q | Q | I | Q | Q | I | I | I | Q | I | I |
| Threonine | Q | R | | | Q | Q | | | | | | | | | | | | | | | | | | | | | |
| Tyrosine | Q | R | R | R | Q | Q | R | R | I | | | | | | | | | | | | | | | I | | | |
| Uracil | | | | | Q | Q | | | | | | | | | | | | | | | | | | I | | | |
| Uridine | | R | | | | | | | | | | | | | | | | | | | | | | | | | |
| Valine | Q | R | | | Q | Q | | | I | | | I | | | | | | | I | | | | | | | I | I |

### 5. Metabolite Quantification with HR-MAS NMR

There are several options for quantitative analysis with HR-MAS NMR spectroscopy, which have certain advantages and disadvantages. As already mentioned above, the method for absolute quantification used in twelve publications to determine metabolite concentrations in breast cancer tissue is indicated in Table 1.

One approach is the quantification with an internal standard, which is at the same time used for calibrating the spectrum [39]. Tetramethylsilane (TMS) is widely used as internal standard in organic chemistry. However, it is insoluble in water and therefore its water-soluble forms, trimethylsilylpropionate (TSP) [73] and sodium 2,2-dimethyl-2-silapentane-5-sulfonate (DSS) are used in HR-MAS NMR [74]. As the weaker acid compared to DSS, TSP is more affected by sample pH. Both TSP and DSS are reported to bind to hydrophobic parts of proteins [75], which partly distorts the quantification. In prostate tissue, it was observed that more than 70% of the added TSP became "NMR-invisible," because the TSP was bound to macromolecules in the tissue [76]. If the TSP signal is subsequently used for quantification, the observed error in the TSP peak area gives rise to a significant over-estimation of the metabolite concentrations, as was also mentioned elsewhere [73]. In their tutorial about NMR metabolomics analysis [77], the authors state that large deviations of the TSP peak area and shape in a series of samples are often due to an insufficient suppression of macromolecules like proteins, as they are partners for non-specific bonding to TSP and DSS. Nowick et al. suggested DSA (4,4-dimethyl-4-silapentane-1-ammonium trifluoroacetate) as a new internal standard that does not suffer from interactions with cationic peptides like DSS [78]. Alum et al. compared DSA to TSP and found that the integral of the DSA signal correlated linearly with its concentration under all pH, whereas no such linear correlation could be found with TSP. The authors suggest DSA to be used both as a universal chemical shift reference and a concentration standard [79]. In Figure 5, examples of free and breast cancer tissue-bound TSP are shown. It is also possible to use the water content of the tissue as an internal standard [19,60], as demonstrated on brain tissue samples [80]. However, it is questionable whether quantification based on water content is well suited in heterogeneous tissue types, such as breast cancer, where fat and therefore also water content varies greatly.

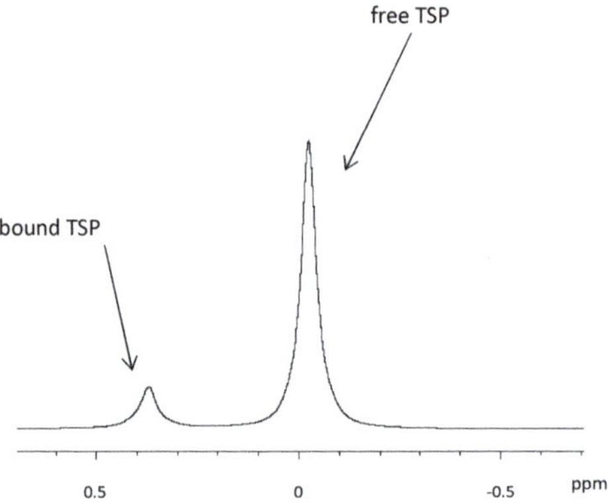


**Figure 5.** Influence of TSP interaction with tissue on the quantification with [1]H HR-MAS NMR spectroscopy.

External standards are also used for metabolite quantification with HR-MAS NMR. A widely used method to quantify metabolites in tissue is ERETIC (Electronic REference To access In vivo Concentrations), an artificially generated radio frequency signal, which is pre-calibrated to a reference sample, such as sucrose, in an independent measurement [81,82]. The advantage of ERETIC is that no standard has to be added to the sample and thus no distortion due to the interaction between the standard and the sample matrix can occur. This was shown for instance by Martinez-Bisbal et al., who compared metabolite concentrations obtained with ERETIC and the internal standard DSS in tissue biopsies from glioblastoma multiforme, observing consistently higher concentrations with DSS [74]. Sitter et al. found a larger relative standard deviation (RSD) with ERETIC (> 6.7%) than with TSP (> 4.4%) for the quantification of serial dilutions of creatine in phosphate buffered saline (1, 5 and 10 mM) [39], which they attributed to radiofrequency inhomogeneity influencing ERETIC more than TSP. Nevertheless, the authors still recommended ERETIC as the better alternative, since this method avoids matrix effects, is stable, accurate and precise and has a reproducible signal area, as reported by Albers et al. [76]. To assess the robustness of ERETIC- and TSP based quantification, Albers et al. studied the stability of their peak areas in solution over time, reporting a long-term RSD of 4.10% for ERETIC and 2.60% for TSP. An external standard other than ERETIC has been used by Taylor et al. [52]. Here, the authors applied a silicone rubber sample to function as an external standard for both frequency reference (0.06 ppm from TMS) and quantification. The rubber sample (approximately 100 µg) was permanently mounted inside a Kel-F spacer in the MAS rotor in such a way that it is not in contact with the sample but is located inside the detection coil. Pulse length-based concentration determination (PULCON) can also be used to measure the concentrations in 1D NMR spectra [83], especially when several (unknown) signals are superimposed. This methodology was originally developed for the analysis of proteins. The sequence consists of first determining the 360° pulse after tuning and matching. After the integration of known signals, the protein concentration can be determined [83]. This is particularly helpful if one or more superimposed signals are unknown. PULCON is rarely used in HR-MAS NMR-based metabolomics and its use in breast cancer metabolomics is reported only once [42]. In view of the high technical reproducibility and the fact that ERETIC is free from matrix effects [76], ERETIC is today generally considered the preferred method for metabolite quantification in tissue by means of HR-MAS NMR spectroscopy. Nevertheless, since the performance of different methods for quantitative determination of metabolites in tissue by means of HR-MAS NMR has not yet been directly compared, a comprehensive study of the above-mentioned methods would be of great advantage.

One potential problem with metabolite quantification in breast cancer tissue with HR-MAS NMR is related to the presence of strong lipid signals that cannot be sufficiently suppressed by the CPMG pulse sequence. This has been reported to particularly affect the quantification of lactate, since the lactate methyl resonance at 1.32 ppm may be masked by a broad lipid signal in lipid-rich environments [38,84]. Even though the quantification of lactate was possible in several other studies [39,50,68,85], the detection of lactate in tissue may require a selective excitation technique, such as the Sel-MQC sequence of He et al., which is a spectral editing sequence that uses multiple-quantum filtration [86]. However, experimental problems, such as $B_1$ inhomogeneity, are a challenge for the reliable quantification of low lactate concentrations [87] and, in addition, severe signal losses can lead to inefficient suppression of unwanted lipid signals. Therefore, an improved spectral editing scheme that is robust to inhomogeneous fields was recently reported and shown to achieve selective excitation of lactate with minimal signal loss [87,88]. In addition, a variant of this pulse sequence [88] provides in-phase magnetization, which can be more accurately quantified than the antiphase magnetization of the pulse sequence [86] and allows the selection of experimental parameters that meet optimal lipid suppression requirements [88]. Lactate editing was originally developed for MRS [88] but it has been demonstrated that it can also be used in HR-MAS $^1$H-NMR to quantify lactate in intact lipid-rich tissue [36]. This approach can, in theory, be extended to other metabolites, that can be edited employing multiple-quantum filtration techniques. This holds especially for alanine and threonine [87]. In summary, sophisticated spectral editing techniques that are based on Optimal Control (OC) theory [89] have become available, that allow for the design of tailored pulses that excite only the signal of a given metabolite. However, these have so far merely been applied to the quantification of lactate and alanine [87].

The problem with superimposed signals is not restricted to that of strong lipid signals. The "add to subtract" approach, described by Ye et al. [90], is one approach to handle superimposed signals. Here, the signals of metabolites which are usually strong like lactate or glucose, are subtracted from the spectrum prior to analysis and small, low-concentration molecules, which may be of biological significance, appear and can be quantified [8]. Another approach to deconvolute superimposed signals, which has so far only been used in in vivo NMR, is the Linear Combination of Model (LCModel) method [91,92]. This method analyses an NMR spectrum as a linear combination of model spectra obtained from individual metabolite solutions, using a constrained regularization method which takes the baseline and the line shape of the spectra into account without employing a restrictive parameterization on the data [91,92]. Using LCModel analysis, a set of in vitro metabolite spectra combined with simulated lipid and protein spectra assists in the analysis of HR-MAS spectra. Application of this LCModel setup to brain tumour biopsy HR-MAS data revealed interactions between metabolites and the macromolecular background via the analysis of small peak shifts [28]. Moreover Opstad et al showed, that LCModel provides a user-independent protocol of analysis of brain tumour HR-MAS spectra [28]. So far, however, no applications of LCModel analysis to breast cancer tissue HR-MAS spectra have been reported.

Finally, a factor which is independent of the quantification method but that may affect the quantitative analysis when metabolite concentrations are determined per gram tissue, as is commonly done with HR-MAS NMR, is related to the cellular composition of the tumour tissue specimen, which might vary within a tumour as well as between tumours. Tumour tissue does not only consist of tumour cells but also of, for example, cancer-associated fibroblasts, endothelial and lymphatic cells and cells of the immune system. Moreover, adipocytes may be present in breast carcinomas (Figure 6) and breast tumour tissues also differ with regard to the amount of tumour cells in relation to the amount of necrosis and extra-cellular matrix-rich stroma. Consequently, tissue samples from two different tumours or samples from two different areas within the same tumour, with the same weight may have different cellular content, which may translate into differences in metabolite content. Several studies assessed the tumour cell content in haematoxylin-eosin stained tissues sections, either prepared after HR-MAS NMR or from an adjacent tumour area, with notable differences between the tumours [84,93]. Some authors reported study inclusion criteria based on the tissue composition; for instance, >30 tumour cells [38] or >5% tumour area [85] were used as an inclusion criterion, while others excluded samples with a high lipid content [43]. The influence of the tissue composition on the observed metabolite levels should be taken into account, since correlations between metabolite concentrations and tissue composition have been reported, including higher levels of glycine, GPC and phosphocholine with higher tumour cell fraction [39], as well as poor SNR in spectra with a high level of connective tissue [93].

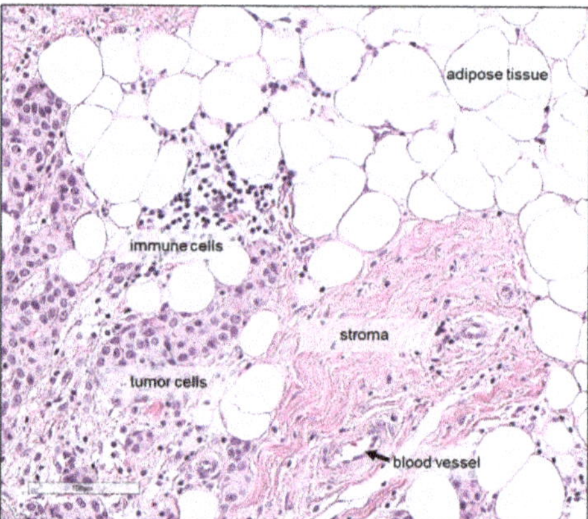


**Figure 6.** Haematoxylin-eosin staining showing the tumour tissue microenvironment of a breast carcinoma and different types of cells, including tumour cells, immune cells and adipocytes—as well as stroma and blood vessel. Scale bar: 100 μm.

Intra-tumour variability of metabolite concentrations may, in addition to differences in the tissue composition, be related to the coexistence of different subpopulations of cancer cells, which differ in their genetic and phenotypic characteristics [94,95]. Therefore, despite the high technical reproducibility of NMR [8], recognizing the potential impact of intra-tumour heterogeneity on metabolite concentrations determined by quantitative HR-MAS NMR is essential. Understanding intra-tumour heterogeneity is also a prerequisite to decide whether analysis of a small number of samples or even a single biopsy, is sufficient to obtain information representative for the entire tumour. Thus far, three studies addressed intra-tumour differences in metabolite concentrations in breast tumour tissue [35,50,96]. Based on the correlation between paired samples from the same tumour compared to random sample pairs, Cao et al. stated that the metabolic profile varied more between the different tumours than within the tumours [96]. This conclusion is supported by a second investigation, where intra-tumour concentration differences were assessed for 32 metabolites, as well as lipid content indicated by the signals at 1.3 ppm and 0.9 ppm, by sampling 8-10 tissue cores (diameter: 2 mm) from resected breast tumour tissue and duplicate cores from additionally 15 breast tumours [35]. A high degree of intra-specimen variability was observed in the tumour tissue (mean RSD: 0.48-0.74) compared to normal liver tissue (mean RSD: 0.16-0.20), which is morphologically more homogeneous. Nevertheless, it was shown that inter-tumour differences were, on average, larger than those observed within a tumour, suggesting that the analysis of one or a few, replicates per tumour might be sufficient. Park et al. used a slightly different approach and considered also the intra-tumoural localization by sampling tissue cores from both the tumour centre and periphery from surgically removed tumour tissue, concluding that the intra-tumoral localization had a limited impact on the observed concentrations of the 34 reported metabolites [50].

## 6. Significant Associations with Clinical Factors

Breast cancer is a heterogeneous disease, with diverse biological features as well as clinical behaviour. To assess the clinical course of the disease and to make decisions about treatment, factors such as tumour stage, tumour grade, oestrogen receptor (ER) and progesterone receptor (PR) status and expression of the human epidermal growth factor receptor (HER2) are considered. Chemotherapy is indicated for tumours that are negative for ER, PR and HER2 (triple-negative) and also represents the only available therapy for this subtype, in addition to surgery [97]. For the HER2 positive subtype, anti-HER2 therapy and chemotherapy are recommended, irrespective of ER status; whereas, hormonal therapy is recommended for ER positive tumours, with chemotherapy additionally administered in case of a high risk of recurrence, as indicated by tumour grade, proliferation or a prognostic gene expression assay [97]. Several studies thus far used quantitative HR-MAS NMR to correlate metabolite levels in intact breast cancer tissue with clinicopathological factors [39–41,47,96], survival [37,38] or treatment response [38,48]. The main findings of these studies are summarized below, focusing on significant differences of metabolite levels between clinically relevant patient subgroups (Figure 7).

Significantly higher choline concentrations in ER- and PR- tumours, as well as higher concentrations of creatine and taurine in PR- tumours, were reported by Choi et al. [47]. An elevated level of choline in the ER- subtype was also reported by Cao et al., together with higher levels of glycine, lactate and glutamate and lower levels of glutamine [96]. In relation to HER2 positivity, significant associations with higher levels of taurine, scyllo-inositol and myo-inositol [47], as well as higher levels of glycine, glutamine, succinate, creatine and lower levels of alanine [96], were described. Triple-negative status was found to be associated with higher choline levels in three separate publications [40,47,96] and with lower creatine levels in two publications [40,96]. Moreover, triple-negativity was associated with higher choline to creatine and total choline to creatine ratios [47], higher levels of glutamate and lower levels of glutamine [96]. Comparing basal-like and luminal-like xenografts, glycerophosphocholine and glycine were higher in the basal-like and phosphocholine was higher in the luminal-like [40].

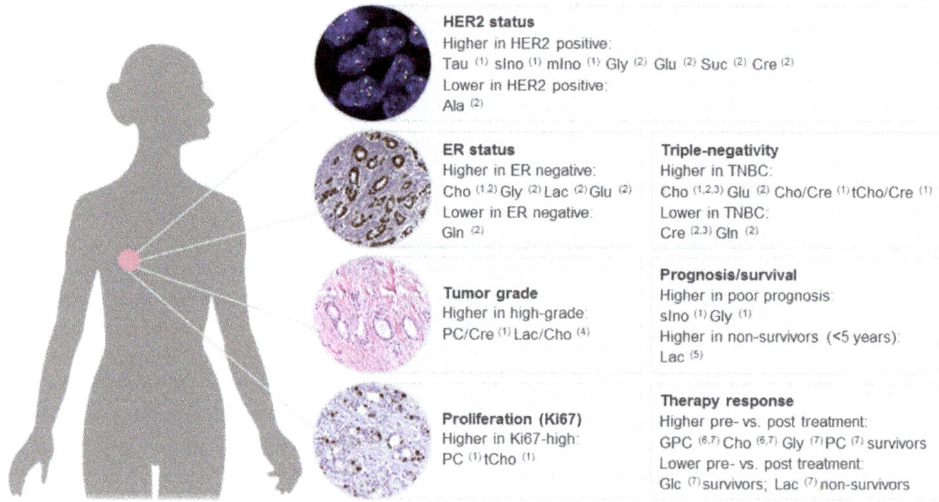


**Figure 7.** Significant associations of metabolite levels determined by HR-MAS NMR with clinicopathological factors in human breast cancer. [1] Choi et al. 2012 [47]; [2] Cao et al. 2014 [96]; [3] Moestue et al. 2010 [40]; [4] Cheng et al. 1998 [49]; [5] Giskeödegård et al. 2012 [85]; [6] Cao et al. 2012b [38]; [7] Cao et al. 2012a [37].

Few studies thus far looked at metabolite concentrations in relation to tumour grade. A significantly increased phosphocholine to creatine ratio in grade III compared to grade I-II tumours was reported by Choi et al. and higher concentrations of phosphocholine and total choline were reported by the same authors in highly proliferative tumours, as assessed by Ki-67 [47]. The association between phosphocholine and tumour grade is supported by findings from the first study of metabolites in breast cancer tissue using HR-MAS NMR, where Cheng et al. showed a higher phosphocholine to choline ratio in high-grade tumours already in 1998, although this difference was not statistically significant [49]. A significantly higher lactate to choline ratio in high-grade tumours was also reported [49].

Sitter et al. observed higher concentrations of choline and glycine in tumours larger than 2 cm [41]. Taking several clinicopathological factors into account and comparing tumours from patients with good prognosis (defined as node-negative, < 2 cm and ER+ and PR+) and poor prognosis (node-positive, > 2 cm or ER- or PR-), significantly higher concentrations of scyllo-inositol and glycine were characteristic for the tumours from patients with poor prognosis [47]. Another study identified no significant differences between prognostic groups, defined similarly, for individual metabolites; however, a trend towards higher concentrations of glycine in tumours from patients with a poor prognosis was observed [39]. Using survival time shorter or longer than five years as the endpoint, higher lactate levels were found in ER+ non-survivors, while higher levels of glycine almost reached statistical significance [85]. The association of lactate and glycine with survival is supported by a second study, although here the reported association were not statistically significant [38]. Whereas we here focused primarily on studies that used quantitative HR-MAS NMR to correlate metabolite concentrations with clinicopathological factors, the use of multivariate modelling for prediction of factors, such as tumour grade, oestrogen receptor status and lymph-node status, has also been pursued [46,98]. Future studies of larger patient cohorts, with accompanying information on metastasis-free and overall survival, are warranted to clarify if metabolic signatures can be used to predict clinically relevant prognostic endpoints, such as identifying patients with a low risk of disease recurrence.

The response to neoadjuvant therapy also provides information on prognosis, since a pathological complete response (pCR) is indicative of a lower risk of recurrent disease [99]. Comparing patients with a complete pathological response (pCR) after neoadjuvant anthracycline and/or taxane-based chemotherapy to patients who did not achieve a pCR, no significant differences in metabolite concentrations were found; only, a trend towards a lower phosphocholine to creatine ratio in patients with pCR was observed [48]. In a recent study by Euceda et al., including 122 breast cancer patients with biopsies taken before, during and after neoadjuvant 5-fluorouracil, epirubicin and cyclophosphamide, followed by taxane-based therapy – where the patients were additionally randomized to receive bevacizumab or not – principal component analysis (PCA) indicated overall changes in the metabolic profile with chemotherapy over time [43]. However, no metabolic differences were found between pre-treatment biopsies from pathological complete responders and non-responders. Moreover, no significant differences in metabolite levels were found between patients treated with chemotherapy only and patients treated with chemotherapy plus bevacizumab, either before, during or after completion of therapy [43]. In a study of triple-negative patient-derived xenografts, the response to everolimus, an inhibitor of mammalian target of rapamycin (mTOR), could not be predicted based on the metabolic profile [44]. The prediction of chemotherapy response prior to treatment thus appears to be difficult based on the metabolic evaluation of pre-treatment biopsies. The observed differences between the metabolic profile before and after treatment can be further explored to reveal mechanisms behind therapy response and resistance but may also reflect differences between tumour and normal breast tissue. Analysis of changes in metabolite levels between 33 paired pre- and post-treatment specimens showed lower glycerophosphocholine and choline levels post-treatment compared to pre-treatment in survivors ($\geq$ 5 years), while no significant differences pre- versus post-treatment were found in non-survivors (< 5 years) [38]. In another study of the same authors, where tumour tissue specimens from 89 breast cancer patients who received neoadjuvant chemotherapy were analysed, survivors showed a significant decrease in the levels of glycine, choline, phosphocholine and glycerophosphocholine and an increase in glucose, post-treatment compared to pre-treatment; whereas, non-survivors displayed an increased level of lactate after treatment [37]. A different metabolic response to PI3K/mTOR inhibition between basal-like and luminal-like patient-derived xenografts was also reported, with lower level of phosphoethanolamine and higher levels of phosphocholine and glycerophosphocholine compared to untreated controls in basal-like but not in luminal-like, xenografts [72]. This study also highlights the potential of using $^{31}$P HR-MAS NMR in biomarker studies to analyse phosphorus-containing metabolites, such as phosphoethanolamine, for which the weak signal may be difficult to detect using $^1$H HR-MAS NMR.

## 7. Summary

This review provides an overview of analytical aspects related to the use of quantitative HR-MAS NMR for metabolic profiling of intact breast tumour tissue, including an overview of the most widely used NMR techniques and briefly summarizes significant findings with regard to metabolite levels and clinically relevant factors.

A robust and standardized protocol is required for reproducible analysis of metabolite concentrations in tumour tissue samples from large patient cohorts. This means that factors such as the time and the temperature of each step — from sample collection during surgery to the actual measurement — must be taken into account. Although tissue freezing can alter the metabolic profile, there is often no better alternative since the logistics needed for the direct analysis of fresh tissue after surgery is missing, making snap-freezing of the tissue, followed by storage at $-80$ °C, indispensable. A simple rule of thumb is to minimize the time period before freezing, followed by storage in a low-temperature freezer. During the measurement, a low temperature, for example, 4 °C, for a time as short as possible, might be preferable but the effect of these factors on the observed metabolite concentrations has not been directly studied in breast cancer tissue thus far. Clearly, a frozen sample would be inappropriate for HR-MAS NMR because of the associated line broadening. A too strong reduction of the measurement time would make the identification of certain metabolites impossible.

Finally, no direct comparison has been made to study the impact of rotation frequency on metabolite concentrations in breast cancer tissue.

After the generation of the HR-MAS NMR spectrum, the correct identification and quantification of metabolites is crucial. The limited resolution and sensitivity of the method is reflected in the number of specific metabolites and in the analytical accuracy, which can be obtained. Typically, approximately 40 metabolites in tissue can be determined by HR-MAS NMR. Problems of spectral overlap can be alleviated by using higher magnetic field strengths, which improves both sensitivity and resolution. Another approach to improve accuracy is to cope with the overlap problem by decoupling $^1$H-$^1$H-scalar interactions via pure shift methods [100] which give peaks free of splittings. These techniques suffer, however, from sensitivity losses and are merely suitable for the profiling of high concentration metabolites. Another boost of accuracy can be obtained from hybrid techniques like the combination of MS/MS and NMR [100]. In the presence of strong peak overlaps and especially if previously unknown metabolites have to be identified, signal assignment may require assistance by 2D NMR techniques. Here, TOCSY is particularly valuable since it shows which peaks belong to a shared spin system and thus to a metabolite. In addition, HSQC provides information about the $^1J$ correlation of $^1$H NMR signals with $^{13}$C nuclei. By combining these two methods, it is possible to assign a particular spin system from the $^1$H NMR spectrum to a metabolite and unambiguously identify it by means of its $^{13}$C NMR shifts. Still, since tissue concentrations of many metabolites are very low, identification may be problematic. This may be compensated by increasing the measurement time in 2D-NMR studies. However, increasing the measurement time will at some point be insufficient, since the slope of the SNR can only be improved by exponentially increasing the number of scans, eventually making further improvements infeasible.

Following metabolite identification, the choice of the quantification method is also important. Nowadays, ERETIC has become established as the method of choice, since it avoids the protein binding observed with internal standards such as TSP. However, for absolute quantification further aspects must be taken into account. In most publications that used HR-MAS NMR for quantification of metabolites in breast cancer tissue, the sum of the acquisition time and the relaxation delay was greater than $5 \times T_{1max}$, where $T_{1max}$ denotes the spin-lattice relaxation time of the slowest-relaxing metabolite. Therefore, a correction of the effects of $T_1$ relaxation is not necessary for quantification. On the other hand, to achieve absolute concentrations, a $T_2$ correction has to be applied to compensate for the signal loss during the extended echo train. Most of the publications listed in Table 1 used spin-echo sequences with approximate total durations of 285 ms but did not report a $T_2$ correction, that is, $T_2$ relaxation losses were deliberately accepted. Since the determination of $T_2$ times is not trivial due to the low intensity of some metabolites and the high fat content in some breast cancer specimens, $T_2$ time determination in HR-MAS NMR based metabolomics is a matter of further investigations. However, for many purposes, such as the comparison of metabolite levels between clinically relevant subgroups, relative concentrations are sufficient.

An important question is why one should use HR-MAS NMR for metabolic profiling of tissue when MS-based methods provide information about a considerably larger number of metabolites. The availability of human tissue for scientific research is limited and its collection is associated with great organizational and time effort. Therefore, in explorative tissue profiling studies, it is natural to opt for an approach that obtains as much information as possible from a limited amount of tissue. The detection and quantification of up to 46 metabolites in breast cancer tissue using HR-MAS NMR has been shown to be feasible by a number of groups (Table 1). Moreover, the non-destructive nature of HR-MAS NMR sets it aside from MS-based methods, since it allows the analysis of metabolites to be combined with other analytical techniques on the same tissue specimen. This may be important when small amounts of tissue are available, such as when working with pre-treatment core needle biopsies or resection specimens from early stage tumours, where most of the tissue is fixed in formalin and used to establish the pathological diagnosis. Histological examination after HR-MAS NMR has for instance been pursued [38,85], providing information about the tissue composition, such as the percentage of tumour cells. It has also been shown that high-quality RNA, with RNA integrity (RIN) values in the range 7–10, can be isolated from tissue after HR-MAS NMR [36,84]. Therefore, a combinatory approach with transcriptomics is feasible. Such a combination has for instance been used to identify subtypes of luminal-A breast cancers [84]; however, these subtypes are yet to be proven to have clinical relevance. A combination of HR-MAS NMR-based metabolomics, transcriptomics and proteomics—although using tissue from different areas of the tumour sample—has also been pursued. A recent study used hierarchical clustering based on metabolite data from a large breast cancer cohort generated by HR-MAS NMR to identify three metabolic clusters, with differences in glycerophospholipid metabolism and glycolysis [45]. Interestingly, these clusters neither overlap with classification based on grading, nodal status, tumour size or hormone receptor status, nor with the gene expression-based PAM50 subtypes. One cluster contained an overrepresentation of tumours with lobular histology as well as all ductal carcinomas in situ and significant differences between clusters were also found for the reactive I and II subtypes (Cancer Genome Atlas Network 2012) defined based on reverse phase protein array analysis (RPPA), as well as for genes related with extracellular matrix, basement membrane and cell adhesion. Whether these clusters are also linked to prognosis remains to be clarified. Generally, it remains to be determined whether metabolic biomarkers quantified in breasts cancer tissue ex vivo by HR-MAS NMR are useful in a clinical setting. Today, clinicopathological parameters are used in the clinic, together with commercially available gene expression-based assays, to predict the risk of recurrence. Several studies used quantitative HR-MAS NMR to study associations between metabolite concentrations in intact human breast cancer tissue with clinicopathologic factors and clinically relevant endpoints, such as survival or therapy response but relatively few statistically significant findings were reported that were additionally validated in independent studies. Small sample sizes and large intra-group variability in some cases likely contribute to the lack of statistical significance. Consistently, higher levels of choline-containing metabolites have been reported to be associated with poor prognostic features, including tumour grade, proliferation and ER negativity [47], triple-negativity [96] and a tumour size larger than 2 cm [41], as well as higher levels of glycine with larger tumour size [41] and survival time shorter than 5 years [85], although the latter did only reach borderline significance. Higher lactate levels in non-survivors were also reported in two separate studies [38,85], although statistical significance was only reached in one study [85].

In conclusion, additional studies with larger numbers of patients are required to establish reliable associations between metabolites and prognosis of breast cancer. Moreover, studies are required to analyse whether metabolites contribute information independent from transcriptomics. Nonetheless, a unique advantage of HR-MAS NMR is that tissue can be analysed in a non-destructive manner which allows combination of this technique with either transcriptomics or with other omics techniques.

**Funding:** Financial support by the Ministerium für Innovation, Wissenschaft und Forschung des Landes Nordrhein-Westfalen, the Senatsverwaltung für Wirtschaft, Technologie und Forschung des Landes Berlin and the Bundesministerium für Bildung und Forschung is gratefully acknowledged.

**Acknowledgments:** Mikheil Gogiashvili dedicates this review to Jan T. Andersson for the invaluable support in completing his doctoral thesis. We thank Michael Weil for his assistance in making Figure 1.

**Conflicts of Interest:** The authors declare no conflict of interest.

## References

1. Perou, C.M.; Sørlie, T.; Eisen, M.B.; van de Rijn, M.; Jeffrey, S.S.; Rees, C.A.; Pollack, J.R.; Ross, D.T.; Johnsen, H.; Akslen, L.A.; et al. Molecular portraits of human breast tumours. *Nature* **2000**, *406*, 747–752. [CrossRef] [PubMed]
2. Sørlie, T.; Perou, C.M.; Tibshirani, R.; Aas, T.; Geisler, S.; Johnsen, H.; Hastie, T.; Eisen, M.B.; van de Rijn, M.; Jeffrey, S.S.; et al. Gene expression patterns of breast carcinomas distinguish tumor subclasses with clinical implications. *Proc. Natl. Acad. Sci. USA* **2001**, *98*, 10869–10874. [CrossRef]

3.  Parker, J.S.; Mullins, M.; Cheang, M.C.U.; Leung, S.; Voduc, D.; Vickery, T.; Davies, S.; Fauron, C.; He, X.; Hu, Z.; et al. Supervised risk predictor of breast cancer based on intrinsic subtypes. *J. Clin. Oncol. Off. J. Am. Soc. Clin. Oncol.* **2009**, *27*, 1160–1167. [CrossRef] [PubMed]

4.  Paik, S.; Shak, S.; Tang, G.; Kim, C.; Baker, J.; Cronin, M.; Baehner, F.L.; Walker, M.G.; Watson, D.; Park, T.; et al. A multigene assay to predict recurrence of tamoxifen-treated, node-negative breast cancer. *New Engl. J. Med.* **2004**, *351*, 2817–2826. [CrossRef] [PubMed]

5.  Filipits, M.; Rudas, M.; Jakesz, R.; Dubsky, P.; Fitzal, F.; Singer, C.F.; Dietze, O.; Greil, R.; Jelen, A.; Sevelda, P.; et al. A new molecular predictor of distant recurrence in ER-positive, HER2-negative breast cancer adds independent information to conventional clinical risk factors. *Clin. Cancer Res. An. Off. J. Am. Assoc. Cancer Res.* **2011**, *17*, 6012–6020. [CrossRef] [PubMed]

6.  Sparano, J.A. Prognostic gene expression assays in breast cancer: Are two better than one? *NPJ Breast Cancer* **2018**, *4*, 11. [CrossRef] [PubMed]

7.  Wishart, D.S. Emerging applications of metabolomics in drug discovery and precision medicine. *Nat. Rev. Drug Discov.* **2016**, *15*, 473–484. [CrossRef]

8.  Nagana Gowda, G.A.; Raftery, D. Can NMR solve some significant challenges in metabolomics? *J. Magn. Reson. (San Diego Calif 1997)* **2015**, *260*, 144–160. [CrossRef] [PubMed]

9.  Maniara, G.; Rajamoorthi, K.; Rajan, S.; Stockton, G.W. Method performance and validation for quantitative analysis by (1)h and (31)p NMR spectroscopy. Applications to analytical standards and agricultural chemicals. *Anal. Chem.* **1998**, *70*, 4921–4928. [CrossRef]

10. Mountford, C.; Ramadan, S.; Stanwell, P.; Malycha, P. Proton MRS of the breast in the clinical setting. *NMR Biomed.* **2009**, *22*, 54–64. [CrossRef] [PubMed]

11. Beckonert, O.; Coen, M.; Keun, H.C.; Wang, Y.; Ebbels, T.M.D.; Holmes, E.; Lindon, J.C.; Nicholson, J.K. High-resolution magic-angle-spinning NMR spectroscopy for metabolic profiling of intact tissues. *Nat. Protoc.* **2010**, *5*, 1019–1032. [CrossRef] [PubMed]

12. Martin, R.W.; Jachmann, R.C.; Sakellariou, D.; Nielsen, U.G.; Pines, A. High-resolution nuclear magnetic resonance spectroscopy of biological tissues using projected magic angle spinning. *Magn. Reson. Med.* **2005**, *54*, 253–257. [CrossRef] [PubMed]

13. Chen, J.-H.; Enloe, B.M.; Xiao, Y.; Cory, D.G.; Singer, S. Isotropic susceptibility shift under MAS: The origin of the split water resonances in 1H MAS NMR spectra of cell suspensions. *Magn. Reson. Med.* **2003**, *50*, 515–521. [CrossRef] [PubMed]

14. Andrew, E.R.; Bradbury, A.; Eades, R.G. Removal of Dipolar Broadening of Nuclear Magnetic Resonance Spectra of Solids by Specimen Rotation. *Nature* **1959**, *183*, 1802–1803. [CrossRef]

15. Lowe, I.J. Free Induction Decays of Rotating Solids. *Phys. Rev. Lett.* **1959**, *2*, 285–287. [CrossRef]

16. Keifer, P.A.; Baltusis, L.; Rice, D.M.; Tymiak, A.A.; Shoolery, J.N. A Comparison of NMR Spectra Obtained for Solid-Phase-Synthesis Resins Using Conventional High-Resolution, Magic-Angle-Spinning, and High-Resolution Magic-Angle-Spinning Probes. *J. Magn. Reson. Ser. A* **1996**, *119*, 65–75. [CrossRef]

17. Millis, K.K.; Maas, W.E.; Cory, D.G.; Singer, S. Gradient, high-resolution, magic-angle spinning nuclear magnetic resonance spectroscopy of human adipocyte tissue. *Magn. Reson. Med.* **1997**, *38*, 399–403. [CrossRef]

18. Millis, K.; Weybright, P.; Campbell, N.; Fletcher, J.A.; Fletcher, C.D.; Cory, D.G.; Singer, S. Classification of human liposarcoma and lipoma using ex vivo proton NMR spectroscopy. *Magn. Reson. Med.* **1999**, *41*, 257–267. [CrossRef]

19. Cheng, L.L.; Ma, M.J.; Becerra, L.; Ptak, T.; Tracey, I.; Lackner, A.; Gonzalez, R.G. Quantitative neuropathology by high resolution magic angle spinning proton magnetic resonance spectroscopy. *Proc. Natl. Acad. Sci.* **1997**, *94*, 6408–6413. [CrossRef]

20. Cheng, L.L.; Lean, C.L.; Bogdanova, A.; Wright, S.C.; Ackerman, J.L.; Brady, T.J.; Garrido, L. Enhanced resolution of proton NMR spectra of malignant lymph nodes using magic-angle spinning. *Magn. Reson. Med.* **1996**, *36*, 653–658. [CrossRef]

21. Doty, F.D.; Entzminger, G.; Yang, Y.A. Magnetism in high-resolution NMR probe design. I: General methods. *Concepts Magn. Reson.* **1998**, *10*, 133–156. [CrossRef]

22. Doty, F.D.; Entzminger, G.; Yang, Y.A. Magnetism in high-resolution NMR probe design. II: HR MAS. *Concepts Magn. Reson.* **1998**, *10*, 239–260. [CrossRef]

23. Tosi, R.; Tugnoli, V. *Nuclear Magnetic Resonance Spectroscopy in the Study of Neoplastic Tissue*; Nova Science: New York, NY, USA, 2005.
24. Sitter, B.; Sonnewald, U.; Spraul, M.; Fjösne, H.E.; Gribbestad, I.S. High-resolution magic angle spinning MRS of breast cancer tissue. *NMR Biomed.* **2002**, *15*, 327–337. [CrossRef] [PubMed]
25. Esteve, V.; Martínez-Granados, B.; Martínez-Bisbal, M.C. Pitfalls to be considered on the metabolomic analysis of biological samples by HR-MAS. *Front. Chem.* **2014**, *2*, 33. [CrossRef]
26. Righi, V.; Schenetti, L.; Maiorana, A.; Libertini, E.; Bettelli, S.; Bonetti, L.R.; Mucci, A. Assessment of freezing effects and diagnostic potential of BioBank healthy and neoplastic breast tissues through HR-MAS NMR spectroscopy. *Metabol. Off. J. Metabol. Soc.* **2015**, *11*, 487–498. [CrossRef]
27. Haukaas, T.H.; Moestue, S.A.; Vettukattil, R.; Sitter, B.; Lamichhane, S.; Segura, R.; Giskeødegård, G.F.; Bathen, T.F. Impact of Freezing Delay Time on Tissue Samples for Metabolomic Studies. *Front. Oncol.* **2016**, *6*, 17. [CrossRef] [PubMed]
28. Opstad, K.S.; Bell, B.A.; Griffiths, J.R.; Howe, F.A. An investigation of human brain tumour lipids by high-resolution magic angle spinning 1H MRS and histological analysis. *NMR Biomed.* **2008**, *21*, 677–685. [CrossRef] [PubMed]
29. Middleton, D.A.; Bradley, D.P.; Connor, S.C.; Mullins, P.G.; Reid, D.G. The effect of sample freezing on proton magic-angle spinning NMR spectra of biological tissue. *Magn. Reson. Med.* **1998**, *40*, 166–169. [CrossRef]
30. Waters, N.J.; Garrod, S.; Farrant, R.D.; Haselden, J.N.; Connor, S.C.; Connelly, J.; Lindon, J.C.; Holmes, E.; Nicholson, J.K. High-resolution magic angle spinning (1)H NMR spectroscopy of intact liver and kidney: Optimization of sample preparation procedures and biochemical stability of tissue during spectral acquisition. *Anal. Biochem.* **2000**, *282*, 16–23. [CrossRef]
31. Shabihkhani, M.; Lucey, G.M.; Wei, B.; Mareninov, S.; Lou, J.J.; Vinters, H.V.; Singer, E.J.; Cloughesy, T.F.; Yong, W.H. The procurement, storage, and quality assurance of frozen blood and tissue biospecimens in pathology, biorepository, and biobank settings. *Clin. Biochem.* **2014**, *47*, 258–266. [CrossRef]
32. Jordan, K.W.; He, W.; Halpern, E.F.; Wu, C.-L.; Cheng, L.L. Evaluation of Tissue Metabolites with High Resolution Magic Angle Spinning MR Spectroscopy Human Prostate Samples after Three-Year Storage at −80 °C. *BiomarkInsights* **2017**, *2*, 117727190700200. [CrossRef]
33. Giskeødegård, G.F.; Cao, M.D.; Bathen, T.F. High-resolution magic-angle-spinning NMR spectroscopy of intact tissue. *Methods Mol. Biol. (Clifton, NJ)* **2015**, *1277*, 37–50. [CrossRef]
34. Bertilsson, H.; Angelsen, A.; Viset, T.; Skogseth, H.; Tessem, M.-B.; Halgunset, J. A new method to provide a fresh frozen prostate slice suitable for gene expression study and MR spectroscopy. *Prostate* **2011**, *71*, 461–469. [CrossRef] [PubMed]
35. Gogiashvili, M.; Horsch, S.; Marchan, R.; Gianmoena, K.; Cadenas, C.; Tanner, B.; Naumann, S.; Ersova, D.; Lippek, F.; Rahnenführer, J.; et al. Impact of intratumoral heterogeneity of breast cancer tissue on quantitative metabolomics using high-resolution magic angle spinning 1 H NMR spectroscopy. *NMR Biomed.* **2018**, *31*. [CrossRef] [PubMed]
36. Gogiashvili, M.; Edlund, K.; Gianmoena, K.; Marchan, R.; Brik, A.; Andersson, J.T.; Lambert, J.; Madjar, K.; Hellwig, B.; Rahnenführer, J.; et al. Metabolic profiling of ob/ob mouse fatty liver using HR-MAS 1H-NMR combined with gene expression analysis reveals alterations in betaine metabolism and the transsulfuration pathway. *Anal. Bioanal. Chem.* **2017**, *409*, 1591–1606. [CrossRef] [PubMed]
37. Cao, M.D.; Giskeødegård, G.F.; Bathen, T.F.; Sitter, B.; Bofin, A.; Lønning, P.E.; Lundgren, S.; Gribbestad, I.S. Prognostic value of metabolic response in breast cancer patients receiving neoadjuvant chemotherapy. *BMC Cancer* **2012**, *12*, 39. [CrossRef]
38. Cao, M.D.; Sitter, B.; Bathen, T.F.; Bofin, A.; Lønning, P.E.; Lundgren, S.; Gribbestad, I.S. Predicting long-term survival and treatment response in breast cancer patients receiving neoadjuvant chemotherapy by MR metabolic profiling. *NMR Biomed.* **2012**, *25*, 369–378. [CrossRef] [PubMed]
39. Sitter, B.; Bathen, T.F.; Singstad, T.E.; Fjøsne, H.E.; Lundgren, S.; Halgunset, J.; Gribbestad, I.S. Quantification of metabolites in breast cancer patients with different clinical prognosis using HR MAS MR spectroscopy. *NMR Biomed.* **2010**, *23*, 424–431. [CrossRef]
40. Moestue, S.A.; Borgan, E.; Huuse, E.M.; Lindholm, E.M.; Sitter, B.; Børresen-Dale, A.-L.; Engebraaten, O.; Maelandsmo, G.M.; Gribbestad, I.S. Distinct choline metabolic profiles are associated with differences in gene expression for basal-like and luminal-like breast cancer xenograft models. *BMC Cancer* **2010**, *10*, 433. [CrossRef] [PubMed]

41. Sitter, B.; Lundgren, S.; Bathen, T.F.; Halgunset, J.; Fjosne, H.E.; Gribbestad, I.S. Comparison of HR MAS MR spectroscopic profiles of breast cancer tissue with clinical parameters. *NMR Biomed.* **2006**, *19*, 30–40. [CrossRef]

42. Grinde, M.T.; Skrbo, N.; Moestue, S.A.; Rødland, E.A.; Borgan, E.; Kristian, A.; Sitter, B.; Bathen, T.F.; Børresen-Dale, A.-L.; Mælandsmo, G.M.; et al. Interplay of choline metabolites and genes in patient-derived breast cancer xenografts. *Breast Cancer Res. BCR* **2014**, *16*, R5. [CrossRef] [PubMed]

43. Euceda, L.R.; Haukaas, T.H.; Giskeødegård, G.F.; Vettukattil, R.; Engel, J.; Silwal-Pandit, L.; Lundgren, S.; Borgen, E.; Garred, Ø.; Postma, G.; et al. Evaluation of metabolomic changes during neoadjuvant chemotherapy combined with bevacizumab in breast cancer using MR spectroscopy. *Metabol. Off. J. Metabol. Soc.* **2017**, *13*, 80. [CrossRef]

44. Euceda, L.R.; Hill, D.K.; Stokke, E.; Hatem, R.; El Botty, R.; Bièche, I.; Marangoni, E.; Bathen, T.F.; Moestue, S.A. Metabolic Response to Everolimus in Patient-Derived Triple-Negative Breast Cancer Xenografts. *J. Proteome Res.* **2017**, *16*, 1868–1879. [CrossRef] [PubMed]

45. Haukaas, T.H.; Euceda, L.R.; Giskeødegård, G.F.; Lamichhane, S.; Krohn, M.; Jernström, S.; Aure, M.R.; Lingjærde, O.C.; Schlichting, E.; Garred, Ø.; et al. Metabolic clusters of breast cancer in relation to gene- and protein expression subtypes. *Cancer Metabol.* **2016**, *4*, 12. [CrossRef] [PubMed]

46. Bathen, T.F.; Jensen, L.R.; Sitter, B.; Fjösne, H.E.; Halgunset, J.; Axelson, D.E.; Gribbestad, I.S.; Lundgren, S. MR-determined metabolic phenotype of breast cancer in prediction of lymphatic spread, grade, and hormone status. *Breast Cancer Res. Treat.* **2007**, *104*, 181–189. [CrossRef] [PubMed]

47. Choi, J.S.; Baek, H.-M.; Kim, S.; Kim, M.J.; Youk, J.H.; Moon, H.J.; Kim, E.-K.; Han, K.H.; Kim, D.-H.; Kim, S.I.; et al. HR-MAS MR spectroscopy of breast cancer tissue obtained with core needle biopsy: Correlation with prognostic factors. *PLoS ONE* **2012**, *7*, e51712. [CrossRef] [PubMed]

48. Choi, J.S.; Baek, H.-M.; Kim, S.; Kim, M.J.; Youk, J.H.; Moon, H.J.; Kim, E.-K.; Nam, Y.K. Magnetic resonance metabolic profiling of breast cancer tissue obtained with core needle biopsy for predicting pathologic response to neoadjuvant chemotherapy. *PLoS ONE* **2013**, *8*, e83866. [CrossRef] [PubMed]

49. Cheng, L.L.; Chang, I.W.; Smith, B.L.; Gonzalez, R.G. Evaluating human breast ductal carcinomas with high-resolution magic-angle spinning proton magnetic resonance spectroscopy. *J. Magn. Reson. (San Diego Calif 1997)* **1998**, *135*, 194–202. [CrossRef]

50. Park, V.Y.; Yoon, D.; Koo, J.S.; Kim, E.-K.; Kim, S.I.; Choi, J.S.; Park, S.; Park, H.S.; Kim, S.; Kim, M.J. Intratumoral Agreement of High-Resolution Magic Angle Spinning Magnetic Resonance Spectroscopic Profiles in the Metabolic Characterization of Breast Cancer. *Medicine* **2016**, *95*, e3398. [CrossRef]

51. Li, M.; Song, Y.; Cho, N.; Chang, J.M.; Koo, H.R.; Yi, A.; Kim, H.; Park, S.; Moon, W.K. An HR-MAS MR metabolomics study on breast tissues obtained with core needle biopsy. *PLoS ONE* **2011**, *6*, e25563. [CrossRef]

52. Taylor, J.L.; Wu, C.-L.; Cory, D.; Gonzalez, R.G.; Bielecki, A.; Cheng, L.L. High-resolution magic angle spinning proton NMR analysis of human prostate tissue with slow spinning rates. *Magn. Reson. Med.* **2003**, *50*, 627–632. [CrossRef] [PubMed]

53. Weybright, P.; Millis, K.; Campbell, N.; Cory, D.G.; Singer, S. Gradient, high-resolution, magic angle spinning1H nuclear magnetic resonance spectroscopy of intact cells. *Magn. Reson. Med.* **1998**, *39*, 337–345. [CrossRef] [PubMed]

54. Aime, S.; Bruno, E.; Cabella, C.; Colombatto, S.; Digilio, G.; Mainero, V. HR-MAS of cells: A "cellular water shift" due to water-protein interactions? *Magn. Reson. Med.* **2005**, *54*, 1547–1552. [CrossRef] [PubMed]

55. André, M.; Dumez, J.-N.; Rezig, L.; Shintu, L.; Piotto, M.; Caldarelli, S. Complete protocol for slow-spinning high-resolution magic-angle spinning NMR analysis of fragile tissues. *Anal. Chem.* **2014**, *86*, 10749–10754. [CrossRef]

56. Chae, E.Y.; Shin, H.J.; Kim, S.; Baek, H.-M.; Yoon, D.; Kim, S.; Shim, Y.E.; Kim, H.H.; Cha, J.H.; Choi, W.J.; et al. The Role of High-Resolution Magic Angle Spinning 1H Nuclear Magnetic Resonance Spectroscopy for Predicting the Invasive Component in Patients with Ductal Carcinoma In Situ Diagnosed on Preoperative Biopsy. *PLoS ONE* **2016**, *11*, e0161038. [CrossRef] [PubMed]

57. Renault, M.; Shintu, L.; Piotto, M.; Caldarelli, S. Slow-spinning low-sideband HR-MAS NMR spectroscopy: Delicate analysis of biological samples. *Sci. Rep.* **2013**, *3*, 3349. [CrossRef] [PubMed]

58. Hoult, D.I. Solvent peak saturation with single phase and quadrature fourier transformation. *J. Magn. Reson. (1969)* **1976**, *21*, 337–347. [CrossRef]

59. Tzika, A.A.; Cheng, L.L.; Goumnerova, L.; Madsen, J.R.; Zurakowski, D.; Astrakas, L.G.; Zarifi, M.K.; Scott, R.M.; Anthony, D.C.; Gonzalez, R.G.; et al. Biochemical characterization of pediatric brain tumors by using in vivo and ex vivo magnetic resonance spectroscopy. *J. Neurosurg.* **2002**, *96*, 1023–1031. [CrossRef]

60. Cheng, L.L.; Chang, I.W.; Louis, D.N.; Gonzalez, R.G. Correlation of high-resolution magic angle spinning proton magnetic resonance spectroscopy with histopathology of intact human brain tumor specimens. *Cancer Res.* **1998**, *58*, 1825–1832.

61. Barton, S.J.; Howe, F.A.; Tomlins, A.M.; Cudlip, S.A.; Nicholson, J.K.; Anthony Bell, B.; Griffiths, J.R. Comparison of in vivo1H MRS of human brain tumours with1H HR-MAS spectroscopy of intact biopsy samples in vitro. *MAGMA* **1999**, *8*, 121–128. [CrossRef]

62. Ludwig, C.; Viant, M.R. Two-dimensional J-resolved NMR spectroscopy: Review of a key methodology in the metabolomics toolbox. *Phytochem. Anal. PCA* **2010**, *21*, 22–32. [CrossRef]

63. Palmer, A.G.; Cavanagh, J.; Wright, P.E.; Rance, M. Sensitivity improvement in proton-detected two-dimensional heteronuclear correlation NMR spectroscopy. *J. Magn. Reson. (1969)* **1991**, *93*, 151–170. [CrossRef]

64. Morris, G.A.; Freeman, R. Enhancement of nuclear magnetic resonance signals by polarization transfer. *J. Am. Chem. Soc.* **1979**, *101*, 760–762. [CrossRef]

65. Ravikumar, M.; Bothner-By, A.A. A two-dimensional NMR experiment for the correlation of spin-locked and free-precession frequencies. *J. Am. Chem. Soc.* **1993**, *115*, 7537–7538. [CrossRef]

66. Kupce, E.; Keifer, P.A.; Delepierre, M. Adiabatic TOCSY MAS in liquids. *J. Magn. Reson. (San Diego Calif 1997)* **2001**, *148*, 115–120. [CrossRef]

67. Wieruszeski, J.M.; Montagne, G.; Chessari, G.; Rousselot-Pailley, P.; Lippens, G. Rotor synchronization of radiofrequency and gradient pulses in high-resolution magic angle spinning NMR. *J. Magn. Reson. (San Diego Calif 1997)* **2001**, *152*, 95–102. [CrossRef]

68. Yoon, H.; Yoon, D.; Yun, M.; Choi, J.S.; Park, V.Y.; Kim, E.-K.; Jeong, J.; Koo, J.S.; Yoon, J.H.; Moon, H.J.; et al. Metabolomics of Breast Cancer Using High-Resolution Magic Angle Spinning Magnetic Resonance Spectroscopy: Correlations with 18F-FDG Positron Emission Tomography-Computed Tomography, Dynamic Contrast-Enhanced and Diffusion-Weighted Imaging MRI. *PLoS ONE* **2016**, *11*, e0159949. [CrossRef]

69. Antzutkin, O.N.; Shekar, S.C.; Levitt, M.H. Two-Dimensional Sideband Separation in Magic-Angle-Spinning NMR. *J. Magn. Reson. Ser. A* **1995**, *115*, 7–19. [CrossRef]

70. Hu, J.Z.; Wang, W.; Liu, F.; Solum, M.S.; Alderman, D.W.; Pugmire, R.J.; Grant, D.M. Magic-Angle-Turning Experiments for Measuring Chemical-Shift-Tensor Principal Values in Powdered Solids. *J. Magn. Reson. Ser. A* **1995**, *113*, 210–222. [CrossRef]

71. Aguilar, J.A.; Nilsson, M.; Bodenhausen, G.; Morris, G.A. Spin echo NMR spectra without J modulation. *Chem. Commun. (Camb. Engl.)* **2012**, *48*, 811–813. [CrossRef] [PubMed]

72. Esmaeili, M.; Bathen, T.F.; Engebråten, O.; Mælandsmo, G.M.; Gribbestad, I.S.; Moestue, S.A. Quantitative (31)P HR-MAS MR spectroscopy for detection of response to PI3K/mTOR inhibition in breast cancer xenografts. *Magn. Reson. Med.* **2014**, *71*, 1973–1981. [CrossRef]

73. Swanson, M.G.; Zektzer, A.S.; Tabatabai, Z.L.; Simko, J.; Jarso, S.; Keshari, K.R.; Schmitt, L.; Carroll, P.R.; Shinohara, K.; Vigneron, D.B.; et al. Quantitative analysis of prostate metabolites using 1H HR-MAS spectroscopy. *Magn. Reson. Med.* **2006**, *55*, 1257–1264. [CrossRef] [PubMed]

74. Martínez-Bisbal, M.C.; Monleon, D.; Assemat, O.; Piotto, M.; Piquer, J.; Llácer, J.L.; Celda, B. Determination of metabolite concentrations in human brain tumour biopsy samples using HR-MAS and ERETIC measurements. *NMR Biomed.* **2009**, *22*, 199–206. [CrossRef] [PubMed]

75. Kriat, M.; Confort-Gouny, S.; Vion-Dury, J.; Sciaky, M.; Viout, P.; Cozzone, P.J. Quantitation of metabolites in human blood serum by proton magnetic resonance spectroscopy. A comparative study of the use of formate and TSP as concentration standards. *NMR Biomed.* **1992**, *5*, 179–184. [CrossRef] [PubMed]

76. Albers, M.J.; Butler, T.N.; Rahwa, I.; Bao, N.; Keshari, K.R.; Swanson, M.G.; Kurhanewicz, J. Evaluation of the ERETIC method as an improved quantitative reference for 1H HR-MAS spectroscopy of prostate tissue. *Magn. Reson. Med.* **2009**, *61*, 525–532. [CrossRef] [PubMed]

77. Kostidis, S.; Addie, R.D.; Morreau, H.; Mayboroda, O.A.; Giera, M. Quantitative NMR analysis of intra- and extracellular metabolism of mammalian cells: A tutorial. *Anal. Chim. Acta* **2017**, *980*, 1–24. [CrossRef] [PubMed]

78. Nowick, J.S.; Khakshoor, O.; Hashemzadeh, M.; Brower, J.O. DSA: A new internal standard for NMR studies in aqueous solution. *Org. Lett.* **2003**, *5*, 3511–3513. [CrossRef] [PubMed]

79. Alum, M.F.; Shaw, P.A.; Sweatman, B.C.; Ubhi, B.K.; Haselden, J.N.; Connor, S.C. 4,4-Dimethyl-4-silapentane-1-ammonium trifluoroacetate (DSA), a promising universal internal standard for NMR-based metabolic profiling studies of biofluids, including blood plasma and serum. *Metabol. Off. J. Metabol. Soc.* **2008**, *4*, 122–127. [CrossRef]

80. Barker, P.B.; Soher, B.J.; Blackband, S.J.; Chatham, J.C.; Mathews, V.P.; Bryan, R.N. Quantitation of proton NMR spectra of the human brain using tissue water as an internal concentration reference. *NMR Biomed.* **1993**, *6*, 89–94. [CrossRef] [PubMed]

81. Barantin, L.; Le Pape, A.; Akoka, S. A new method for absolute quantitation MRS metabolites. *Magn. Reson. Med.* **1997**, *38*, 179–182. [CrossRef]

82. Akoka, S.; Barantin, L.; Trierweiler, M. Concentration Measurement by Proton NMR Using the ERETIC Method. *Anal. Chem.* **1999**, *71*, 2554–2557. [CrossRef] [PubMed]

83. Wider, G.; Dreier, L. Measuring protein concentrations by NMR spectroscopy. *J. Am. Chem. Soc.* **2006**, *128*, 2571–2576. [CrossRef] [PubMed]

84. Borgan, E.; Sitter, B.; Lingjærde, O.C.; Johnsen, H.; Lundgren, S.; Bathen, T.F.; Sørlie, T.; Børresen-Dale, A.-L.; Gribbestad, I.S. Merging transcriptomics and metabolomics—Advances in breast cancer profiling. *BMC Cancer* **2010**, *10*, 628. [CrossRef] [PubMed]

85. Giskeødegård, G.F.; Lundgren, S.; Sitter, B.; Fjøsne, H.E.; Postma, G.; Buydens, L.M.C.; Gribbestad, I.S.; Bathen, T.F. Lactate and glycine-potential MR biomarkers of prognosis in estrogen receptor-positive breast cancers. *NMR Biomed.* **2012**, *25*, 1271–1279. [CrossRef] [PubMed]

86. He, Q.H.; Shungu, D.C.; Vanzijl, P.C.M.; Bhujwalla, Z.M.; Glickson, J.D. Single-Scan in Vivo Lactate Editing with Complete Lipid and Water Suppression by Selective Multiple-Quantum-Coherence Transfer (Sel-MQC) with Application to Tumors. *J. Magn. Reson. Ser. B* **1995**, *106*, 203–211. [CrossRef]

87. Holbach, M.; Lambert, J.; Suter, D. Optimized multiple-quantum filter for robust selective excitation of metabolite signals. *J. Magn. Reson. (San Diego Calif 1997)* **2014**, *243*, 8–16. [CrossRef]

88. Holbach, M.; Lambert, J.; Johst, S.; Ladd, M.E.; Suter, D. Optimized selective lactate excitation with a refocused multiple-quantum filter. *J. Magn. Reson. (San Diego Calif 1997)* **2015**, *255*, 34–38. [CrossRef]

89. Maximov, I.I.; Tosner, Z.; Nielsen, N.C. Optimal control design of NMR and dynamic nuclear polarization experiments using monotonically convergent algorithms. *J. Chem. Phys.* **2008**, *128*, 184505. [CrossRef]

90. Ye, T.; Zheng, C.; Zhang, S.; Gowda, G.A.N.; Vitek, O.; Raftery, D. "Add to subtract": A simple method to remove complex background signals from the 1H nuclear magnetic resonance spectra of mixtures. *Anal. Chem.* **2012**, *84*, 994–1002. [CrossRef]

91. Provencher, S.W. Estimation of metabolite concentrations from localizedin vivo proton NMR spectra. *Magn. Reson. Med.* **1993**, *30*, 672–679. [CrossRef]

92. Provencher, S.W. Automatic quantitation of localized in vivo 1H spectra with LCModel. *NMR Biomed.* **2001**, *14*, 260–264. [CrossRef] [PubMed]

93. Bathen, T.F.; Geurts, B.; Sitter, B.; Fjøsne, H.E.; Lundgren, S.; Buydens, L.M.; Gribbestad, I.S.; Postma, G.; Giskeødegård, G.F. Feasibility of MR metabolomics for immediate analysis of resection margins during breast cancer surgery. *PLoS ONE* **2013**, *8*, e61578. [CrossRef] [PubMed]

94. Martelotto, L.G.; Ng, C.K.Y.; Piscuoglio, S.; Weigelt, B.; Reis-Filho, J.S. Breast cancer intra-tumor heterogeneity. *Breast Cancer Res. BCR* **2014**, *16*, 210. [CrossRef] [PubMed]

95. Ng, C.K.Y.; Pemberton, H.N.; Reis-Filho, J.S. Breast cancer intratumor genetic heterogeneity: Causes and implications. *Expert Rev. Anticancer Ther.* **2012**, *12*, 1021–1032. [CrossRef] [PubMed]

96. Cao, M.D.; Lamichhane, S.; Lundgren, S.; Bofin, A.; Fjøsne, H.; Giskeødegård, G.F.; Bathen, T.F. Metabolic characterization of triple negative breast cancer. *BMC Cancer* **2014**, *14*, 941. [CrossRef] [PubMed]

97. Curigliano, G.; Burstein, H.J.; P Winer, E.; Gnant, M.; Dubsky, P.; Loibl, S.; Colleoni, M.; Regan, M.M.; Piccart-Gebhart, M.; Senn, H.-J.; et al. De-escalating and escalating treatments for early-stage breast cancer: The St. Gallen International Expert Consensus Conference on the Primary Therapy of Early Breast Cancer 2017. *Ann. Oncol. Off. J. Eur. Soc. Med. Oncol.* **2017**, *28*, 1700–1712. [CrossRef] [PubMed]

98. Giskeødegård, G.F.; Grinde, M.T.; Sitter, B.; Axelson, D.E.; Lundgren, S.; Fjøsne, H.E.; Dahl, S.; Gribbestad, I.S.; Bathen, T.F. Multivariate modeling and prediction of breast cancer prognostic factors using MR metabolomics. *J. Proteome Res.* **2010**, *9*, 972–979. [CrossRef]

99.  Cortazar, P.; Zhang, L.; Untch, M.; Mehta, K.; Costantino, J.P.; Wolmark, N.; Bonnefoi, H.; Cameron, D.; Gianni, L.; Valagussa, P.; et al. Pathological complete response and long-term clinical benefit in breast cancer: The CTNeoBC pooled analysis. *Lancet* **2014**, *384*, 164–172. [CrossRef]
100. Bingol, K. Recent Advances in Targeted and Untargeted Metabolomics by NMR and MS/NMR Methods. *High.-Throughput* **2018**, *7*. [CrossRef]

 *metabolites*

*Article*

# Dynamic Metabolic Response to Adriamycin-Induced Senescence in Breast Cancer Cells

**Rong You [1,2], Jin Dai [2], Ping Zhang [2,3], Gregory A. Barding Jr. [2,4] and Daniel Raftery [2,5,*]**

[1] College of Life Sciences, South China Normal University, 55 Zhongshan Avenue West, Guangzhou 510631, China; 20021141@m.scnu.edu.cn

[2] Mitochondria and Metabolism Center, Department of Anesthesiology and Pain Medicine, University of Washington, 850 Republican Street, Seattle, WA 98109, USA; daij@uw.edu (J.D.); ping17028@swu.edu.cn (P.Z.); gabarding@cpp.edu (G.A.B.J.)

[3] College of Plant Protection, Southwest University, 2 Tiansheng Road, Chongqing 400715, China

[4] Chemistry and Biochemistry Department, California State Polytechnic University, Pomona, CA 91768, USA

[5] Fred Hutchinson Cancer Research Center, 1100 Fairview Avenue N., Seattle, WA 98109, USA

* Correspondence: draftery@uw.edu; Tel.: +1-(206)-543-9709; Fax: +1-(206)-616-4819

Received: 31 October 2018; Accepted: 11 December 2018; Published: 15 December 2018

**Abstract:** Cellular senescence displays a heterogeneous set of phenotypes linked to tumor suppression; however, after drug treatment, senescence may also be involved in stable or recurrent cancer. Metabolic changes during senescence can provide detailed information on cellular status and may also have implications for the development of effective treatment strategies. The metabolic response to Adriamycin (ADR) treatment, which causes senescence as well as cell death, was obtained with the aid of metabolic profiling and isotope tracing in two human breast cancer cell lines, MCF7 and MDA-MB-231. After 5 days of ADR treatment, more than 60% of remaining, intact cells entered into a senescent state, characterized by enlarged and flattened morphology and positive blue staining using SA-β-gal. Metabolic trajectory analysis showed that the two cell lines' responses were significantly different and were divided into two distinct stages. The metabolic shift from the first stage to the second was reflected by a partial recovery of the TCA cycle, as well as amino acid and lipid metabolisms. Isotope tracing analysis indicated that the higher level of glutamine metabolism helped maintain senescence. The results suggest that the dynamic changes during senescence indicate a multi-step process involving important metabolic pathways which might allow breast cancer cells to adapt to persistent ADR treatment, while the higher level of anapleurosis may be important for maintaining the senescent state. Ultimately, a better understanding of metabolic changes during senescence might provide targets for cancer therapy and tumor eradication.

**Keywords:** senescence MCF7; MDA-MB-231; metabolomics; isotope tracing analysis; gas chromatography–mass spectrometry (GC–MS)

## 1. Introduction

Cellular senescence was initially identified as cell cycle arrest from the limited replicative capacity in normal human diploid fibroblasts (HDFs); it was termed "replicative senescence" and associated with telomere shortening or dysfunction [1]. More recently, it has been recognized that a number of different stressors including oncogenic mutations, chemotherapeutic drugs and oxidative stress can cause telomere-independent cellular senescence, which is termed premature senescence [2]. Senescent cells display heterogeneous phenotypes including enlarged cell size, flattened cell morphology, an inability to synthesize DNA, formation of senescence-associated heterochromatin foci (SAHF) and expression of an endogenous senescence-associated β-galactosidase activity (SA-β-gal) [3]. Premature senescence in cancer therapy serves as an effective tumor-suppressor by preventing cancer cell proliferation or blocking the acquisition of tumor transformation [4]. However, the senescence-associated secretory phenotype during cancer therapy may influence tissue microenvironments, and even stimulate tumorgenesis and metastasis in vitro and in vivo [5,6]. Moreover, premature senescent cancer cells have the capability to escape growth arrest and re-enter the cell cycle, leading to tumor relapse [7,8]. Thus, the process of tumor cells undergoing cell senescence after drug treatment results in stable disease rather than regression of the tumor, which represents a non-optimal outcome and significant health risk in cancer therapy.

Cellular survival and growth require specific metabolic reprogramming to adapt to genetic or environmental stresses because of the need for metabolic pathways to continue to produce energy, precursors and substrates for macromolecular synthesis and/or cell signaling. One of the hallmarks in cancer biology is the higher levels of glucose uptake and glycolysis during tumor growth, commonly known as the Warburg effect, as well as the importance of glutamine as an anapleurotic substrate for the TCA cycle [9]. It has been noted that metabolic reprogramming might be exploited therapeutically for cancer therapy. With the aim of providing further insights into cancer biology and new targets for cancer therapy, metabolic phenotypes associated with cell senescence have been studied over the last few years. Higher utilization of glucose and higher ATP production were observed in therapy induced senescent (TIS)-competent lymphomas [10]. Consequently, TIS lymphomas were sensitive to blocking glucose utilization, which led to their selective eradication. The mitochondrial gatekeeper pyruvate dehydrogenase (PDH) was found to be a crucial mediator of oncogene-induced senescence (OIS) by BRAF$^{V600E}$, which was accompanied by increased pyruvate oxidation and mitochondrial oxidative phosphorylation [11]. A lower level of deoxyribonucleotide triphosphate was also found in oncogene-induced senescence, caused by oncogene-induced repression of ribonucleotide reductase subunit M2 [12]. These studies revealed that senescent cells display metabolically active and context-dependent phenotypes. It was also suggested that understanding the metabolic changes during cell senescence may have implications for the development of new and effective strategies to treat cancer.

Although significant progress towards understanding the senescence-associated phenotype and its underlying molecular mechanisms have been made recently, a global metabolic view at the systems level is still lacking. Metabolomics provides a comprehensive characterization of the metabolic changes in biological systems that occur in response to different stimuli, and their interpretation in terms of metabolic pathway changes allows an understanding of the physiological variation [13,14]. Metabolomics has found widespread applications in cellular metabolism for elucidating perturbed cellular homeostasis, cell transformation and stem cell differentiation [15–17] and providing an understanding of the metabolic control of cell fate. In addition, stable isotope tracing has become an important and complementary tool in metabolomics, allowing one to track individual atoms and determine the fate of individual metabolites, through which metabolic network and flux changes can be obtained to facilitate the identification of altered pathways [18,19]. Here, we use gas chromatography–mass spectrometry (GC–MS) based global metabolomics and isotope tracer analysis to identify the metabolic changes during the progression of senescence in two breast cancer cell lines induced by Adriamycin (ADR) treatment to better understand how senescent cells can maintain their metabolic activity.

## 2. Results

*Morphological characteristics of ADR-induced cell senescence.* ADR treatment enabled the preparation of stable and repeatable senescent cell samples in MCF7 and MDA-MB-231 breast cancer cell lines. ADR treated cells became flattened and enlarged (Figure 1), and the effects of ADR treatment on the breast cancer cells were dose and time dependent (data not shown). When the cells were treated with 0.04 µg/mL ADR for 5 days more than 60% of the remaining, adherent cells became senescent, according to SA-β-gal analysis, indicating that ADR treatment caused very significant cell senescence in MCF-7 and MDA 231 cells. Higher dosages of ADR in preliminary experiments results in significant cell death and thus were not useful to study senescence. By comparison, less than 4% of cells were senescent after 7 days of culture without ADR treatment.

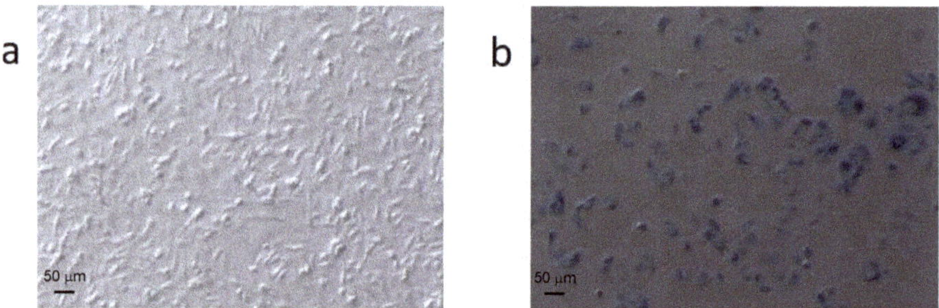

**Figure 1.** Comparison of (**a**) normal and (**b**) Adriamycin (ADR)-treated MCF7 cells stained for SA-β-gal activity. ADR-treated cells were larger and had a flatter morphology than untreated cells. Cell counting (250 cells per condition) showed that the number of SA-β-gal-positive (blue) cells versus the number of total cells was approximately 4% in the untreated cells and ~60% in the treated cells.

*GC–MS based metabolomics characterization of cellular metabolic changes during ADR-induced senescence.* Multivariate analysis of the global metabolic changes at 0, 1, 3, and 5 days of ADR treatment was performed using principal component analysis (PCA) (Figure 2) and showed large changes over the treatment time. The metabolic trajectories during ADR-induced cell senescence indicated that responses of these two cell lines to ADR-induced senescence were different. The two types of cell line samples were clustered together in the absence of ADR treatment. Distinct responses of MCF7 and MDA-MB-231 cells became evident after 3 days of ADR treatment (see Figure 2b), with the sample trajectory of MCF7 moving away from the initial pre-treatment cluster along PC2 while that of MDA-MB-231 moved away along PC1. However, after 5 days of ADR treatment, at a time when the two cell lines showed more than 60% cell senescence based on the SA-β-gal analysis, both cell lines moved partly back towards the initial clusters.

Individual metabolites perturbed by ADR treatment were identified using the Student's *t*-test. Since the 3-day ADR treatment caused the most obvious responses for both MCF7 and MDA-MB-231 as indicated by PCA, a direct comparison of each metabolite for the 3-day ADR treatment time point was performed with those samples prior to ADR treatment. A further comparison was carried out between the 3- and 5-day treatment periods to better understand the recovery trend. Metabolites with significantly altered levels are listed in Table 1 for MCF7 and Table 2 for MDA-MB-231.

For MCF7, 2-keto-3-methylvaleric acid, an intermediate metabolite of branched-chain amino acid (BCAA) metabolism was elevated after ADR treatment. TCA cycle metabolites including fumaric acid, malic acid and citric acid decreased after 3 days of ADR treatment. However, these metabolites were significantly increased when MCF7 cells entered into senescence after 5 days of treatment. Meanwhile, amino acids associated with the TCA cycle, aspartic acid and glutamic acid, which are transformed from oxaloacetate and α-ketoglutarate, respectively, were also tracked with the TCA intermediates. The level of serine, which is transformed from intermediates in the glycolysis pathway was decreased after 3 days but recovered when cells entered into more than 60% senescence. Moreover, fatty acids in MCF7 including heptadecanoic acid, linoleic acid, oleic acid and stearic acid also initially decreased, followed by recovery from ADR-induced DNA damage after 5 days of ADR treatment. In all, the characteristics of these significantly changed amino acids, fatty acids and the intermediates of TCA cycles showed that ADR treatment caused obvious effects on MCF7 cell metabolism, most of which showed their lowest levels after 3 days of ADR treatment but which then later recovered after 5 days.

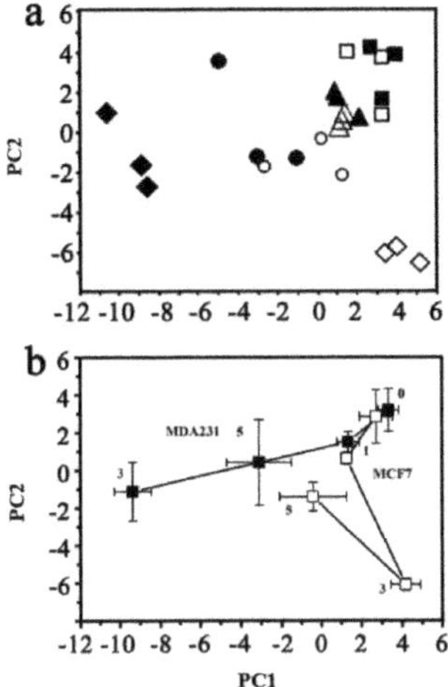

**Figure 2.** The overall metabolic response to ADR damage visualized using (**a**) principal component analysis (PCA) of all cell samples ($R^2$ = 0.547 and $Q^2$ = 0.367). MDA-MB-231 cells with ADR treatment for: ■ 0 days; ▲ 1 day; ◆ 3 days; ● 5 days. MCF7 cells with ADR treatment for: □ 0 days; △ 1 day; ◇ 3 days; ○ 5 days. (**b**) The centroided metabolic trajectories for the two cell lines during ADR treatment: ■ MDA-MB-231; □ MCF7. PC1 loading: 36.3%; PC2 loading: 18.4%.

**Table 1.** Metabolites showing significantly altered levels during cell senescence in the MCF7 cell line and their related metabolic pathways.

| | MCF7 Cells | | | | |
|---|---|---|---|---|---|
| | 3 Day versus 0 Day | | 5 Day versus 3 Day | | |
| **Metabolite** | **Fold Change** | **p-Value** | **Fold Change** | **p-Value** | **Involved Pathway** |
| Malic acid | 0.42 | 0.0016 | 1.85 | 0.0039 | TCA cycle |
| Fumaric acid | 0.58 | 0.0071 | 1.69 | 0.021 | |
| Citric acid | 0.21 | 0.00024 | 3.69 | 0.05 | |
| Valine | 0.57 | 0.0019 | 1.76 | 0.048 | Amino acid metabolism |
| Leucine | 0.56 | 0.0036 | 1.73 | 0.028 | |
| Isoleucine | 0.59 | 0.012 | 1.66 | 0.043 | |
| Proline | 2.51 | 0.012 | 1.39 | 0.029 | |
| Serine | 0.19 | 0.0061 | 1.79 | 0.017 | |
| Aspartic acid | 0.34 | 0.00081 | 1.81 | 0.0015 | |
| Glutamine | 0.3 | 0.0084 | 1.82 | 0.092 | |
| Lauric acid | 0.25 | 0.0039 | 2.6 | 0.0022 | Fatty acid metabolism |
| Palmitic acid | 0.28 | 0.00084 | 2.57 | 0.0014 | |
| Linoleic acid | 0.37 | 0.00098 | 2.02 | 0.0021 | |
| Heptadecanoic acid | 0.067 | 0.0019 | 2.73 | 0.013 | |
| Oleic acid | 0.35 | 0.0042 | 2.07 | 0.0086 | |
| Stearic acid | 0.5 | 0.0042 | 2.63 | 0.019 | |
| 2-Hydrloxyethyl palmitate | 0.26 | 0.0017 | 2.62 | 0.0061 | |
| 2-Ketobutyric acid, enol | 2.12 | 0.035 | 1.55 | 0.093 | Branched-chain amino acid (BCAA) metabolism |
| 2-Keto-3-methylvaleric acid | 2.94 | 0.037 | 2.87 | 0.061 | |
| Creatinine | 0.045 | 0.016 | 1.58 | 0.0028 | Nucleic acid metabolism |
| Glycerol-3-phosphate | 0.074 | 0.0056 | 1.55 | 0.03 | |
| Cytindine-5′-monophosphate | 0.26 | 0.049 | 2.41 | 0.0096 | |
| Uridine-5-monophosphate | 0.0066 | 0.019 | 1.45 | 0.31 | |
| Pantothenic acid | 0.34 | 0.026 | 1.28 | 0.069 | Others |
| Cholest-8(14)-en-3-ol | 0.031 | 0.0031 | 4.79 | 0.018 | |

**Table 2.** Metabolites showing significantly altered levels during cell senescence in the MDA-MB-231 cell line and their related metabolic pathways.

| | MDA-MB-231 Cells | | | | |
| Metabolite | 3 Day versus 0 Day | | 5 Day versus 3 Day | | Involved Pathway |
| | Fold Change | *p*-Value | Fold Change | *p*-Value | |
|---|---|---|---|---|---|
| Malic acid | 0.33 | 0.00045 | 0.91 | 0.61 | TCA cycle |
| Fumaric acid | 0.44 | 0.0019 | 0.88 | 0.48 | |
| Citric acid | 0.2 | 0.019 | 2.68 | 0.1 | |
| Alanine | 17.62 | 0.019 | 0.47 | 0.042 | Amino acid metabolism |
| Leucine | 2.21 | 0.00073 | 0.66 | 0.0038 | |
| Isoleucine | 2.6 | 0.002 | 0.59 | 0.00035 | |
| Proline | 3.42 | 0.01 | 0.43 | 0.01 | |
| Glycine | 2.2 | 0.0029 | 0.47 | 0.02 | |
| Threonine | 2.69 | 0.0053 | 0.47 | 0.0036 | |
| Glutamic acid | 1.97 | 0.0056 | 0.43 | 0.00249 | |
| Phenylalanine | 4.34 | 0.014 | 0.46 | 0.026 | |
| 2-Aminoadipic acid | 1.83 | 0.012 | 0.42 | 0.0045 | |
| Tyrosine | 2.96 | 0.0028 | 0.85 | 0.021 | |
| L-Tryptophan | 5.61 | 0.029 | 0.48 | 0.05 | |
| Heptadecanoic acid | 20.42 | 0.00043 | 1.55 | 0.17 | Fatty acid metabolism |
| Oleic acid | 1.27 | 0.04 | 0.9 | 0.2 | |
| Stearic acid | 0.58 | 0.0029 | 1.48 | 0.19 | |
| 2-Keto-3-methylvaleric acid | 2.98 | 0.03 | 1.12 | 0.57 | BCAA metabolism |
| Isobutyric acid | 4.86 | 0.023 | 1.13 | 0.43 | |
| Glycerol-3-phosphate | 21.28 | 0.0065 | 1.046 | 0.81 | Nucleic acid metabolism |
| Creatinine | 10.81 | 0.026 | 0.75 | 0.23 | |
| 5-Methylthioadenosine | 3.7 | 0.0085 | 0.65 | 0.045 | |
| Phosphorylethanolamine | 63.71 | 0.035 | 0.78 | 0.52 | Others |
| Pantothenic acid | 2.45 | 0.0075 | 0.94 | 0.73 | |
| Scyllo-inositol | 0.69 | 0.023 | 1.54 | 0.26 | |
| D-Myo-inositol | 1.57 | 0.0037 | 0.68 | 0.13 | |
| Cellobiose | 3.7 | 0.0085 | 0.65 | 0.045 | |
| Cholest-8(14)-en-3-ol | 19.94 | 0.00052 | 0.61 | 0.044 | |

For MDA-MB-231, changes in 2-keto-3-methylvaleric acid showed the same trend as for MCF7, which was elevated after ADR treatment. ADR treatment also caused obvious effects on the TCA cycle, lipid metabolism and amino acid metabolism (Table 2). The same downregulation of TCA intermediates, including fumaric acid, malic acid and citric acid occurred after 3 days of ADR treatment. However, a few amino acid and fatty acid changes in MDA-MB-231 cells were opposite to those observed in MCF7 cells. Specifically, alanine, glutamic and other amino acids were upregulated after 3 days to their highest levels, and then were reduced after 5 days of ADR treatment. These results suggest very different responses to ADR treatment based on cell type, even though both cell types showed the same morphological and SA-β-gal detected changes.

*GC–MS based isotope tracing analysis.* Stable isotope tracing analysis using [U-$^{13}$C]-labeled glucose was also employed to compare the cellular metabolism of ADR treated and non-treated MDF7 and MDA-231 cancer cells to help identify how senescent cells maintain an active metabolic phenotype after 5 days of ADR treatment. From the results of tracing experiments, we observed that the levels of m + 3 isotopologues of lactate and pyruvate derived from labeled glucose in these two cell lines did not change in an obvious manner based on an analysis of the isotopologue distribution of metabolites (Figure 3a). However, the level of the m + 3 isotopologue of alanine in MCF7 was lower than that for non-senescent cells (Figure 3b), indicating that in senescent cells a lower flux of pyruvate to alanine occurred. Meanwhile, the m + 0 level of the essential amino acid threonine was lower in senescent MCF7 (Figure 3c), which might be related with higher threonine degradation (to α-ketoglutaric acid) in senescent MCF7 cells.

The results also showed that the mean enrichments of TCA cycle intermediates derived from labeled glucose for senescent MCF7 and MDA131 were lower than those for the non-senescent cells;

less glucose entered into the TCA cycle in senescent cells (Figure 4a,b). The doubly $^{13}$C-labeled isotopologues (m + 2) of citric, α-ketoglutaric, succinic, fumaric and malic acids in the TCA cycle significantly increased in the senescent cells (Figure 5a,b). However, lower levels of (m + 4), (m + 6) isotopologues of citric and malic acids appeared in senescent MCF7 (Figure 5c), which was also related to lower glucose entering the TCA cycle. It is also notable that a large fraction of these TCA metabolites (m + 0) were higher than those of the untreated cells (Figure 6a,b).

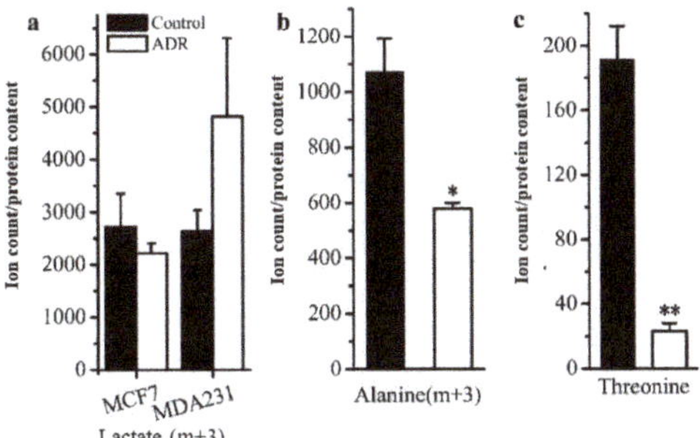

**Figure 3.** Comparison of cell metabolism between the ADR-treated senescent and non-treated cancer cells measured by isotope tracing analysis: (**a**) levels of the m + 3 isotopologue of lactate in MCF7 and MDA-MB-231; (**b**) levels of the m + 3 isotopologue of alanine in MCF7. (**c**) Measure of threonine levels in MCF7. * signifies $p < 0.05$ and ** signifies $p < 0.005$.

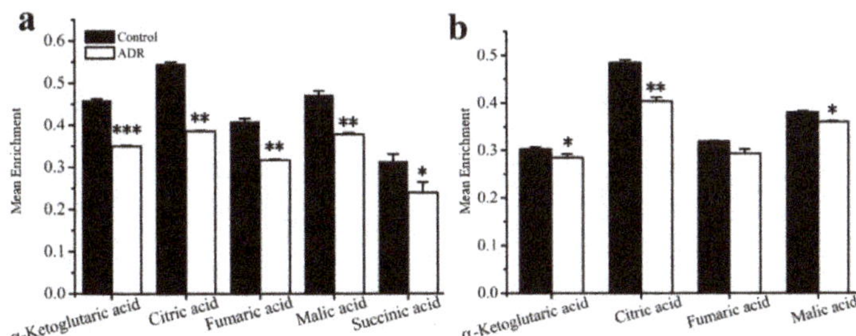

**Figure 4.** Mean enrichment of TCA cycle intermediates for non-treated cells and ADR-treated cells by isotope tracing analysis: (**a**) MCF7, (**b**) MDA-MB-231. * signifies $p < 0.05$; ** signifies $p < 0.005$ and *** signifies $p < 0.0001$.

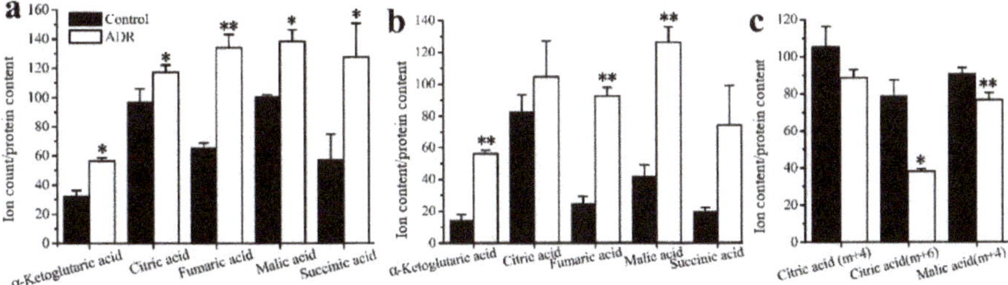

**Figure 5.** Distribution of the isotopologues of TCA cycle intermediates for non-treated cells and ADR-treated cells (**a**) (m + 2) levels of intermediates in MCF7 cells; (**b**) (m + 2) levels intermediates in MDA-MB-231 cells; (**c**) (m + 4) and (m + 6) isotopologues of citrate and malate. * signifies $p < 0.05$ and ** signifies $p < 0.005$.

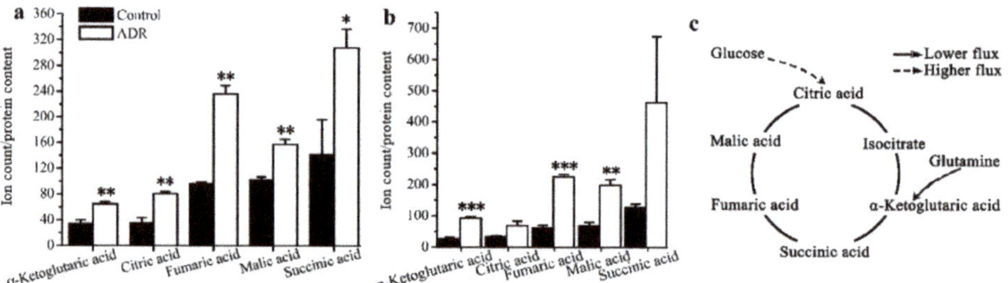

**Figure 6.** Levels of the (m + 0) isotopologue of TCA cycle intermediates for non-treated cells and ADR-treated cells: (**a**) MCF7; (**b**) MDA-MB-231. (**c**) A schematic characterization of TCA metabolites and how they might maintain cell senescence based on the isotopologue (m + 0) levels of TCA intermediates. * signifies $p < 0.05$; ** signifies $p < 0.005$ and *** signifies $p < 0.0001$.

## 3. Discussion

DNA-damaging agents can induce premature senescence in cancer cells and, therefore, could provide an effective method to limit tumor progression by preventing cancer cell proliferation [20]. However, increasing evidence shows that prematurely senescent cells maintain their metabolic activity and have the ability to escape growth arrest to re-enter into the cell cycle [12,21]. Thus, tumor cells undergoing cell senescence after drug treatment results in stable disease rather than regression of the tumor and, therefore, represents a potential lurking hazard in cancer therapy that might lead to tumor relapse. In fact, the physiological consequences of senescent cancer cells still remain elusive. Only a few studies that address the potentially altered metabolism occurring under cell senescence have been performed [22,23]. In the recent work by Wu et al., senescence and apoptosis were contrasted using a combination of metabolomics and proteomics. The study revealed that senescent MCF7 cells underwent metabolic reprogramming to survive by facilitating reactive oxygen species (ROS) elimination and DNA damage repair, while the metabolism of apoptotic MCF7 cells was downregulated when faced with irreparable DNA damage. Nevertheless, the metabolic changes associated with the duration of stress-induced senescence are not sufficiently clear, nor how senescent cells maintain their metabolic activity. Understanding the metabolic processes important in cell senescence might have profound implications for the development of effective strategies to treat cancer and for finding related biomarkers to measure cellular responses.

The main focus in our study was to identify underlying metabolic alterations in ADR-induced breast cancer cell senescence and describe the metabolic reprogramming that occurs. With the aid of GC–MS based metabolomics and isotope tracing methods, we obtained detailed metabolic information associated with breast cancer cell senescence. Two cell human breast cell lines: MCF7 (p53 wild-type, estrogen receptor, ER+) and MDA-MB-231 (p53 mutant, estrogen receptor, ER−) were treated by moderate dosage of the anticancer drug ADR for 5 days to induce senescence, by which time more than 60% of the remaining, adherent breast cancer cells entered into cell senescence. Morphological characterization and staining analysis suggested that ADR-induced senescent cells represented reliable models for studying the metabolic events associated with response to chemotherapy-induced senescence. Dramatic metabolic changes in response to ADR-induced cell senescence were observed clearly using GC–MS based metabolomics.

Metabolic responses were dependent on treatment time, indicating that these two cell lines might be very sensitive to the duration of ADR stress. Thus, the metabolic changes across multiple time points were needed to reflect changes during the progression of cell senescence. PCA analysis indicated that the two cell lines underwent different metabolic trajectories after ADR treatment. This result might be due in part to their different genetic backgrounds, which could contribute to the different metabolic phenotypes. In fact, cell senescence is not characterized by a stable and homogeneous status but rather a heterogeneous phenotype depending on cell types and diverse stimuli [24]. Moreover, one of the most striking features observed in our study was that the metabolic responses of MCF7 and MDA-MB-231 to ADR treatment were divided into different stages. For both cell lines, the metabolic profiles after 3 days of treatment were very different compared to their initial profiles. Unexpectedly, after 5 days of ADR treatment both cell lines exhibited some recovery as they entered into cell senescence. The dynamic nature and the two-stage development of the metabolic profiles might be correlated with a DNA damage response and metabolic adaption, respectively. Genotoxic stresses initiate related signaling pathways to repair DNA damage through the DNA damage response (DDR) machinery [25]. However, damaged cells can also be induced to undergo senescence with persistent DNA damage [26]. Ultimately, there may be a molecular switch that regulates cellular metabolic responses to genotoxic stresses.

In MCF7 cells, the downregulation of TCA cycle metabolism was observed after the first stage. It has been reported that mitochondria are the primary target involved in ADR treatment [27], so the observed lowered metabolism may represent mitochondrial dysfunction and energy imbalance during the first stage. In addition, amino acid and fatty acid metabolism in MCF7 cells was also reduced during the first stage. The TCA cycle coordinates energy production and biosynthesis as well as the exit of intermediates such as malic, alpha-ketoglutaric and citric acids from the cycle to supply various biosynthetic pathways including amino acid and fatty acid synthesis. Thus, ADR treatment can induce decreases in TCA cycle metabolism concomitant with decreases in amino acid and fatty acid metabolism.

An interesting finding in our study was the modulation of the concentrations of heptadecanoic acid in MCF7 cells. While heptadecanoic acid and other odd-chain fatty acids are normally considered of exogenous origin or produced from propionate, a recent study found that the cytidine-5-monophosphate/heptadecanoic acid metabolic ratio can be used as a powerful biomarker of breast cancer, implying that fatty acid synthesis is potentially regulated in breast cancer [28]. The regulation of heptadecanoic acid here might be related to fatty acid synthesis and membrane lipid metabolism changes. Even larger initial changes in heptadecaonic acid were observed for MDA-MB-231 cells.

The metabolic recovery during the second stage (day 3–5) was substantiated by the synchronous upregulation of TCA cycle, amino acid and fatty acid metabolisms as the MCF7 cells entered into senescence. We can conclude that in MCF7, TCA cycle metabolism is strongly coupled to ADR-induced mitochondrial changes during cell senescence. Previously, it was shown that iron accumulation in mitochondria occurs as a result of ADR and causes cardiotoxicity [29]. However, the metabolic response to DNA damage in MDA-MB-231 was more complicated and sometimes in contrast to changes observed in MCF7. Changes in the TCA cycle showed an unanticipated inverse relationship with amino acid and fatty acid metabolisms in the first stage of treatment. During the second stage, the recovery of citric acid levels, concomitant with the decrease in amino acid and fatty acid metabolisms also showed metabolic recovery in MDA-MB-231 after ADR treatment. However, the synchronization between TCA cycle metabolism and amino and fatty acids did not occur in MDA-MB-231 after ADR treatment. Instead, TCA cycle metabolism appears to be inversely correlated with amino acid and fatty acids metabolism. Our data suggest that MCF7 cells appear to have a higher level of basal respiration and an improved ability to sustain TCA cycle metabolite levels with loss of glucose oxidation after ADR treatment, while MDA-MB-231 cells were required to continue using amino acids as fuel, leading to their depletion and less anapleurotic recovery.

Overall, the metabolic alteration of both MCF7 and MDA-MB-231 cells undergoing ADR-induced senescence passed through two different stages. The metabolic shift observed from the first stage to the second was likely caused by the partial adaptation of the cells to persistent ADR stress such that they could enter into senescence [30]. A number of signaling pathways would likely become activated in response to the metabolic imbalance during the periods of ADR treatment, including mTORC1, AMPK, and NFKB [31]. Therefore, such metabolic shifts might be explored as therapeutic targets to treat cancers to avoid senescence and possible relapse. In addition, we note that the level of 2-keto-3-methylvaleric acid, an intermediate of the BCAA catabolism pathway, increased for both

MCF7 and MDA-MB-231. The observed change might be related to DNA protective effects, as higher levels of BCAAs have been suggested to elevate glutathione-S-transferase (GST) activities against DNA damage [32]. In addition, BCAAs can feed into the TCA cycle through the formation of acetyl-CoA or succinyl-CoA, and thus the elevation of 2-keto-3-methylvaleric acid might indicate the activation of a metabolic restorative mechanism. Considering ADR treatment caused obvious effects on TCA cycle metabolism, the higher level of BCAA metabolism also suggested a metabolic shift toward oxidative metabolism using non-glucose substrates (2-keto-3-methylvaleric acid) due to the inactivation of the TCA cycle.

Based on isotope-tracing analysis, both MCF7 and MDA-MB-231 exhibited almost the same metabolic characteristics after 5 days of ADR treatment. We observed that the glycolysis pathway in both ADR-induced senescent cell lines functioned similarly under senescence. There were no obvious distinctions in the levels of m + 3 isotopologues of lactate and pyruvate derived from labeled glucose between these two cell lines. This result suggests that the energy demands in senescent cells may not represent a prerequisite for maintaining the senescent status. Furthermore, the lower mean enrichment of TCA cycle metabolites in senescent MCF7 and MDA-MB-231 cells showed that the glucose flux into the TCA cycle was lower compared to pretreated cells, which might be related to mitochondrial damage and the activation of the tumor suppressor p53 upon ADR-induced cell senescence; p53 can down-regulate the expression of glucose transporters GLUT1 and GLUT4 [33]. Nevertheless, the isotopologue m + 2 levels of fumaric, malic, citric and alpha-ketoglutaric acids were increased, suggesting that flux through the first turn of the TCA cycle was higher in senescent cells. Since senescent cells appeared to develop a highly active secretory phenotype characterized by robust production of various inflammatory cytokines, the TCA cycle intermediates might need to exit the cycle to supply substrate for various biosynthetic pathways. If true, higher levels of m + 2 isotopologues might be related with the senescence associated secretory phenotype (SASP). The higher levels of m + 0 isotopologues of TCA cycle intermediates were observed in both senescent cell types, most of which could be supplied by glutamine anapleurosis or threonine in MCF7. Senescent cells may need a higher level of glutamine anapleurosis to survive DNA damage and maintain senescence. A number of studies have reported on glutamine's ability to support the TCA cycle to produce energy and provide precursors for macromolecular synthesis for cell survival [34].

## 4. Materials and Methods

***Chemicals.*** MSTFA+1%TMCS (N,O-Bis(trimethylsilyl)trifluoroacetamide with 1% (vol/vol) Trimethylchlorosilane) (Thermo Fisher Scientific, Waltham, MA, USA), MTBSTFA (N-(*tert*-butyldimethylsilyl)-N-methyltrifluoroacetamide with 1% (*v/v*) (Sigma-Aldrich, St. Louis, MO, USA), and TBDMCS (*tert*-butyldimethylsilyl chloride) (Sigma-Aldrich) were purchased for metabolite derivatization. The FAME (fatty acid methyl-ester) library of compounds with different carbon chain lengths for retention index (RI) calculations was purchased from Agilent (Santa Clara, CA, USA). Methoxyamine hydrochloride (MeOX), chloroform, and pyridine were purchased from Sigma-Aldrich. Methanol (high-performance liquid chromatography (HPLC) grade) was purchased from Thermo Fisher Scientific. [U-$^{13}$C] labeled glucose was purchased from Cambridge Isotope Laboratory (Tewksbury, MA, USA). DMEM medium was purchased from Gibco cell culture (Los Angeles, CA, USA).

***Cell culture and ADR treatment.*** Human breast cancer cell lines MCF7 and MDA-MB-231 were supplied by ATCC (Manassas, VA, USA) and cultured in a medium containing DMEM, 10% fetal calf serum, 2 mM glutamine, 1% penicillin-streptomycin (Gibco, Los Angeles, CA, USA) at 37 °C with 5% $CO_2$. Cell viability was assessed using the Vybrant MTT Cell Proliferation Assay (ThermoFisher, Waltham, MA, USA). Senescence-associated over expression and accumulation of β-galactosidase (SA-β-gal) was analyzed using kits obtained from Cell Biolabs (Atlanta, GA, USA). Protein content was determined using the Pierce bicinchoninic acid (BCA) assay obtained from ThermoFisher. For GC–MS based metabolomic profiling experiments, both cell lines were seeded using approximately 500,000 cells onto 10 cm petri dishes in triplicate. Cells were cultured in Dulbecco modified Eagle's medium (DMEM) containing 0.04 µg/mL ADR for 0, 1, 3, and 5 days, with the media replaced every other day during which time any non-adherent dead cells removed. The cells were then washed twice with ice cold PBS (pH 7.4) for 3 s, and then again with ice cold de-ionized (DI) water for 1 s. Cold extraction solvent (methanol:chloroform, 9:1) was added to quench the cellular metabolism. The cells were detached using a cell scraper and cell suspensions were then transferred into Eppendorf tubes and centrifuged. Supernatants were collected and dried under vacuum using an Eppendorf Concentrator plus (Eppendorf, Hamburg, Germany).

For isotope tracing experiments, the cells were prepared as described above in 10 cm petri dishes and were grown to approximately 90% confluence. Both cell lines were then cultured in medium containing 0.04 µg/mL ADR for 5 days. At the end of ADR treatment, the media were replaced with new medium containing 2 mM [U-$^{13}$C]-glucose for 24 h. Cell metabolite extraction was carried out as described above.

*Sample preparation.* After drying, cell sample extracts were again dissolved in 200 µL methanol:choloroform (9:1) and centrifuged for 10 min. A 150 µL aliquot of the supernatant was transferred to an Eppendorf tube containing a glass insert and centrifuged under vacuum using an Eppendorf Concentrator plus. Dried extracts were stored at −20 °C until derivatization was performed for GC–MS analysis.

The dried extracts were treated first with 25 µL MeOX (20 mg/mL) reagent at 37 °C for 90 min, followed by derivatization using 75 µL MSTFA with 1% TCMS for 30 min at 37 °C. Finally, 2 µL FAME solution was added to the mixture and vortexed in preparation for GC–MS analysis.

For isotope tracing analysis, the dried extracts were treated first with 30 µL MeOX (20 mg/mL) reagent at 37 °C for 90 min, followed by derivatization with 70 µL MTBSTFA with 1% TBDMCS for 30 min at 70 °C. Finally, 2 µL FAME solution was added to the mixture and vortexed prior to GC–MS analysis.

*Protein content.* Protein concentrations were determined using a BCA assay according to the manufacturer's instructions.

*GC–MS analysis.* All samples were analyzed using an Agilent 7890 GC instrument equipped with a 5975 mass selective detector (MSD), employing an HP-5 ms GC column (30 m length, 0.25 mm i.d., 0.20 µm film thickness). The sample injection (1 µL) was performed in splitless mode at an injection temperature of 250 °C. Helium carrier gas flow was 1.0 mL/min. The ion source temperature was 250 °C. The temperature gradient for GC separation was initially 60 °C for 1 min, then increased from 60 °C to 325 °C at 10 °C/min, where it remained for 10 min.

*Data preprocessing.* Retention indices (RIs) were calculated for each sample using AMDIS software (National Institute of Standards and Technology (NIST), Gaithersburg, MD). RI information was subsequently applied to the chromatographic analysis of each sample. The NIST-08 mass spectral and the Agilent Fiehn Metabolomics Retention Time Lock (RTL) Libraries were used to match both mass spectra and RIs to identify metabolites ($\Delta$RI < 2%). For isotope tracing analysis, metabolites were identified based on their RI and mass fragmentation patterns by comparison with metabolite standard compounds.

Collected GC–MS data were converted from the Agilent ChemStation to MassHunter formats using the Agilent GC/MSD translator. Processing of converted data was performed using Agilent Mass Hunter Quantitative Analysis. First, a batch library was created containing the names of identified compounds, retention times, *m/z* values, and their tolerances. Based on the existing library, peak integration and deconvolution were then performed with Mass Hunter using the batch method. A minimum absolute abundance of 1000 counts was used to filter the data. The extracted data were exported from Mass Hunter in "tsv" format, which can be viewed using Microsoft Excel software.

*Data analysis.* The extracted data for each cell sample were normalized to protein content and introduced into SIMCA-P software v11.5 (Umetrics, Malmö, Sweden) for PCA to identify outliers and to visualize general data clustering and trends among the observations [35]. The overall metabolic trajectory plots of two cell lines were prepared with Origin 8.5 (OriginLab, Northampton, MA, USA) using the centroids determined from PCA analysis. *p* values comparing metabolites that contribute to discrimination were calculated in Excel using a two-tailed Student's *t*-test. For isotope tracing analysis, the mean isotope enrichment and isotopotologue distribution of labeled metabolites in the TCA cycle and glycolysis pathway were calculated using IsoCor (http://metasys.insa-toulouse.fr/software/IsoCor/) analysis [36].

## 5. Conclusions

In sum, relatively little is known about the metabolic changes that accompany and maintain cell senescence. Here, we performed metabolomic analyses of ADR-induced senescence in two well-known breast cancer cell lines. Our metabolomics results indicate large and dynamic metabolic changes during ADR-induced cell senescence. Based on metabolic trajectory and univariate analysis, we conclude that the metabolic changes of MCF7 and MDA-MB-231 cells subjected to ADR treatment were characterized by a two-stage process. Moreover, most of the significantly regulated metabolites in the first stage exhibited partial metabolic recovery during the second stage.

Our work demonstrated that ADR treatment induced a number of important metabolic responses in MCF7 and MDA-MB-231 cells, which are illustrated in Figure 7. The metabolic changes observed at earlier treatment times caused by a moderate dosage of ADR included energy, amino acid and lipid metabolisms as well as others. Persistent genotoxic stresses activated cell metabolic recovery as cells went into senescence, a process that might be related to maintaining senescence. Isotope incorporation analysis results suggested a lower glucose flux into the TCA cycle, and glutamine anapleurosis might be a key component to maintain cell senescence in MCF7 (Figure 6c). This pathway potentially can be explored as a therapeutic target to treat senescent cells, with the goal of increasing the vulnerability of cells after mild and repairable genotoxic stress [37].

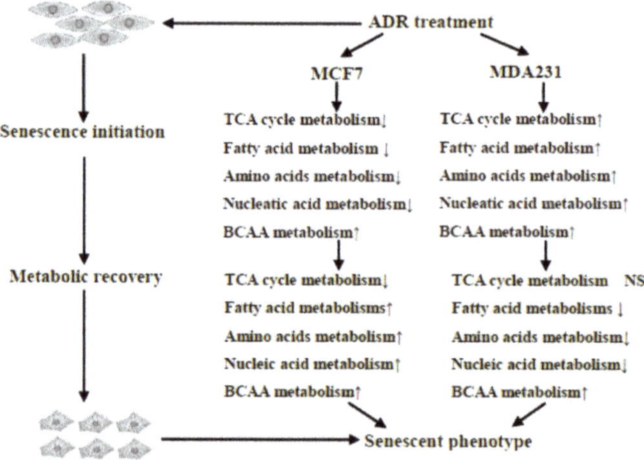


**Figure 7.** Overview of the dynamic metabolic changes during ADR-induced cell senescence.

**Author Contributions:** Formal analysis, P.Z., G.A.B.J. and D.R.; Investigation, R.Y.; Methodology, R.Y., J.D. and G.A.B.J.; Project administration, D.R.; Resources, D.R.; Visualization, P.Z.; Writing—original draft, R.Y.; Writing—review and editing, G.A.B.J. and D.R.

**Funding:** We gratefully acknowledge financial support from the Chinese National Scholarship Council, the NIH (R01GM085291, P30CA015704) and the University of Washington.

**Acknowledgments:** We thank David Hockenbery for his insightful comments on our manuscript.

**Conflicts of Interest:** D.R. serves an executive officer for and holds equity in Matrix-Bio, Inc., a biomarker discovery company. The other authors report they have no conflicts of interest.

## References


1. Hayflick, L. The limited in vitro lifetime of human diploid cell strains. *Exp. Cell Res.* **1965**, *37*, 614–636. [CrossRef]
2. Gey, C.; Seeger, K. Metabolic changes during cellular senescence investigated by proton nmr-spectroscopy. *Mech. Ageing Dev.* **2013**, *134*, 130–138. [CrossRef] [PubMed]
3. Roninson, I.B. Tumor cell senescence in cancer treatment. *Cancer Res.* **2003**, *63*, 2705–2715. [PubMed]
4. Collado, M.; Blasco, M.A.; Serrano, M. Cellular senescence in cancer and aging. *Cell* **2007**, *130*, 223–233. [CrossRef] [PubMed]
5. Davalos, A.R.; Coppe, J.J.; Desprez, P.Y. Senescent cells as a source of inflammatory factors for tumor progression. *Cancer Metast. Rev.* **2010**, *29*, 273–283. [CrossRef]
6. Judith, C.; Andersen, J.K.; Pankaj, K.; Simon, M. Cellular senescence: A link between cancer and age-related degenerative disease? *Semin. Cancer Biol.* **2011**, *21*, 354–359.
7. Elmore, L.W.; Di, X.; Dumur, C.; Holt, S.E.; Gewirtz, D.A. Evasion of a single-step, chemotherapy-induced senescence in breast cancer cells: Implications for treatment response. *Clin. Cancer Res.* **2005**, *11*, 2637–2643. [CrossRef]
8. Evan, G.I.; di Fagagna, F.d.A. Cellular senescence: Hot or what? *Curr. Opin. Genet. Dev.* **2009**, *19*, 25–31. [CrossRef]
9. Vander Heiden, M.G.; Cantley, L.C.; Thompson, C.B. Understanding the warburg effect: The metabolic requirements of cell proliferation. *Science* **2009**, *324*, 1029–1033. [CrossRef]
10. Dörr, J.R.; Yu, Y.; Milanovic, M.; Beuster, G.; Zasada, C.; Däbritz, J.H.M.; Lisec, J.; Lenze, D.; Gerhardt, A.; Schleicher, K. Synthetic lethal metabolic targeting of cellular senescence in cancer therapy. *Nature* **2013**, *501*, 421–425. [CrossRef]
11. Kaplon, J.; Zheng, L.; Meissl, K.; Chaneton, B.; Selivanov, V.A.; Mackay, G.; van der Burg, S.H.; Verdegaal, E.M.; Cascante, M.; Shlomi, T. A key role for mitochondrial gatekeeper pyruvate dehydrogenase in oncogene-induced senescence. *Nature* **2013**, *498*, 109–112. [CrossRef]
12. Aird, K.M.; Zhang, G.; Li, H.; Tu, Z.; Bitler, B.G.; Garipov, A.; Wu, H.; Wei, Z.; Wagner, S.N.; Herlyn, M. Suppression of nucleotide metabolism underlies the establishment and maintenance of oncogene-induced senescence. *Cell Rep.* **2013**, *3*, 1252–1265. [CrossRef] [PubMed]
13. Nicholson, J.K.; Lindon, J.C. Systems biology: Metabonomics. *Nature* **2008**, *455*, 1054–1056. [CrossRef] [PubMed]
14. Gowda, G.A.N.; Zhang, S.C.; Gu, H.W.; Asiago, V.; Shanaiah, N.; Raftery, D. Metabolomics-based methods for early disease diagnostics. *Expert Rev. Mol. Diagn.* **2008**, *8*, 617–633. [CrossRef]
15. Madhu, B.; Narita, M.; Jauhiainen, A.; Menon, S.; Stubbs, M.; Tavaré, S.; Narita, M.; Griffiths, J.R. Metabolomic changes during cellular transformation monitored by metabolite–metabolite correlation analysis and correlated with gene expression. *Metabolomics* **2015**, *11*, 1848–1863. [CrossRef] [PubMed]
16. Oblong, J.E. The evolving role of the NAD+/nicotinamide metabolome in skin homeostasis, cellular bioenergetics, and aging. *DNA Repair* **2014**, *23*, 59–63. [CrossRef]
17. Klontzas, M.E.; Vernardis, S.I.; Heliotis, M.; Tsiridis, E.; Mantalaris, A. Metabolomics analysis of the osteogenic differentiation of umbilical cord blood mesenchymal stem cells reveals differential sensitivity to osteogenic agents. *Stem Cells Dev.* **2017**, *26*, 723–733. [CrossRef]
18. Fan, T.W.; Lorkiewicz, P.K.; Sellers, K.; Moseley, H.N.; Higashi, R.M.; Lane, A.N. Stable isotope-resolved metabolomics and applications for drug development. *Pharmacol. Ther.* **2012**, *133*, 366–391. [CrossRef]
19. Lane, A.N.; Tan, J.; Wang, Y.; Yan, J.; Higashi, R.M.; Fan, T.W. Probing the metabolic phenotype of breast cancer cells by multiple tracer stable isotope resolved metabolomics. *Metab. Eng.* **2017**, *43*, 125–136. [CrossRef]
20. Sawyers, C. Targeted cancer therapy. *Nature* **2004**, *432*, 294–297. [CrossRef]
21. Back, J.H.; Rezvani, H.R.; Zhu, Y.; Guyonnet-Duperat, V.; Athar, M.; Ratner, D.; Kim, A.L. Cancer cell survival following DNA damage-mediated premature senescence is regulated by mammalian target of rapamycin (mtor)-dependent inhibition of sirtuin 1. *J. Biol. Chem.* **2011**, *286*, 19100–19108. [CrossRef] [PubMed]
22. Wiley, C.D.; Campisi, J. From ancient pathways to aging cells-connecting metabolism and cellular senescence. *Cell Metab.* **2016**, *23*, 1013–1021. [CrossRef] [PubMed]

23. Wu, M.Q.; Ye, H.; Shao, C.; Zheng, X.; Li, Q.R.; Wang, L.; Zhao, M.; Lu, G.Y.; Chen, B.Q.; Zhang, J.; et al. Metabolomics-proteomics combined approach identifies differential metabolism-associated molecular events between senescence and apoptosis. *J. Proteome Res.* **2017**, *16*, 2250–2261. [CrossRef] [PubMed]

24. Salama, R.; Sadaie, M.; Hoare, M.; Narita, M. Cellular senescence and its effector programs. *Genes Dev.* **2014**, *28*, 99–114. [CrossRef] [PubMed]

25. Chou, W.C.; Hu, L.Y.; Hsiung, C.N.; Shen, C.Y. Initiation of the atm-chk2 DNA damage response through the base excision repair pathway. *Carcinogenesis* **2015**, *36*, 832–840. [CrossRef] [PubMed]

26. Mckenna, E.; Traganos, F.; Zhao, H.; Darzynkiewicz, Z. Persistent DNA damage caused by low levels of mitomycin c induces irreversible cell senescence. *Cell Cycle* **2012**, *11*, 3132–3140. [CrossRef] [PubMed]

27. Berthiaume, J.; Wallace, K. Adriamycin-induced oxidative mitochondrial cardiotoxicity. *Cell Biol. Toxicol.* **2007**, *23*, 15–25. [CrossRef] [PubMed]

28. Budczies, J.; Denkert, C.; Muller, B.M.; Brockmoller, S.F.; Klauschen, F.; Gyorffy, B.; Dietel, M.; Richter-Ehrenstein, C.; Marten, U.; Salek, R.M.; et al. Remodeling of central metabolism in invasive breast cancer compared to normal breast tissue—A gc-tofms based metabolomics study. *BMC Genom.* **2012**, *13*, 334. [CrossRef]

29. Ichikawa, Y.; Ghanefar, M.; Bayeva, M.; Wu, R.; Khechaduri, A.; Prasad, S.V.N.; Mutharasan, R.K.; Naik, T.J.; Ardehali, H. Cardiotoxicity of doxorubicin is mediated through mitochondrial iron accumulation. *J. Clin. Investig.* **2014**, *124*, 617–630. [CrossRef]

30. Bartek, J.; Lukas, J. DNA damage checkpoints: From initiation to recovery or adaptation. *Curr. Opin. Cell Biol.* **2007**, *19*, 238–245. [CrossRef]

31. Tacar, O.; Sriamornsak, P.; Dass, C.R. Doxorubicin: An update on anticancer molecular action, toxicity and novel drug delivery systems. *J. Pharm. Pharmacol.* **2013**, *65*, 157–170. [CrossRef] [PubMed]

32. Nakano, M.; Nakashima, A.; Nagano, T.; Ishikawa, S.; Kikkawa, U.; Kamada, S. Branched-chain amino acids enhance premature senescence through mammalian target of rapamycin complex i-mediated upregulation of p21 protein. *PLoS ONE* **2013**, *8*, e80411. [CrossRef] [PubMed]

33. Fabiana, S.B.Y.; Michal, A.; Eddy, K. The tumor suppressor p53 down-regulates glucose transporters glut1 and glut4 gene expression. *Cancer Res.* **2004**, *64*, 2627–2633.

34. Yang, C.; Ko, B.; Hensley, C.; Jiang, L.; Wasti, A.; Kim, J.; Sudderth, J.; Calvaruso, M.A.; Lumata, L.; Mitsche, M. Glutamine oxidation maintains the tca cycle and cell survival during impaired mitochondrial pyruvate transport. *Mol. Cell* **2014**, *56*, 414–424. [CrossRef] [PubMed]

35. Ivosev, G.; Burton, L.; Bonner, R. Dimensionality reduction and visualization in principal component analysis. *Anal. Chem.* **2008**, *80*, 4933–4944. [CrossRef] [PubMed]

36. Millard, P.; Letisse, F.; Sokol, S.; Portais, J.-C. Isocor: Correcting ms data in isotope labeling experiments. *Bioinformatics* **2012**, *28*, 1294–1296. [CrossRef] [PubMed]

37. Sawako, S.; Tomoaki, T.; Poyurovsky, M.V.; Hidekazu, N.; Takafumi, M.; Shuichi, O.; Maria, L.; Hiroyuki, H.; Toshinori, N.; Yutaka, S. Phosphate-activated glutaminase (gls2), a p53-inducible regulator of glutamine metabolism and reactive oxygen species. *Proc. Natl. Acad. Sci. USA* **2010**, *107*, 7461–7466.

*Article*

# GC-MS Metabolomics Reveals Distinct Profiles of Low- and High-Grade Bladder Cancer Cultured Cells

Daniela Rodrigues [1,*], Joana Pinto [1], Ana Margarida Araújo [1], Carmen Jerónimo [2,3], Rui Henrique [2,3,4], Maria de Lourdes Bastos [1], Paula Guedes de Pinho [1] and Márcia Carvalho [1,5,*]

[1] UCIBIO/REQUIMTE, Department of Biological Sciences, Laboratory of Toxicology, Faculty of Pharmacy, University of Porto, 4050-313 Porto, Portugal; jipinto@ff.up.pt (J.P.); ana.margarida.c.araujo@gmail.com (A.M.A.); mlbastos@ff.up.pt (M.d.L.B.); pguedes@ff.up.pt (P.G.d.P.)

[2] Cancer Biology & Epigenetics Group, Research Center (CI-IPOP) Portuguese Oncology Institute of Porto (IPO Porto), 4200-072 Porto, Portugal; carmenjeronimo@ipoporto.min-saude.pt (C.J.); rmhenrique@icbas.up.pt (R.H.)

[3] Department of Pathology and Molecular Immunology-Biomedical Sciences Institute (ICBAS), University of Porto, 4050-313 Porto, Portugal

[4] Department of Pathology, Portuguese Oncology Institute of Porto (IPO Porto), 4200-072 Porto, Portugal

[5] UFP Energy, Environment and Health Research Unit (FP-ENAS), University Fernando Pessoa, 4249-004 Porto, Portugal

* Correspondence: daniela.fgrodrigues@gmail.com (D.R.); mcarv@ufp.edu.pt (M.C.); Tel.: +310-655-956-721 (D.R.); +351-225-071-300 (M.C.)

Received: 26 December 2018; Accepted: 15 January 2019; Published: 18 January 2019

**Abstract:** Previous studies have shown that metabolomics can be a useful tool to better understand the mechanisms of carcinogenesis; however, alterations in biochemical pathways that lead to bladder cancer (BC) development have hitherto not been fully investigated. In this study, gas chromatography-mass spectrometry (GC-MS)-based metabolomics was applied to unveil the metabolic alterations between low-grade and high-grade BC cultured cell lines. Multivariable analysis revealed a panel of metabolites responsible for the separation between the two tumorigenic cell lines. Significantly lower levels of fatty acids, including myristic, palmitic, and palmitoleic acids, were found in high-grade versus low-grade BC cells. Furthermore, significantly altered levels of some amino acids were observed between low- and high-grade BC, namely glycine, leucine, methionine, valine, and aspartic acid. This study successfully demonstrated the potential of metabolomic analysis to discriminate BC cells according to tumor aggressiveness. Moreover, these findings suggest that bladder tumorigenic cell lines of different grades disclose distinct metabolic profiles, mainly affecting fatty acid biosynthesis and amino acid metabolism to compensate for higher energetic needs.

**Keywords:** bladder cancer; cancer progression; in vitro; metabolomic signatures; endometabolome; GC-MS; metabolic pathways

---

## 1. Introduction

Bladder cancer (BC) is the second most common genitourinary malignancy and one of the deadliest cancers worldwide [1]. BC can be classified as low-grade (LG) or high-grade (HG) according to histopathological characteristics [2,3], with low-grade meaning that the differentiation of the tumor is more similar to normal than high-grade. Importantly, HG BC is more aggressive and prone to invasion than LG [2]. Although significant progresses in unveiling new diagnostic strategies based on molecular markers have been made [4,5], their high cost and flaw in detecting early BC do not offer advantage over classical ones [6], thus hindering their clinical application [5]. Therefore, there is an urgent need for discovering early, specific, cost-effective, and non-invasive diagnosis methods, so that therapeutics can be more effective.

Metabolomics has proven to be a promising and alternative tool for early cancer detection, through the comprehensive analysis of alterations in metabolite levels that can be translated into biomarkers. These metabolite signatures reflect the biological activity of each cancer cell type, which brings the possibility of distinguishing unique dysregulations in metabolic pathways characteristic of different cancer subtypes [7]. This approach has been already applied to several cancers including those of breast [8], ovary [9], kidney [10], colorectum [11], and hepatocellular carcinomas [12], through the screening and detection of metabolite levels in human urine, feces, or biofluids, which represents an advantage over classical invasive diagnosis methods [13]. Nevertheless, discrepancies among studies that aimed to investigate cancer biomarkers for early detection through metabolomic analysis are compelling, hampering the development of a diagnostic tool based on such molecules. This particularly applies to the case of BC, for which studies based on metabolomic diagnosis biomarkers are controversial and limited [3]. For these reasons, translatability to clinical settings has been hindered. Therefore, more studies focusing on early detection and the progression of BC are urgently needed.

Recently, studies on metabolomics have also focused on in vitro approaches to obtain further information about metabolic pathways that lead to BC progression [3]. In vitro cell culture systems represent the least complex disease model and have a number of advantages over tissue or biofluid analysis, including simpler and controllable experimental settings, less variability among samples, reduction in animal testing, and the provision of better insight into metabolic changes [14]. Nevertheless, this study model has only comprised immortalized cell lines and primary cell cultures so far, which leads to some limitations mainly related to extrapolation to in vivo systems [15], the inability to mimic the environment and communication between surrounding cells [16], the need of routine evaluation, and a careful interpretation of results, since metabolic perturbations can be caused by changes in the culture cell medium rather than to the disease itself [14].

The application of metabolomic approaches, using different disease models, has led to the discovery of several metabolites whose levels are altered in BC cells (see review by Rodrigues et al. [3]) which are involved in important biochemical pathways that produce energy, including glycolysis, tricarboxylic acid (TCA) cycle, fatty acid β-oxidation, and amino acid metabolism. Nevertheless, there is a clear gap in the search for metabolites that can be used to distinguish different grades of BC, particularly in in vitro studies, since past research has mostly focused on distinguishing normal and benign cancer samples. There are very few studies conducted using a metabolomics approach to distinguish LG from HG BC, all of which applied to either human plasma/serum or urine [17–19]. However, to the best of our knowledge, no study has yet investigated the endometabolome signatures of BC cultured cell lines with different tumor grades, making our in vitro study a pivotal one in searching for metabolic differences between LG and HG transitional cell carcinoma (TCC) of the bladder. In this study, we applied gas chromatography-mass spectrometry (GC-MS)-based metabolomics to determine the endometabolome signatures of two BC cell lines of different tumor grades. This approach not only allowed for a demonstration of the potential of metabolomics analysis to discriminate BC cells according to tumor aggressiveness but also extended the knowledge on the metabolic alterations between LG and HG bladder TCC, an evaluation which is lacking among published in vitro studies.

## 2. Material and Methods

### 2.1. Chemicals

Eagle's minimum essential medium (MEM) supplemented with L-glutamine, *N*-trimethylsilyl-*N*-methyl trifluoroacetamide (MSTFA, ≥98.5%), desmosterol, and L-norvaline was purchased from Sigma-Aldrich (St. Louis, MO, USA). Penicillin, streptomycin, trypsin, and fetal bovine serum (FBS) were purchased from Invitrogen (Karlsruhe, Germany). Phosphate buffer solution (PBS) was purchased from Biochrom (Merck, Berlin, Germany); methanol (99.9%) and dichloromethane (99%) were purchased from VWR (Leuven, Belgium). All chemicals were of analytical grade and were dissolved in ultrapure water unless otherwise indicated.

## 2.2. Cell Lines and Culture Conditions

The BC cell lines 5637 and J82 were obtained from American Type Culture Collection (ATCC; Manassas, VA, USA). Both cell lines were derived from transitional cell carcinoma of the human urinary bladder, with 5637 being classified as grade II and J82 as grade III (stage pT3) [20]. BC cell lines were cultured in MEM to ensure optimal cell growth and maintenance of epithelial cell characteristics. The culture medium was prepared as indicated by the manufacturer and supplemented with 10% FBS and 100 units·mL$^{-1}$ penicillin/100 µg·mL$^{-1}$ streptomycin. All cell lines were routinely tested for *Mycoplasma* spp. contamination (PCR Mycoplasma Detection Set, Clontech Laboratories).

## 2.3. Sample Collection

The experiments were carried out during five passages, with triplicates for each passage, after an adaptation stage of at least three passages for both cell lines. Cells were grown for 48 h in T75 culture flasks to near confluence. The culture medium was discarded and cells were gently washed with 2 mL PBS to remove all medium. Cold methanol (3 mL) was then added to each flask to effectively quench the metabolism of cells. Subsequently, cells were scraped off the flasks on ice, transferred to a falcon tube and centrifuged for 10 min at 3000 *g* at 4 °C. The supernatant was collected and stored at −80 °C until analysis.

## 2.4. Sample Preparation for GC-MS Analysis

Each sample (1 mL) was transferred into a glass vial, followed by the addition of 10 µL desmosterol (1 mg/mL) and norvaline (1 mg/mL) as internal standards. These compounds, which were added in equal amounts to all samples, were used to monitor the performance of the of GC-MS acquisition (injection issues and retention time deviation). Subsequently, the mixture was vortex-mixed at high speed for 1 minute and carefully evaporated to dryness at room temperature under a gentle stream of nitrogen gas. The derivatization process was adapted from a previous work performed in our laboratory [21,22], which was carried out by adding 50 µL MSTFA and 50 µL dichloromethane to the dried extract, followed by vortex-mixing at high speed for 1 minute and incubation for 30 min at 80 °C. Then, the derivatized solution was cooled and transferred into screw-top autosampler vials for subsequent GC-MS analysis. In addition, quality control samples (QCs) were prepared as a pool of all samples in the study and divided into aliquots to avoid the repeated freeze–thaw cycles. QCs were derivatized using the same protocol applied for samples.

## 2.5. GC-MS Analysis: Equipment and Conditions

The endometabolome profiles of J82 and 5637 cells were obtained using an EVOQ 436 GC system (Bruker Daltonics, Fremont, CA) coupled to a SCION TQ mass detector, a Bruker Daltonics MS workstation software (version 8.2), and a Combi-PAL autosampler (Varian Pal Autosampler, Switzerland), as described previously [22]. Briefly, random injection of all samples was performed and a QC sample was also injected in every four samples, under the same conditions, for a total of seven QCs. Chromatographic separation was obtained by using a GC fused silica capillary column BR-5ms (5% phenyl, 95% dimethyl polysiloxane, 30 m × 0.25 mm × 0.25 µm) (Bruker Daltonics, Freemont, CA, USA). The injector temperature was 250 °C and samples (1 µL) were introduced in split mode with a 1:5 ratio. Helium C-60 (Gasin, Portugal) was the carrier gas with a flow rate of 1.0 mL/min. The program set for the column temperature was as follows: initial temperature at 70 °C held for 2 min, then ramped at 15 °C/min to 250 °C, held for 2 min, and finally increased at 10 °C/min to 300 °C, then held for 8 min, having a total duration of 29 min per run. The MS detector was functioning in Electron Impact (EI) mode. The transfer line temperature was 230 °C, the manifold temperature was 40 °C, and the ion source temperature was 250 °C. The mass range selected was 50–600 *m*/*z*, with a scan rate of 6 scans per second. The resolution and intensity of the chromatographic peaks among all samples were monitored using norvaline and desmosterol as internal standards.

### 2.6. GC-MS Data Pre-Processing

All raw data files obtained from GC-TQ-MS were exported as Computable Document Format (CDF) files and pre-processed to convert instrumental data sets into a manageable format for data analysis and remove any biases such as background, noise, and retention time (RT) fluctuations over a set of samples. Data pre-processing was performed using the software MZmine 2.21 [23]. The parameters used in these steps were set as follows: RT range 2.8–29.0 min; $m/z$ range 50–600; MS data noise level $2.5 \times 10^4$; $m/z$ tolerance 0.5; chromatogram baseline level $2.0 \times 10^4$; peak duration range 0.05–0.50 min. Chromatographic peaks regarded as trash or irrelevant were manually removed from the data matrix, as well as all peaks with a relative standard deviation (RSD) $\geq$30% across all QCs. Normalization of the data was performed by determining the mean of the chromatogram's area of each set of triplicates, which was subsequently divided by the total area. Ultimately, statistical analysis was performed with the resulting normalized peak areas.

### 2.7. Metabolite Identification

The identification of metabolites in GC-TQ-MS chromatograms was performed by comparing the retention indices (RI) and mass spectra fragmentation patterns of each compound with the RI and mass spectra present in the National Institute of Standards and Technology (NIST) spectral library version 14 (Gaithersburg, MA, USA). The reverse match obtained by NIST 14 was also considered and the identification was obtained when a value of 700 or above was achieved. Compounds for which no satisfactory match was found were listed as "unknown i" ($i = 1,2,3 \dots$) according to their RT in ascending order. When possible, the compound identification was confirmed through comparison of their RTs and mass spectra with that obtained from pure standards. Pathway and metabolite information were extracted from the Human Metabolome Database (HMDB) [24] and the Kyoto Encyclopedia of Genes and Genomes (KEGG) [25].

### 2.8. Metabolic Pathway Analysis

To explore which metabolic pathways were altered and contributed the most to the separation between HG J82 and LG 5637 cell lines, Metabolite Set Enrichment Analysis (MSEA) [26], specifically the pathway over-representation analysis, was performed using the freely available online software Metaboanalyst 4.0 [27], where biologically meaningful patterns were investigated using the set of significantly different metabolites (compound names).

### 2.9. Statistical Analysis

Principal component analysis (PCA) and partial least squares discriminant analysis (PLS-DA) were applied to the final matrix scaled to pareto (Par) [28], using SIMCA-P 13.0.3 (Umetrics, Umea, Sweden). The $R^2X$ (variance explained by the $X$ matrix), $R^2Y$ (variance explained by the $Y$ matrix), and $Q^2$ (goodness of prediction or prediction power) parameters, obtained by 7-fold cross validation, were used to evaluate the model robustness (SIMCA-P 13.0.3). The PLS-DA loadings plot and the variable importance to the projection (VIP > 1) of each variable were used to assess the variables ($m/z$-RT pairs) responsible for group separation. For subsequent analyses, the peak area of all detected derivatives from the same metabolite were summed, as recommended in the literature [29].

The statistical significance of relevant compounds identified in the PLS-DA loadings plots was assessed using the unpaired Mann–Whitney test (non-normally distributed data), in GraphPad Prism version 6 (GraphPad Software, San Diego, CA, USA). This software was also used to determine the area under the curve (AUC) for each metabolite. For each model, the discriminative compounds were considered statistically significant when $p < 0.05$ (confidence level 95%). Effect size [30] was computed for compounds with $p < 0.05$ and corrected for small sample sizes.

## 3. Results and Discussion

In this study, the endometabolome profiles of the LG BC cell line 5637 and the HG BC cell line J82 were analyzed to search for metabolites with the potential to assess tumor aggressiveness. Representative GC-MS chromatograms of the in vitro tumorigenic cells are shown in Supplementary Figure S1. A total of 37 metabolites were identified across all samples, except for one metabolite which remained unknown (Supplementary Table S1). Most of the identified metabolites belong to the classes of amino acids, fatty acids, and organic acids or derivatives. Less represented classes included amino alcohols, monosaccharides, sterols, and sugar alcohols. The general characteristics of the metabolites detected in the endometabolome of BC cell lines, such as RT, RI, characteristic ions ($m/z$), metabolite identification method, HMDB identification, and the main pathways in which they are involved, are summarized in Supplementary Table S1. Whenever available, BC studies in which those metabolites have been previously found are also described.

PCA was first performed to explore the metabolic differences between LG BC and HG BC cell endometabolomes, showing a tendency to separate both ($R^2X = 0.690$, Supplementary Figure S2). PLS-DA confirmed a clear separation between LG BC and HG BC (Figure 1a) with a good prediction power ($Q^2 = 0.870$). Eight metabolites were identified as important for discriminating between LG BC and HG BC cell lines, as represented in the Volcano plot (Table 1, Figure 1b). Glycine, myristic, palmitic, and palmitoleic acids were found to be significantly decreased in HG J82 cells, whereas leucine, methionine, valine, and aspartic acid were found to be significantly increased in the HG compared to the LG cell line.

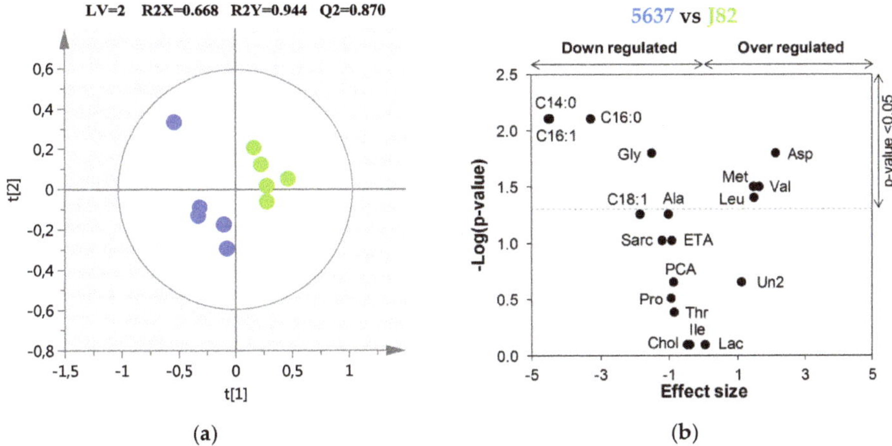



**Figure 1.** (**a**) Partial least squares discriminant analysis (PLS-DA) scores scatter plot obtained for the cancer cell lines 5637 ($n = 5$, low-grade, (•) and J82 ($n = 5$, high-grade, (•). The ellipses indicate the 95% confidence limit of the model; (**b**) Volcano plot showing the metabolites contributing to the discrimination between high-grade (HG) J82 and low-grade (LG) 5637 cells. Horizontal dashed line indicates the significance level ($p < 0.05$). Three-letter codes are used for amino acids. C14:0, myristic acid; C16:0, palmitic acid; C16:1, palmitoleic acid; Chol, cholesterol; ETA, ethanolamine; Lac, lactic acid; PCA, pyroglutamic acid; Sarc, sarcosine; Un2, unknown 2.

Boxplots of two of the most significantly altered metabolites between the two cancer cell lines are represented in Figure 2a (aspartic acid) and Figure 2b (myristic acid). Figure 3 shows the metabolic pathways that are significantly disturbed between the two BC cell lines, which will be discussed in detail below.

**Table 1.** List of metabolites selected as important for the discrimination between HG (J82) and LG (5637) bladder cancer cells. Values for ES, ES$_{SE}$, *p*, and AUC are indicated for each metabolite.

| Metabolite | HG J82 vs LG 5637 | | |
|---|---|---|---|
| | ES (±ES$_{SE}$) [a] | *p*-Value [b] | AUC |
| **Amino acids and derivatives** | | | |
| Glycine | ↓  −1.51 (±1.30) | $1.59 \times 10^{-2}$ | 0.960 |
| Aspartic acid | ↑  2.13 (±1.46) | $1.59 \times 10^{-2}$ | 0.960 |
| Leucine | ↑  1.48 (±1.29) | $3.97 \times 10^{-2}$ | 0.880 |
| Methionine | ↑  1.46 (±1.29) | $3.17 \times 10^{-2}$ | 0.920 |
| Valine | ↑  1.63 (±1.33) | $3.17 \times 10^{-2}$ | 0.920 |
| **Fatty Acids** | | | |
| Myristic acid | ↓  −4.50 (±2.27) | $7.90 \times 10^{-3}$ | 1.000 |
| Palmitic acid | ↓  −3.28 (±1.82) | $7.90 \times 10^{-3}$ | 1.000 |
| Palmitoleic acid | ↓  −4.46 (±2.25) | $7.90 \times 10^{-3}$ | 1.000 |

AUC, area under the curve; ES, effect size; ↑, metabolites increased; ↓, metabolites decreased in the endometabolome of HG vs. LG cancer cells. [a] ES determined as described in Reference [30]; [b] 95% significance level ($p < 0.05$).

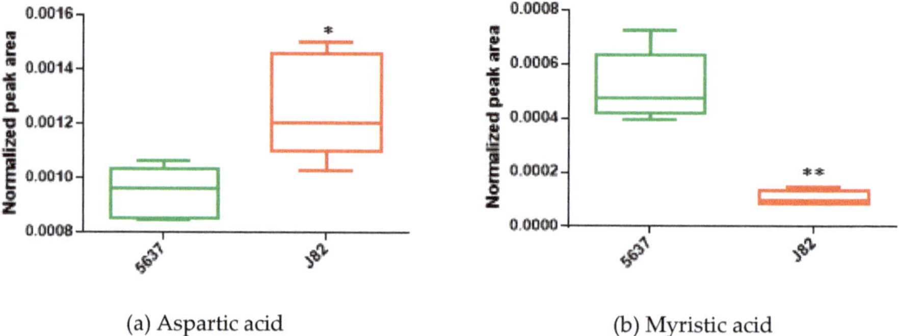

(a) Aspartic acid           (b) Myristic acid

**Figure 2.** Boxplots of (**a**) aspartic acid and (**b**) myristic acid, two metabolites found to be significantly altered between the LG bladder cancer cell line 5637 ($n = 5$) and the HG bladder cancer cell line J82 ($n = 5$). * $p < 0.05$; ** $p < 0.01$.

The levels of myristic, palmitic, and palmitoleic acids were significantly decreased in the higher tumor grade cell line, which is concordant with the well-known fact that cancer cells, as they grow into more advanced stages, change their energetic requirements to keep on growing and proliferating. This is because several important biological processes, such as the synthesis of DNA and proteins or the production of new cellular components, need to be maintained and enhanced as the tumor becomes more aggressive [31,32]. Fatty acids (FAs), when metabolized by cells, yield great amounts of energy that serve as fuel for several cellular processes such as the TCA cycle and β-oxidation [33]. Moreover, FAs are also involved in other cellular processes, particularly cell signaling and cell membrane integrity. To the best of our knowledge, no previous studies have reported alterations in the levels of these three FAs according to BC grading. Therefore, our results offer an important insight into the metabolic differences between tumor grades and clearly demonstrate that LG and HG BC cells generate a different in vitro signature that reflects the distinct metabolic needs of those cancer cells, with advanced grades requiring more FA consumption for survival and continuous growth and proliferation [3].

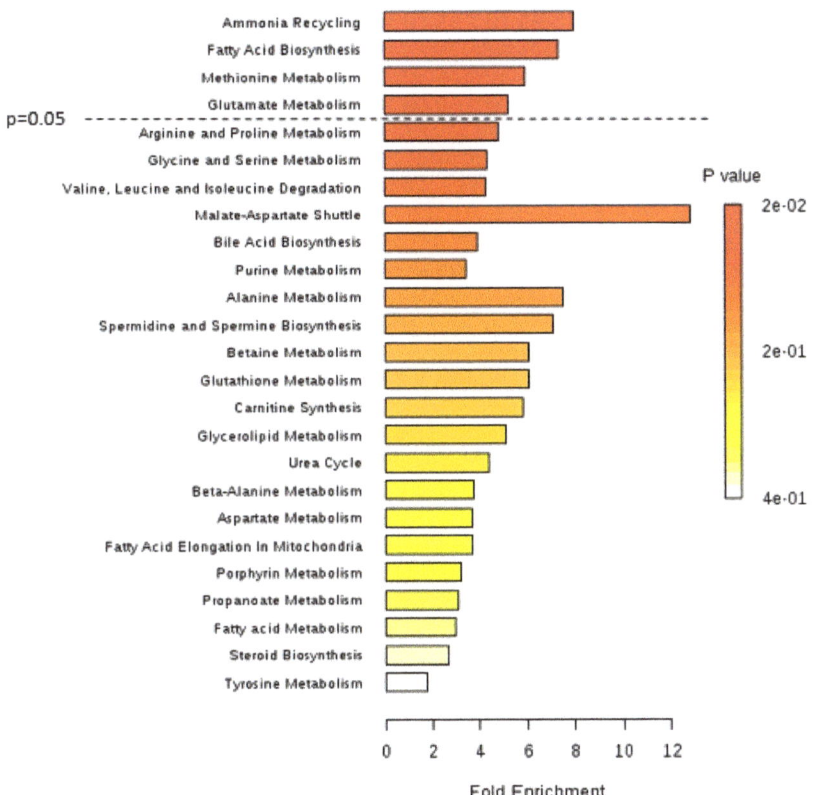



**Figure 3.** Summary plot for metabolite set enrichment analysis (MSEA) performed for the HG bladder cancer (BC) cell line (J82) versus the LG BC cell line (5637), where metabolite sets are ranked according to the Holm *p*-value. The horizontal bar graph summarizes metabolic pathways that were different between HG and LG bladder cancer cells. Significantly different pathways include ammonia recycling, fatty acid biosynthesis, and methionine and glutamate metabolisms.

Referring to amino acids, glycine was found to be decreased in HG BC cells (J82) compared to LG BC cells (5637). Corroborating our findings, previous studies by Dettmer et al. [34] and Cao et al. [35] observed lower levels of glycine in the serum of patients with advanced BC compared to patients with LG BC. Glycine may play a critical role in tumor progression to higher grades [36], being involved not only in amino acid metabolism but also in purine and glutathione metabolism [3]. This suggests that these two pathways might be augmented in HG cancer cells and glycine may be highly used in order to either produce more DNA/RNA components, proteins, or antioxidant metabolites. Glycine is also involved in ammonia recycling (Figure 3), being used to give rise to ammonia through the action of the glycine cleavage system [37]. Upregulation of this ammonia recycling-related pathway might explain the decrease in glycine levels as BC progresses, but further studies are required to confirm this hypothesis.

Apart from glycine, four other amino acids were significantly altered between the two BC cell lines, namely aspartic acid, leucine, valine, and methionine, which may indicate alterations in amino acid metabolism (Figure 3) and consequently in the pathways in which these amino acids are involved, namely protein biosynthesis and the TCA cycle. The potential increase in protein biosynthesis combined with the downregulation of the TCA cycle, which also corresponds with the upregulation of FA biosynthesis, is crucial for cancer cells to maintain their proliferative demands as proteins and FA components are necessary during cell division [38]. Furthermore, TCA cycle downregulation may show that cancer cells, as they develop to more advanced stages, increasingly rely on glycolytic pathways over oxidative phosphorylation [3]. These alterations in amino acid levels and their respective pathways were previously found in other BC metabolomic studies in which different tumor grades were compared in vitro and in human biofluids, namely urine and blood [34,35,39–47].

Noticeably, methionine has not been previously reported as a differentially altered amino acid between BC grades. In this study, we observed that methionine metabolism was found to be one of the most important discriminatory pathways comparing HG and LG BC cells (Figure 3). This demonstrates that this pathway is altered towards the production of methionine, which is an essential amino acid that functions as an intermediate for protein synthesis, in transmethylation reactions as a methyl donor to S-adenosyl methionine (SAM) and in detoxifying processes, as well as an important compound for angiogenesis [48]. Moreover, the increase in methionine levels in HG cancer cells may also be associated with the downregulation of the TCA cycle. This is a novel finding that should be considered in future validation and translational studies with human samples.

## 4. Conclusions

In this study, we used an in vitro metabolomics approach to evaluate alterations in the endometabolome signatures of bladder cancer cells with distinct tumor grades. The bottom-up approach was selected due to its easily controllable setup and lower complexity compared to other disease models, therefore providing better insight into the metabolic changes between the two types of BC cells. The results obtained in this study regarding the endometabolome analysis of HG and LG cancer cells allow us to propose some metabolic alterations occurring during tumor progression in the bladder. Major changes were found in energy-related metabolic pathways, namely FA biosynthesis, amino acid metabolism, and ammonia recycling. Differences in FA levels between HG and LG cancer cell lines may reflect the distinct reliability in β-oxidation to generate energy. The same applies to amino acid metabolism along with ammonia recycling reflecting alterations in protein biosynthesis and the TCA cycle towards the support of tumor cell growth and proliferation. Furthermore, methionine stands out as an amino acid that, to our best knowledge, has never been reported before in in vitro studies comparing HG and LG bladder cancer, making this metabolite a novel finding to be validated. Overall, the results from our study may contribute to the discovery of promising biomarkers applicable to the categorization of the LG and HG forms of BC and the development of new therapeutic approaches, pending further investigation and validation.

**Supplementary Materials:** The following are available online at http://www.mdpi.com/2218-1989/9/1/18/s1, Figure S1: Representative full scan of GC-MS chromatograms obtained from the endometabolome analysis of (A) 5637 and (B) J82 cell lines., Figure S2: PCA scores scatter plot obtained for the GC-MS chromatograms obtained from the endometabolome analysis of the cancer cell lines 5637 (*n* = 5, low-grade, TCC) and J82 (*n* = 5, high-grade, TCC). The ellipses indicate the 95% confidence limit of the model., Table S1: List of all metabolites identified in the endometabolomes of the cancer cell lines 5637 (LG) and J82 (HG). The identification of the metabolites is based on the NIST (2014) and standards. RT, characteristic ions (*m/z*), identification method with reverse % of match from NIST/standards when used, CAS registry number and HMDB code (when available) are indicated for each compound, as well as the main pathways in which they are involved and BC studies where they were previously found (when reported).

**Author Contributions:** D.R. was responsible for the execution of the experimental work and data analysis. A.M.A. supported cell culture and data analysis. J.P. helped with the statistical analysis of the data. C.J. and R.H. kindly provided the cell lines used in the study and gave conceptual advice. P.G.d.P., M.C., and M.d.L.B. contributed to the conception, design, and management of the study. D.R. wrote the manuscript with input from M.C. All authors critically commented on and approved the final submitted version of the paper.

**Funding:** This work received financial support from the European Union (FEDER funds POCI/01/0145/ FEDER/07728) and national funds (FCT/MEC, Fundação para a Ciência e a Tecnologia and Ministério da Educação e Ciência) under the Partnership Agreement PT2020 UID/MULTI/04378/2013. The study is a result of the project NORTE-01-0145-FEDER-000024, supported by the Norte Portugal Regional Operational Programme (NORTE 2020) under the PORTUGAL 2020 Partnership Agreement (DESignBIOtecHealth—New Technologies for three Health Challenges of Modern Societies: Diabetes, Drug Abuse and Kidney Diseases), through the European Regional

Development Fund (ERDF). A.M. Araújo thanks the FCT for her PhD fellowship (SFRH/BD/107708/2015) and M. Carvalho also acknowledges FCT through the UID/MULTI/04546/2016 project.

**Conflicts of Interest:** All authors have read the journal's policy on the disclosure of potential conflicts of interest and have none to declare.

## References


1. Burger, M.; Catto, J.W.; Dalbagni, G.; Grossman, H.B.; Herr, H.; Karakiewicz, P.; Kassouf, W.; Kiemeney, L.A.; La Vecchia, C.; Shariat, S.; et al. Epidemiology and risk factors of urothelial bladder cancer. *Eur. Urol.* **2013**, *63*, 234–241. [CrossRef] [PubMed]

2. Kirkali, Z.; Chan, T.; Manoharan, M.; Algaba, F.; Busch, C.; Cheng, L.; Kiemeney, L.; Kriegmair, M.; Montironi, R.; Murphy, W.M.; et al. Bladder cancer: Epidemiology, staging and grading, and diagnosis. *Urology* **2005**, *66*, 4–34. [CrossRef] [PubMed]

3. Rodrigues, D.; Jeronimo, C.; Henrique, R.; Belo, L.; de Lourdes Bastos, M.; de Pinho, P.G.; Carvalho, M. Biomarkers in bladder cancer: A metabolomic approach using in vitro and ex vivo model systems. *Int. J. Cancer* **2016**, *139*, 256–268. [CrossRef]

4. Chan, E.C.; Pasikanti, K.K.; Hong, Y.; Ho, P.C.; Mahendran, R.; Raman Nee Mani, L.; Chiong, E.; Esuvaranathan, K. Metabonomic profiling of bladder cancer. *J. Proteome Res.* **2015**, *14*, 587–602. [CrossRef] [PubMed]

5. Miremami, J.; Kyprianou, N. The Promise of Novel Molecular Markers in Bladder Cancer. *Int. J. Mol. Sci.* **2014**, *15*, 23897–23908. [CrossRef]

6. Ku, J.H.; Godoy, G.; Amiel, G.E.; Lerner, S.P. Urine survivin as a diagnostic biomarker for bladder cancer: A systematic review. *BJU Int.* **2012**, *110*, 630–636. [CrossRef] [PubMed]

7. Sahu, D.; Lotan, Y.; Wittmann, B.; Neri, B.; Hansel, D.E. Metabolomics analysis reveals distinct profiles of nonmuscle-invasive and muscle-invasive bladder cancer. *Cancer Med.* **2017**, *6*, 2106–2120. [CrossRef]

8. Nam, H.; Chung, B.C.; Kim, Y.; Lee, K.; Lee, D. Combining tissue transcriptomics and urine metabolomics for breast cancer biomarker identification. *Bioinformatics* **2009**, *25*, 3151–3157. [CrossRef]

9. Denkert, C.; Budczies, J.; Kind, T.; Weichert, W.; Tablack, P.; Sehouli, J.; Niesporek, S.; Konsgen, D.; Dietel, M.; Fiehn, O. Mass spectrometry-based metabolic profiling reveals different metabolite patterns in invasive ovarian carcinomas and ovarian borderline tumors. *Cancer Res.* **2006**, *66*, 10795–10804. [CrossRef]

10. Catchpole, G.; Platzer, A.; Weikert, C.; Kempkensteffen, C.; Johannsen, M.; Krause, H.; Jung, K.; Miller, K.; Willmitzer, L.; Selbig, J.; et al. Metabolic profiling reveals key metabolic features of renal cell carcinoma. *J. Cell. Mol. Med.* **2011**, *15*, 109–118. [CrossRef]

11. Cheng, Y.; Xie, G.; Chen, T.; Qiu, Y.; Zou, X.; Zheng, M.; Tan, B.; Feng, B.; Dong, T.; He, P.; et al. Distinct urinary metabolic profile of human colorectal cancer. *J. Proteome Res.* **2012**, *11*, 1354–1363. [CrossRef] [PubMed]

12. Huang, Q.; Tan, Y.; Yin, P.; Ye, G.; Gao, P.; Lu, X.; Wang, H.; Xu, G. Metabolic characterization of hepatocellular carcinoma using nontargeted tissue metabolomics. *Cancer Res.* **2013**, *73*, 4992–5002. [CrossRef] [PubMed]

13. Aboud, O.A.; Weiss, R.H. New opportunities from the cancer metabolome. *Clin. Chem.* **2013**, *59*, 138–146. [CrossRef] [PubMed]

14. Ramirez, T.; Daneshian, M.; Kamp, H.; Bois, F.Y.; Clench, M.R.; Coen, M.; Donley, B.; Fischer, S.M.; Ekman, D.R.; Fabian, E.; et al. Metabolomics in Toxicology and Preclinical Research. *Altex* **2013**, *30*, 209–225. [CrossRef] [PubMed]

15. Yoon, M.; Campbell, J.L.; Andersen, M.E.; Clewell, H.J. Quantitative in vitro to in vivo extrapolation of cell-based toxicity assay results. *Crit. Rev. Toxicol.* **2012**, *42*, 633–652. [CrossRef]

16. Rodrigues, D.; Monteiro, M.; Jeronimo, C.; Henrique, R.; Belo, L.; Bastos, M.L.; Carvalho, M. Renal cell carcinoma: A critical analysis of metabolomic biomarkers emerging from current model systems. *Transl. Res.* **2016**. [CrossRef] [PubMed]

17. Bansal, N.; Gupta, A.; Mitash, N.; Shakya, P.S.; Mandhani, A.; Mahdi, A.A.; Sankhwar, S.N.; Mandal, S.K. Low- and high-grade bladder cancer determination via human serum-based metabolomics approach. *J. Proteome Res.* **2013**, *12*, 5839–5850. [CrossRef] [PubMed]

18. Wittmann, B.M.; Stirdivant, S.M.; Mitchell, M.W.; Wulff, J.E.; McDunn, J.E.; Li, Z.; Dennis-Barrie, A.; Neri, B.P.; Milburn, M.V.; Lotan, Y.; et al. Bladder cancer biomarker discovery using global metabolomic profiling of urine. *PLoS ONE* **2014**, *9*, e115870. [CrossRef]

19. Tan, G.; Wang, H.; Yuan, J.; Qin, W.; Dong, X.; Wu, H.; Meng, P. Three serum metabolite signatures for diagnosing low-grade and high-grade bladder cancer. *Sci. Rep.* **2017**, *7*, 46176. [CrossRef]

20. Zuiverloon, T.C.M.; de Jong, F.C.; Costello, J.C.; Theodorescu, D. Systematic review: Characteristics and preclinical uses of bladder cancer cell lines. *Bladder Cancer* **2018**, *4*, 169–183. [CrossRef]

21. Pereira, D.M.; Vinholes, J.; de Pinho, P.G.; Valentao, P.; Mouga, T.; Teixeira, N.; Andrade, P.B. A gas chromatography-mass spectrometry multi-target method for the simultaneous analysis of three classes of metabolites in marine organisms. *Talanta* **2012**, *100*, 391–400. [CrossRef] [PubMed]

22. Lima, A.R.; Araujo, A.M.; Pinto, J.; Jeronimo, C.; Henrique, R.; Bastos, M.L.; Carvalho, M.; Guedes de Pinho, P. GC-MS-Based Endometabolome Analysis Differentiates Prostate Cancer from Normal Prostate Cells. *Metabolites* **2018**, *8*. [CrossRef] [PubMed]

23. Pluskal, T.; Castillo, S.; Villar-Briones, A.; Oresic, M. MZmine 2: Modular framework for processing, visualizing, and analyzing mass spectrometry-based molecular profile data. *BMC Bioinform.* **2010**, *11*, 395. [CrossRef] [PubMed]

24. Wishart, D.S.; Tzur, D.; Knox, C.; Eisner, R.; Guo, A.C.; Young, N.; Cheng, D.; Jewell, K.; Arndt, D.; Sawhney, S.; et al. HMDB: The Human Metabolome Database. *Nucleic Acids Res.* **2007**, *35*, D521–D526. [CrossRef] [PubMed]

25. Ogata, H.; Goto, S.; Sato, K.; Fujibuchi, W.; Bono, H.; Kanehisa, M. KEGG: Kyoto Encyclopedia of Genes and Genomes. *Nucleic Acids Res.* **1999**, *27*, 29–34. [CrossRef] [PubMed]

26. Xia, J.; Wishart, D.S. MSEA: A web-based tool to identify biologically meaningful patterns in quantitative metabolomic data. *Nucleic Acids Res.* **2010**, *38*, W71–W77. [CrossRef] [PubMed]

27. Chong, J.; Soufan, O.; Li, C.; Caraus, I.; Li, S.; Bourque, G.; Wishart, D.S.; Xia, J. MetaboAnalyst 4.0: Towards more transparent and integrative metabolomics analysis. *Nucleic Acids Res.* **2018**, *46*, W486–W494. [CrossRef] [PubMed]

28. Worley, B.; Powers, R. Multivariate Analysis in Metabolomics. *Curr. Metab.* **2013**, *1*, 92–107. [CrossRef]

29. Mastrangelo, A.; Ferrarini, A.; Rey-Stolle, F.; Garcia, A.; Barbas, C. From sample treatment to biomarker discovery: A tutorial for untargeted metabolomics based on GC-(EI)-Q-MS. *Anal. Chim. Acta* **2015**, *900*, 21–35. [CrossRef]

30. Berben, L.; Sereika, S.M.; Engberg, S. Effect size estimation: Methods and examples. *Int. J. Nurs. Stud.* **2012**, *49*, 1039–1047. [CrossRef]

31. Hanahan, D.; Weinberg, R.A. Hallmarks of cancer: The next generation. *Cell* **2011**, *144*, 646–674. [CrossRef] [PubMed]

32. Putluri, N.; Shojaie, A.; Vasu, V.T.; Vareed, S.K.; Nalluri, S.; Putluri, V.; Thangjam, G.S.; Panzitt, K.; Tallman, C.T.; Butler, C.; et al. Metabolomic profiling reveals potential markers and bioprocesses altered in bladder cancer progression. *Cancer Res.* **2011**, *71*, 7376–7386. [CrossRef]

33. Liu, Y. Fatty acid oxidation is a dominant bioenergetic pathway in prostate cancer. *Prostate Cancer Prostatic Dis.* **2006**, *9*, 230–234. [CrossRef]

34. Dettmer, K.; Vogl, F.C.; Ritter, A.P.; Zhu, W.; Nurnberger, N.; Kreutz, M.; Oefner, P.J.; Gronwald, W.; Gottfried, E. Distinct metabolic differences between various human cancer and primary cells. *Electrophoresis* **2013**, *34*, 2836–2847. [CrossRef] [PubMed]

35. Cao, M.; Zhao, L.; Chen, H.; Xue, W.; Lin, D. NMR-based metabolomic analysis of human bladder cancer. *Anal. Sci.* **2012**, *28*, 451–456. [CrossRef] [PubMed]

36. Letellier, S.; Garnier, J.P.; Spy, J.; Bousquet, B. Determination of the L-DOPA/L-tyrosine ratio in human plasma by high-performance liquid chromatography. Usefulness as a marker in metastatic malignant melanoma. *J. Chromatogr. B* **1997**, *696*, 9–17. [CrossRef]

37. Ananieva, E. Targeting amino acid metabolism in cancer growth and anti-tumor immune response. *World J. Biol. Chem.* **2015**, *6*, 281–289. [CrossRef] [PubMed]

38. Al-Zoughbi, W.; Huang, J.; Paramasivan, G.S.; Till, H.; Pichler, M.; Guertl-Lackner, B.; Hoefler, G. Tumor macroenvironment and metabolism. *Semin. Oncol.* **2014**, *41*, 281–295. [CrossRef]

39. Alberice, J.V.; Amaral, A.F.; Armitage, E.G.; Lorente, J.A.; Algaba, F.; Carrilho, E.; Marquez, M.; Garcia, A.; Malats, N.; Barbas, C. Searching for urine biomarkers of bladder cancer recurrence using a liquid chromatography-mass spectrometry and capillary electrophoresis-mass spectrometry metabolomics approach. *J. Chromatogr. A* **2013**, *1318*, 163–170. [CrossRef]

40. Pasikanti, K.K.; Norasmara, J.; Cai, S.; Mahendran, R.; Esuvaranathan, K.; Ho, P.C.; Chan, E.C. Metabolic footprinting of tumorigenic and nontumorigenic uroepithelial cells using two-dimensional gas chromatography time-of-flight mass spectrometry. *Anal. Bioanal. Chem.* **2010**, *398*, 1285–1293. [CrossRef]

41. Tripathi, P.; Somashekar, B.S.; Ponnusamy, M.; Gursky, A.; Dailey, S.; Kunju, P.; Lee, C.T.; Chinnaiyan, A.M.; Rajendiran, T.M.; Ramamoorthy, A. HR-MAS NMR tissue metabolomic signatures cross-validated by mass spectrometry distinguish bladder cancer from benign disease. *J. Proteome Res.* **2013**, *12*, 3519–3528. [CrossRef] [PubMed]

42. Conde, V.R.; Oliveira, P.F.; Nunes, A.R.; Rocha, C.S.; Ramalhosa, E.; Pereira, J.A.; Alves, M.G.; Silva, B.M. The progression from a lower to a higher invasive stage of bladder cancer is associated with severe alterations in glucose and pyruvate metabolism. *Exp. Cell Res.* **2015**, *335*, 91–98. [CrossRef] [PubMed]

43. Srivastava, S.; Roy, R.; Singh, S.; Kumar, P.; Dalela, D.; Sankhwar, S.N.; Goel, A.; Sonkar, A.A. Taurine—A possible fingerprint biomarker in non-muscle invasive bladder cancer: A pilot study by 1H NMR spectroscopy. *Cancer Biomark.* **2010**, *6*, 11–20. [CrossRef] [PubMed]

44. Gamagedara, S.; Shi, H.; Ma, Y. Quantitative determination of taurine and related biomarkers in urine by liquid chromatography-tandem mass spectrometry. *Anal. Bioanal. Chem.* **2012**, *402*, 763–770. [CrossRef] [PubMed]

45. Pasikanti, K.K.; Esuvaranathan, K.; Ho, P.C.; Mahendran, R.; Kamaraj, R.; Wu, Q.H.; Chiong, E.; Chan, E.C. Noninvasive urinary metabonomic diagnosis of human bladder cancer. *J. Proteome Res.* **2010**, *9*, 2988–2995. [CrossRef] [PubMed]

46. Pasikanti, K.K.; Esuvaranathan, K.; Hong, Y.; Ho, P.C.; Mahendran, R.; Raman Nee Mani, L.; Chiong, E.; Chan, E.C. Urinary metabotyping of bladder cancer using two-dimensional gas chromatography time-of-flight mass spectrometry. *J. Proteome Res.* **2013**, *12*, 3865–3873. [CrossRef]

47. Jin, X.; Yun, S.J.; Jeong, P.; Kim, I.Y.; Kim, W.; Park, S. Diagnosis of bladder cancer and prediction of survival by urinary metabolomics. *Oncotarget* **2014**, *5*, 1635–1645. [CrossRef] [PubMed]

48. Cavuoto, P.; Fenech, M.F. A review of methionine dependency and the role of methionine restriction in cancer growth control and life-span extension. *Cancer Treat. Rev.* **2012**, *38*, 726–736. [CrossRef]

MDPI

St. Alban-Anlage 66

4052 Basel

Switzerland

Tel. +41 61 683 77 34

Fax +41 61 302 89 18

www.mdpi.com

*Metabolites* Editorial Office

E-mail: metabolites@mdpi.com

www.mdpi.com/journal/metabolites